AF736445

Atlas of Chinese Medicinal Plant Diseases

Tielin Wang • Luqi Huang • Lanping Guo

Atlas of Chinese Medicinal Plant Diseases

Springer

中国农业出版社
CHINA AGRICULTURE PRESS

Tielin Wang
National Resource Center for Chinese Materia Medica
China Academy of Chinese Medical Sciences
Beijing, China

Luqi Huang
National Resource Center for Chinese Materia Medica
China Academy of Chinese Medical Sciences
Beijing, China

Lanping Guo
China Academy of Chinese Medical Sciences
Beijing, China

ISBN 978-981-95-6325-8 ISBN 978-981-95-6326-5 (eBook)
https://doi.org/10.1007/978-981-95-6326-5

© China Agriculture Press Co., Ltd. 2026

Jointly published with China Agriculture Press

This work is subject to copyright. All rights are solely and exclusively licensed by the Publisher, whether the whole or part of the material is concerned, specifically the rights of translation, reprinting, reuse of illustrations, recitation, broadcasting, reproduction on microfilms or in any other physical way, and transmission or information storage and retrieval, electronic adaptation, computer software, or by similar or dissimilar methodology now known or hereafter developed.
The use of general descriptive names, registered names, trademarks, service marks, etc. in this publication does not imply, even in the absence of a specific statement, that such names are exempt from the relevant protective laws and regulations and therefore free for general use.
The publishers, the authors, and the editors are safe to assume that the advice and information in this book are believed to be true and accurate at the date of publication. Neither the publishers nor the authors or the editors give a warranty, express or implied, with respect to the material contained herein or for any errors or omissions that may have been made. The publishers remain neutral with regard to jurisdictional claims in published maps and institutional affiliations.

This Springer imprint is published by the registered company Springer Nature Singapore Pte Ltd.
The registered company address is: 152 Beach Road, #21-01/04 Gateway East, Singapore 189721, Singapore

If disposing of this product, please recycle the paper.

Atlas of Chinese Medicinal Plant Diseases

Editorial Board

Editor-in-Chief: Wang Tielin, Huang Luqi, Guo Lanping

Associate Editors: Zhou Rujun, Guan Wei, Zhao Mei, Kong Zhiqiang, Miao Yuhuan

Contributors: Bai Qingrong, Bao Qfiangguo, Chen Chen, Chen Meilan, Chen Zilin, Chi Xiulian, Dai Wei, Du Yongxi, Duan Yuxi, Fan Susu, Ge Yang, He Zhen, Jiang Jingyi, Kang Chuanzhi, Li Jiajun, Li Mingxuan, Li Xiaolin, Li Ying, Li Zibo, Liu Dahui, Liu Li, Liu Juan, Lu Baohui, Lu Heng, Lyu Chaogeng, Luo Bi, Ma Weisi, Ma Xiaojing, Maihemutijiang·Mijiti, Sun Baixin, Sun Jiahui, Sun Kai, Tao Shan, Tang Xiuli, Wan Xiufu, Wang Yiheng, Wang Ping, Wang Rong, Wang Ruishan, Wang Sheng, Wang Yan,Wang Zengxia, Xiao Chenghong, Xu Yang, Yan Honghai, Yan Binbin, Yan Wanrong, Yang Jian, Yang Ye, Yang Yuwen, Yuan Qingjun, Zhan Zhilai, Zhang Cenrong, Zhang Chao, Zhang Xiaobo, Zhang Yan, Zhang Zhihui, Zhao Dan, Zhou Tao, Zhou Xiuteng.

Foreword

Driven by the advancement of Traditional Chinese Medicine (TCM) and the implementation of the Healthy China strategy, demand for both the yield and quality of Chinese medicinal materials has surged. Wild-harvested resources can no longer meet market needs, driving a significant expansion in cultivated medicinal plant varieties and cultivation areas. As cultivation scales expand, disease and pest pressures have intensified considerably. Frequent outbreaks have substantially reduced yields and, in severe cases, caused total crop losses, thereby dampening the enthusiasm of medicinal plant growers. Concurrently, the lack of standardized disease diagnostics and integrated pest management (IPM) in practices has led to widespread misuse of agrochemicals, prioritizing short-term yield over long-term quality, compromising public medication safety. Therefore, precise disease diagnosis and comprehensive management are urgently needed to ensure sustainable medicinal plant cultivation.

Research on medicinal plant diseases remains relatively underdeveloped, with a wide variety of pathogens and unclear distribution and occurrence patterns. Cultivators generally lack expertise in plant protection, rendering disease identification and monitoring particularly challenging. Consequently, there is an urgent need in both production and scientific research for a systematic, comprehensive, and visually informative atlas of medicinal plant diseases. Such an atlas would serve as an essential practical tool to ensure the production of high-quality, safe medicinal materials and foster the green and healthy development of the medicinal industry. To address these challenges, in 2019, a team led by Academician Huang Luqi from the China Academy of Chinese Medical Sciences, leveraging the national technical system for medicinal materials, collaborated with experts in plant protection and the medicinal materials field to conduct the "First National Survey of Medicinal Plant Diseases." This pioneering survey undertook a comprehensive assessment of disease occurrences across the nation's primary production regions and culminated in the compilation of the *Atlas of Medicinal Plants Diseases in China.*

This book is based on the findings from the "First National Survey of Medicinal Plant Diseases," utilizing extensive field data and combined with laboratory scientific identifications. It provides a systematic and objective exposition of the types of diseases prevalent nationwide among major medicinal plants, detailing their pathogenic characteristics and control methods. The atlas is both academic and practical, offering authoritative references for the rapid diagnosis and scientific control of medicinal plant diseases. It plays an essential role in guiding production practices and steadily improving the quality of medicinal materials.

The scope of this book encompasses a broad spectrum of medicinal plants, indicative of both diversity and commonality; it covers diseases across all national production areas, integrating regional specifics with a comprehensive perspective. It provides detailed and timely information on a full range of diseases, including both infectious and noninfectious types.

As a product of interdisciplinary efforts, this atlas will support the monitoring, early warning, accurate diagnosis, and scientific control of medicinal plant diseases. It offers theoretical backing for green development techniques and precision medication applications, providing robust scientific support for the ecological cultivation of medicinal materials. The book is of critical importance for the sustainable development of the medicinal materials industry and will serve as a reference in the fields of disease diagnosis and integrated control for medicinal plants. It aims to offer invaluable guidance and assistance to farmers engaged in the production of medicinal plants, researchers, and industry professionals. Moreover, the publication of this atlas is expected to advance research and application in the prevention and control of medicinal plant diseases, contributing significantly to the sustainable use of medicinal plant resources and ensuring human health.

Jiang Youxu
Academician of the Chinese Academy of Sciences (CAS), Forest Ecologist
Honorary Director & Research Professor, Research Institute of Forest Ecology, Environment and Protection, Chinese Academy of Forestry (CAF)
Professor, College of Urban and Environmental Sciences, Peking University
Honorary Dean, College of Forestry, Beijing Forestry University
August 2023

Preface

Diagnosis and integrated control of medicinal plant diseases are the most challenging technical aspects in the modernization of Chinese medicinal materials. The diversity of diseases, weak research foundation, and management complexity make these conditions difficult to study. The multitude of symptoms and pathogen types, along with intricate patterns of occurrence, pose substantial challenges to the identification, diagnosis, and monitoring of these diseases. Most medicinal plant cultivators lack adequate training in disease recognition, hindering accurate diagnosis and effective control implementation, ultimately compromising the quality of medicinal materials.

To address this, a nationwide survey was initiated to ascertain the prevalence of diseases in the cultivation of medicinal plants and to provide scientific guidelines for their diagnosis and integrated control. Led by Academician Huang Luqi, president of the China Academy of Chinese Medical Sciences, and supported by the national technical system for medicinal materials and a national expert group on Chinese medicinal resources, this effort involved collaborations among multiple research institutions. Participants included the Plant Protection Institute of the Chinese Academy of Agricultural Sciences, the Institute of Medicinal Plant Development at the Chinese Academy of Medical Sciences, China Agricultural University, Shenyang Agricultural University, Qingdao Agricultural University, Yangzhou University, and Hubei University of Chinese Medicine. This collaboration led to the "First National Survey of Medicinal Plant Diseases," focusing on more than 40 medicinal plants to systematically investigate the occurrence, symptoms, pathogens, and patterns of both infectious and noninfectious diseases affecting these plants. The findings were compiled into the *Atlas of Medicinal Plants Diseases in China*.

To enhance the depiction of field symptoms within ecological environments, team members traversed the expanse of the country over four years. Through rigorous field surveys and careful selection of firsthand photographs, they vividly captured the field symptoms of each disease. Additionally, detailed studies and laboratory identifications of each pathogen were conducted, presenting comprehensive visual and textual information in the atlas. It is important to note that all morphological data related to the pathogens presented in this book were obtained through direct measurements, offering new insights to plant protection workers and researchers alike. We extend our heartfelt gratitude to the project team members for their dedicated efforts in sampling and compiling this book.

The creation of this book was a substantial and demanding task, necessitating field collection and scientific laboratory validation for each disease, an endeavor too vast for any single team. We are profoundly grateful for the assistance and contributions from several expert teams within the national technical system for medicinal materials. Our thanks go particularly to Professor Lu Guozhong for his review of the scientific names and taxonomic statuses of pathogens, to Academician Jiang Youxu for his foreword, and to the many other teachers and students who contributed to this publication.

This book summarizes the outcomes of a series of projects, supported by various programs including the National Key Research and Development Program of China (2023YFC3503800, 2024YFD1600600), Scientific and technological innovation project of China Academy of Chinese Medical Sciences (CI2026A039), National Natural Science Foundation of China (82104341, 82574533, 82404809), Technological Innovation Project of China Academy of Chinese Medical Sciences (CI2024E003, CI2024C010YNL), Fundamental Research Funds for the Central public welfare research institutes (ZXKT25043), Project of Hubei Province Traditional Chinese Medicine Innovation Team (ZY2025J002), among others. The publication of this book also received funding from the National Publishing Fund, and we extend our sincere thanks for this support.

China boasts a wealth of medicinal plant resources. This book focuses on the findings from the “First National Survey of Medicinal Plant Diseases” concerning common medicinal plants and aims to be updated and revised as further survey work is conducted, thus benefiting an even broader readership. While this book strives for scientific accuracy and authenticity, any oversight or errors are unintended, and we warmly welcome corrections from our readers to facilitate further improvements.

Beijing, China

Tielin Wang
Luqi Huang
Lanping Guo

Contents

1 The Diseases of Lily ... 1
1.1 Lily Wilt Disease ... 1
1.2 Lily Anthracnose ... 5
1.3 Lily Black Spot Disease ... 7
1.4 Lily Gray Mold Disease ... 9
1.5 Lily Virus Diseases ... 11

2 The Diseases of Fritillaria ... 15
2.1 *Fritillaria maximowiczii* Rust Disease ... 15
2.2 Fritillaria Gray Mold Disease ... 17
2.3 Fritillaria Root Rot ... 18

3 The Diseases of Solomon's Seal ... 23
3.1 Rhizome Rot of *Polygonatum* ... 24
3.2 Anthracnose of *Polygonatum sibiricum* ... 27
3.3 Gray Blotch Disease of *P. sibiricum* ... 30
3.4 Sclerotinia Disease of *Polygonatum* ... 32
3.5 Miscellaneous Diseases of *Polygonatum* ... 34

4 Diseases of *Ophiopogon japonicus* ... 37
4.1 Maidong Leaf Blight ... 38
4.2 Maidong Black Spot Disease ... 39
4.3 Maidong Anthracnose Disease ... 42

5 Diseases of *Anemarrhena asphodeloides* ... 45
5.1 *Anemarrhena asphodeloides* Anthracnose ... 45
5.2 Occasional Diseases of *Anemarrhena asphodeloides* ... 48

6 Diseases of *Paris polyphylla* ... 49
6.1 Leaf Spot Disease of *Paris polyphylla* ... 49
6.2 Anthracnose Disease of *Paris polyphylla* ... 52
6.3 Gray Mold Disease of *Paris polyphylla* ... 54

6.4 Soft Rot Disease of *Paris polyphylla* 56
6.5 Occasional Diseases of *Paris polyphylla* 58

7 Diseases of *Mentha canadensis* 59
7.1 Powdery Mildew of Mint 59
7.2 Gray Mold Disease of Mint 61
7.3 Occasional Diseases of Mint 64

8 Diseases of *Salvia miltiorrhiza* 67
8.1 Root Rot Disease of Danshen 67
8.2 Southern Blight of Danshen 70
8.3 Root-Knot Nematode Disease of Danshen 71

9 Diseases of Patchouli 75
9.1 Bacterial Wilt of Patchouli 75
9.2 Root Rot of Patchouli 78
9.3 Patchouli Occasional Diseases 82

10 Diseases of Schizonepeta 83
10.1 Stem Blight of Schizonepeta 83
10.2 Occasional Diseases of Schizonepeta 86

11 The Diseases of Licorice 87
11.1 Licorice Root Rot 87
11.2 Licorice Leaf Spot Disease 91
11.3 Licorice Rust Disease 95
11.4 Licorice Powdery Mildew 97
11.5 Licorice Viral Diseases 100

12 The Diseases of Kudzu 103
12.1 Kudzu Pseudo-Rust Disease 103
12.2 Kudzu Rust Disease 105
12.3 Kudzu Root Rot Disease 107
12.4 Kudzu Bacterial Leaf Spot Disease 109

13 The Diseases of Astragalus (Huangqi) 113
13.1 Astragalus Root Rot Disease 113
13.2 Powdery Mildew Disease in Astragalus 117
13.3 Downy Mildew Disease in Astragalus 119

14 The Diseases of Alpinia (Yizhi) 123
14.1 Ring Spot Leaf Blight of Alpinia 123
14.2 Anthracnose of Alpinia 125
14.3 Root-Knot Nematode Disease of Alpinia 128
14.4 Fusarium Wilt Disease in Alpinia 130
14.5 Viral Disease in Alpinia 133
14.6 Occasional Diseases of Alpinia Oxyphylla 134

15 The Diseases of *Codonopsis* 135
15.1 Rust of *Codonopsis* 136
15.2 Gray Mold of *Codonopsis* 138
15.3 Root Rot Disease in *Codonopsis* 140

16 The Diseases of Platycodon 145
16.1 Root Rot Disease in *Platycodon* 145
16.2 Root-Knot Nematode Disease of *Platycodon* 148
16.3 Dodder Infestation in *Platycodon* 150
References 152

17 The Diseases of *Artemisia argyi* 153
17.1 Powdery Mildew of *A. argyi* 153
17.2 Sooty Mold Disease of *A. argyi* 155
17.3 Leaf Spot Disease of *A. argyi* 157

18 The Diseases of Atractylodes (Bai Zhu) 161
18.1 Sclerotial Blight of Bai Zhu 161
18.2 Root Rot of Bai Zhu 164
18.3 Damping-Off Disease in Bai Zhu 167
18.4 Atractylodes Leaf Spot Disease 169
18.5 Occasional Disease of Bai Zhu 172

19 The Diseases of Atractylodes 173
19.1 Root Rot of *Atractylodes* 173
19.2 Leaf Spot Disease of Atractylodes 176
19.3 *Atractylodes Lancea* Anthracnose 178
19.4 Sclerotinia Disease of Atractylodes 180
19.5 *Atractylodes Macrocephala* White Scourge Disease 182
19.6 Black Spot Disease of Atractylodes 185
19.7 Atractylodes Rust Disease 186
19.8 Atractylodes Virus Disease 188

20 The Diseases of *Erigeron breviscapus* 193
20.1 Root Rot Disease in *Erigeron breviscapus* 194
20.2 Rust Disease of *Erigeron breviscapus* 197
20.3 Section Three: Downy Mildew of *Erigeron breviscapus* 199
20.4 Root-Knot Nematode Disease in *Erigeron breviscapus* 201
20.5 Major Diseases of *Erigeron breviscapus* 204

21 The Diseases of *Chrysanthemum morifolium* 205
21.1 Wilt Disease of *C. morifolium* 205
21.2 Root Rot Disease of *C. morifolium* 208
21.3 Southern Blight Disease of *C. morifolium* 210
21.4 Black Spot Disease of *C. morifolium* 212
21.5 Viral Diseases of *C. morifolium* 214
21.6 Leaf Blight Disease of *C. morifolium* 217
21.7 Brown Spot Disease of *C. morifolium* 219

21.8 Yellow Spot Disease of *C. morifolium* 220
21.9 Anthracnose of *C. morifolium* 222
21.10 Downy Mildew of *C. morifolium* 223
21.11 Occasional Diseases of *C. morifolium* 225

22 The Diseases of *Andrographis paniculata* 227
22.1 Wilt Disease of *Andrographis paniculata* 227
22.2 Southern Blight of *Andrographis paniculata* 229
22.3 Black Stem Disease of *Andrographis paniculata* 232
22.4 Phytophthora Blight of *Andrographis paniculata* 234

23 The Diseases of Bletilla Species 237
23.1 Bletilla Rust Disease 237
23.2 Anthracnose of Bletilla 239
23.3 Bletilla Leaf Spot Disease 241
23.4 Bletilla Root Rot Disease 241
23.5 Bletilla Opportunistic Diseases 244

24 The Diseases of *Dendrobium* 245
24.1 White Silk Disease of Dendrobium 245
24.2 Leaf Spot Disease in Dendrobium 248
24.3 Viral Disease of Dendrobium 249
24.4 Occasional Diseases in Dendrobium 251

25 The Diseases of Asarum 253
25.1 Leaf Blight in Asarum 253
25.2 Sclerotinia Disease of Asarum 256
25.3 Phytophthora Blight in Asarum 258
25.4 Rust Disease in Asarum 259

26 The Diseases of Aconitum 263
26.1 Root Rot in *A. kusnezoffii* 263
26.2 Southern Blight of *Aconitum kusnezoffii* 266
26.3 Downy Mildew in Aconitum 268
26.4 Damping-Off Disease of *Aconitum kusnezoffii* 270
26.5 Leaf Spot Disease in *A. kusnezoffii* 273
26.6 Leaf Blight Disease of *Aconitum kusnezoffii* 275
26.7 Soft Rot Disease in Aconitum 277
26.8 Virus Disease of Aconitum 279

27 The Diseases of Coptis (Chinese Goldthread) 283
27.1 Root Rot Disease in Coptis 283
27.2 White Rot Disease of Coptis (Huanglian) 286
27.3 Leaf Spot Disease of Coptis 288
27.4 Powdery Mildew of Coptis (Huanglian) 289
27.5 Coptis Anthracnose Disease 291

28 The Diseases of *Paeonia* x *suffruticosa* Andr. 295
28.1 Peony Leaf Mold Disease 296
28.2 Peony Black Spot Disease 299
28.3 Root Rot Disease of Peony 301
28.4 Peony Gray Mold Disease 304
28.5 Brown Spot Disease of Peony 306
28.6 Peony Powdery Mildew 308
28.7 Occasional Parasitic Diseases of Peony 310

29 The Diseases of Schisandra 315
29.1 *Schisandra chinensis* Damping-Off Disease 316
29.2 *Schisandra chinensis* Seedling Damping-Off Disease 318
29.3 *Schisandra chinensis* Leaf Blight 320
29.4 *Schisandra chinensis* Stem Base Rot 322
29.5 *Schisandra chinensis* Powdery Mildew 326
29.6 *Schisandra chinensis* Cluster Rot Disease 328
29.7 *Schisandra chinensis* Frost Injury 330
29.8 *Schisandra chinensis* Sunscald Disease 332

30 The Diseases of Honeysuckle 335
30.1 Honeysuckle Root Rot Disease 335
30.2 Honeysuckle Brown Spot Disease 338
30.3 Honeysuckle Powdery Mildew 341
30.4 Occasional Diseases of Honeysuckle 344

31 The Diseases of *Ligusticum chuanxiong* 345
31.1 Root Rot in *Ligusticum chuanxiong* 345
31.2 Blight Spot Disease of *Ligusticum chuanxiong* 347
31.3 Root-Knot Nematode Disease in *Ligusticum chuanxiong* 350
31.4 Occasional Diseases of *Ligusticum chuanxiong* 352

32 The Diseases of *Angelica sinensis* 355
32.1 Angelica Root Rot Nematode Disease 355
32.2 Angelica Root Rot 359
32.3 Angelica Water Rot Disease 361
32.4 Angelica Leaf Spot Disease 363
32.5 Angelica Powdery Mildew 365
32.6 Angelica Anthracnose Disease 367
32.7 Angelica Dwarf Virus Disease 369
32.8 Occasional Diseases of Angelica 371

33 The Diseases of *Saposhnikovia divaricata* 373
33.1 Powdery Mildew of *Saposhnikovia divaricata* 373
33.2 *Saposhnikovia divaricate* Spot Blight 375
33.3 Gray Mold of *Saposhnikovia divaricata* 377
33.4 *Saposhnikovia divaricata* Root Rot Disease 379
33.5 *Saposhnikovia divaricata* Dodder Disease 382

34 The Diseases of *Paeonia lactiflora* Pall. 385
34.1 Peony Leaf Spot Disease 385
34.2 Peony Anthracnose Disease 388
34.3 Southern Blight of Peony 391
34.4 Peony Viral Disease 394
34.5 Peony Powdery Mildew 396
34.6 Phytophthora Blight of Peony 399
34.7 Common Diseases of Peony 401

35 The Diseases of Woad 405
35.1 Downy Mildew of Woad 405
35.2 Woad Root Rot Disease 408
35.3 Woad Black Spot Disease 410
35.4 Woad Viral Disease 412
35.5 Woad Fungal Diseases 413

36 The Diseases of Prince Ginseng 417
36.1 Leaf Spot Disease of *Pseudostellaria heterophylla* 417
36.2 Tai Zi Shen Root Rot Disease 419
36.3 Tai Zi Shen Violet Root Rot Disease 422
36.4 Tai Zi Shen Virus Disease 423
36.5 Occasional Diseases of Tai Zi Shen 426

37 The Diseases of *Pinellia ternata* 427
37.1 Leaf Spot Disease of *Pinellia ternata* 427
37.2 Southern Blight Disease of *P. ternata* 430
37.3 Soft Rot Disease of *Pinellia ternata* 432
37.4 Viral Disease of Pinellia ternata 435
37.5 Occasional Diseases of *Pinellia ternata* 437

38 The Diseases of Eleuthero (*Eleutherococcus senticosus*) 439
38.1 Eleuthero Gray Mold 439
38.2 Acanthopanax Leaf Spot Disease 441

39 The Diseases of Ginseng 445
39.1 Ginseng Gray Mold 445
39.2 Ginseng Black Spot Disease 447
39.3 Anthracnose of Ginseng 450
39.4 Ginseng Damping-Off Disease 452
39.5 Damping-Off Disease of Ginseng 454
39.6 Ginseng Rust Root Disease 456
39.7 Phytophthora Blight of Ginseng 459

40 The Diseases of Rehmannia 463
40.1 Ring Spot Disease of Rehmannia 463
40.2 Rehmannia Root Cyst Nematode Disease 465

41 The Diseases of Corydalis 469
41.1 Downy Mildew of *Corydalis yuanhusuo* 469
41.2 *Corydalis Sclerotium* Disease 471
41.3 *Corydalis* Rust Disease 473

42 The Diseases of *Belamcanda chinensis* 475
42.1 Southern Blight of *Belamcanda chinensis* 475
42.2 Soft Rot in *Belamcanda chinensis* 478
42.3 Leaf Spot Disease of *Belamcanda chinensis* 480
42.4 Rust Disease of *Belamcanda chinensis* 482
42.5 Occasional Diseases of *Belamcanda chinensis* 485

References 487

Chapter 1
The Diseases of Lily

The Chinese medicinal material "Lily" refers to the dried fleshy scales of plants from the genus Lilium within the Liliaceae family. Common species include Tiger Lily (*Lilium lancifolium* Thunb.), Brown's Lily (*L. brownii* var. viridulum Baker), and Coral Lily (*L. pumilum* DC.).

Lilies are perennial herbs found in Hebei, Shanxi, Henan, Shaanxi, Hubei, Hunan, Jiangxi, Anhui, and Zhejiang. They are revered for their properties of nourishing yin, moisturizing the lungs, clearing the heart, and calming the spirit.

Major diseases affecting lilies include wilt, anthracnose, black spot, gray mold, and viral diseases.

1.1 Lily Wilt Disease

Also known as lily stem rot or root rot, lily wilt disease is a severe soil-borne disease prevalent across national lily cultivation areas. The incidence rate generally ranges from 10% to 30%, with some regions experiencing rates above 80%.

Symptoms

The pathogen enters through wounds at the base of the lily's fleshy roots or scales, causing the roots and base to turn brown and rot, eventually spreading upward. This leads to the development of brown, sunken spots on the scales, which subsequently detach and fall off from the base as the disease progresses. Plants grown from affected bulbs are noticeably stunted, and one side of the plant displays water--soaked spots on the leaves from the bottom up, which then wilt and droop. As the condition worsens, the leaves may yellow or turn brown and purple, ultimately withering and dying under sun exposure. Longitudinal cuts of the diseased stem reveal browned vascular bundles. Severely affected plants have constricted stem bases that are easily broken. The disease can continue to develop during the storage

© China Agriculture Press Co., Ltd. 2026

T. Wang et al., *Atlas of Chinese Medicinal Plant Diseases*,
https://doi.org/10.1007/978-981-95-6326-5_1

and transportation of lilies, causing significant bulb rot. In humid conditions, a pink or white mold layer may be visible on affected parts (Fig. 1.1).

Pathogen

The lily wilt disease is caused by fungi from the genus Fusarium (Fusarium spp.), including *Fusarium oxysporum* Schltdl. ex Snyder et Hansen f. sp. lilii Imle, which specializes in lilies, as well as *Fusarium solani* [(Mart.) Appel et Wollenw. ex Snyder et Hansen] and *Fusarium fragrans* (Wollenw.) among others.

F. solani, when cultured on PDA at 25 °C for 5 days, exhibits sparse aerial mycelium, which is yellow-white to light gray, with a concentric ring pattern, and a soil--yellow sticky mass in the center. Large macroconidia are falcate, slightly curved, with both ends slightly rounded and a less conspicuous foot cell, hyaline, three to five septate, measuring (11–40) × (2–6) μm; small macroconidia are oval, unicellular, measuring (5–14) × (2–6) μm; chlamydospores are nearly spherical, unicellular, measuring (7–28) × (5–10) μm (Fig. 1.2).

F. oxysporum f. sp. lilii, when cultured on PDA at 25 °C for 5 days, produces a cottony, dense, and multicolored mycelium—white, dark red, purple, pink, and wax yellow. Aerial mycelial balls may appear in colonies during later stages. Conidiophores arise laterally from aerial mycelium or on short branches forming sporodochia. Small macroconidia are more numerous, oval or ovoid, generally one septate, measuring (5.65–9.03) × (1.30–1.89) μm; large macroconidia are falcate or oval, centrally swollen, tapering at both ends, with a conspicuous or inconspicuous foot cell, one to four septate, measuring (5.64–23.29) × (1.69–3.66) μm (Fig. 1.3).

F. fragrans, cultured on PDA at 25 °C for 5 days, forms a cobweb-like, pale gray, or green colony without altering the color of the medium. Small macroconidia are mostly ovoid, kidney-shaped, zero to one septate, measuring (4.93–10.78) × (1.70–2.85) μm; large macroconidia are broader than those of *F. oxysporum*, with an inconspicuous foot cell, three to five septate, measuring (6.97–25.77) × (1.78–3.07) μm (Fig. 1.4).

Occurrence Pattern

The primary inoculum for lily wilt disease is the mycelium that overwinters on plant debris, serving as the main source of initial infection the following year. In spring, under favorable conditions, the pathogen becomes more active. Infected plants grow slowly and fail to transport nutrients upward. Generally, disease onset begins in mid to late April, with a sharp rise in cases by early to mid-May, reaching a peak in mid--June. From June to late July, a large number of plants wilt and die. Even harvested lily bulbs can continue to exhibit symptoms. Lilies, which prefer light and are drought-tolerant, often experience more severe disease in years with plentiful rainfall. Conditions such as high temperatures and humidity, poor ventilation, inadequate drainage, and acidic soils also favor the occurrence of the disease.

Control Measures

1. Cultural Control: Choose well-drained planting sites, implement crop rotation, reduce nitrogen fertilizer use, and increase organic fertilizer application. Upon detecting diseased plants in fields, remove them immediately, cease watering, and disinfect the surrounding soil.

Fig. 1.1 Lily Wilt Disease Symptoms. (**a–e**) Field symptoms in Lanzhou lilies. (**f**) Rhizome symptoms in Lanzhou lilies. (**g**, **h**) Field symptoms in Dragon's Tooth lilies. (**i**) Bulb symptoms in Dragon's Tooth lilies

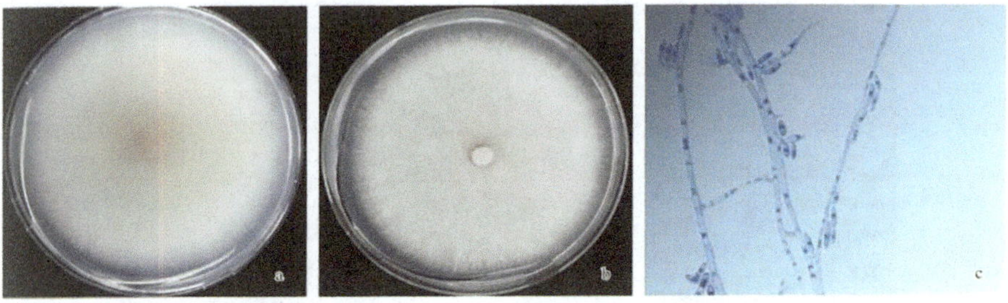

Fig. 1.2 Morphology of *Fusarium solani*. (**a**, **b**) Cultural characteristics. (**c**) Macroconidia

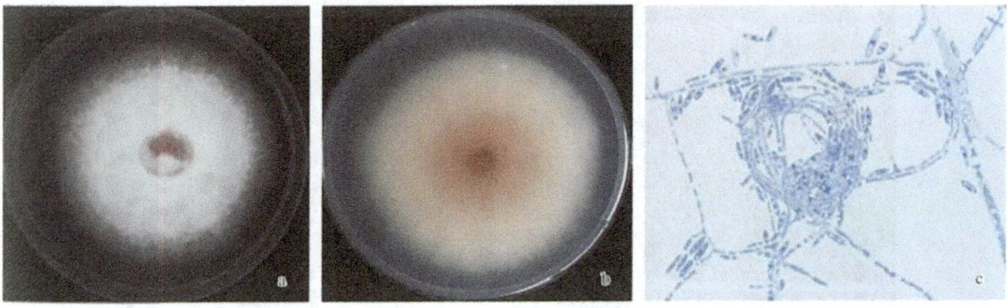

Fig. 1.3 Morphology of *Fusarium oxysporum* f. sp. lilii. (**a**, **b**) Cultural characteristics. (**c**) Macroconidia

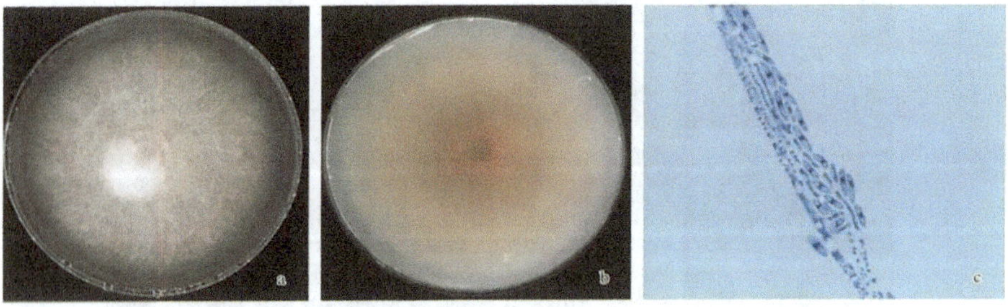

Fig. 1.4 Morphology of *Fusarium fragrans*. (**a**, **b**) Cultural characteristics. (**c**) Macroconidia

2. Chemical Control: Before planting, treat lily bulbs with hot water at 39 °C for 2 h, then immerse them in a 2000-fold dilution of 50% thiophanate-methyl wettable powder (WP) for 30 min. At the onset of the disease, apply treatments such as a 500-fold dilution of 50% carbendazim WP, an 800-fold dilution of 70% thiophanate-methyl WP, an 800-fold dilution of 50% mancozeb WP, a 100-fold dilution of 50% metalaxyl-M WP, or an 800-fold dilution of 30% prochloraz WP. Alternate these treatments and water (or spray) the roots every 7 days, continuing for 3–4 cycles.

1.2 Lily Anthracnose

Lily anthracnose is a significant fungal disease in lily cultivation, occurring both during the growing season and in the storage period of bulbs. This disease is widespread with a high incidence rate, typically resulting in a yield loss of 10% to 20% in general affected areas, and even exceeding 40% in severely affected regions.

Symptoms
Lily anthracnose affects bulbs, stems, leaves, flower stalks, and flowers. The disease first manifests on the outer scales of bulbs, starting with pale red, irregular lesions that clearly demarcate healthy tissue. These lesions deepen to dark brown and harden, eventually causing the entire scale to shrink and turn black-brown. On stems, long, black-brown lesions appear, which may produce tiny black dots in high humidity. Leaves develop irregular or oval lesions, gray-brown in the center and slightly indented. In humid weather, many black, scattered specks grow on the lesions. In severe cases, diseased leaves dry up and fall off. Infected flower stalks show brown soft rot, and infected petals develop several oval or irregularly shaped pale brown lesions, which further expand, causing the flower tissues to thin and decay (Fig. 1.5).

Pathogen
The causative agents of lily anthracnose belong to the genus *Colletotrichum* (*Colletotrichum* spp.), including *Colletotrichum lilii* [*C. lilii* Plakidas ex Boerema et Hamers], *Colletotrichum incanum* (H. C. Yang, J. S. Haudenshield, et G. L. Hartman), and *Colletotrichum spaethianum* [*C. spaethianum* (Allesch.) Damm, P. F. Cannon et Crous]. When cultured on PDA medium, colonies initially appear pale yellow, later turning light brown to dark brown. The mycelium is velvety, gray-white. Conidia are short cylindrical or falcate, smooth-surfaced, thin-walled, hyaline, sometimes containing oil droplets, with blunt rounded tips. Appressoria are pale brown to brown, flattened, with smooth margins (Fig. 1.6).

Fig. 1.5 Symptoms of Anthracnose in Lily

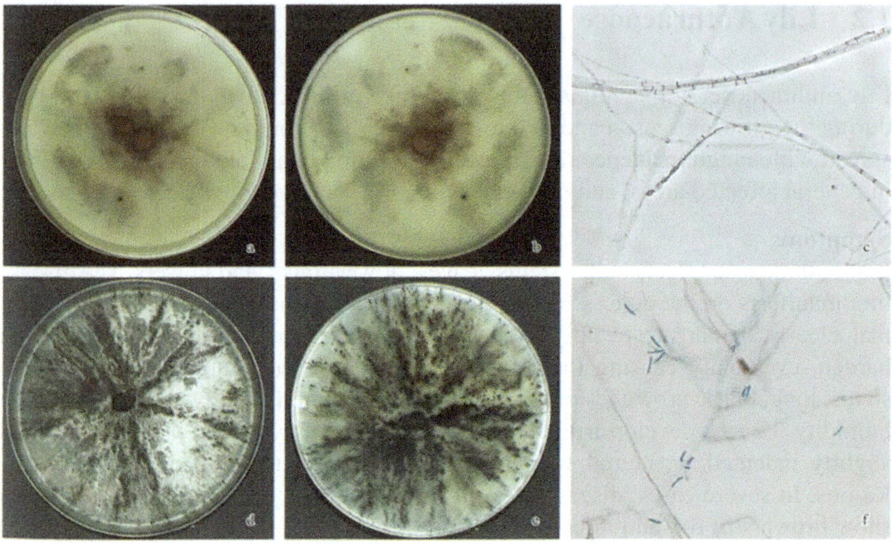

Fig. 1.6 Morphology of Lily Anthracnose pathogen. (**a**, **b**) Cultural characteristics of *Colletotrichum incanum*. (**c**) Mycelium of *Colletotrichum incanum*. (**d**, **e**) Cultural characteristics of *Colletotrichum spaethianum*. (**f**) Mycelium of Colletotrichum spaethianum

Occurrence Pattern

The pathogen survives as mycelia or as sclerotia on diseased plant remnants in the soil, with infected bulbs also serving as a significant source of infection. The pathogen can survive on plant debris for 10–15 months, producing conidia under suitable conditions. These conidia are spread by wind and rain, infecting through wounds or natural openings, or directly via appressoria. During the high incidence period in May and June, conidia formed on diseased tissues can reinfect. Key factors for epidemic development are temperature, humidity, and light conditions; high humidity and continuous light exposure exacerbate infections.

Control Measures

1. Seedling Disinfection: Use sterile bulbs and ensure they are properly disinfected.
2. Field Management: Promptly remove and destroy plant debris to reduce the initial sources of infection. Implementing crop rotation with nonhost crops can reduce the likelihood of pathogen infection.
3. Chemical Control: At the onset of disease, apply treatments such as a 1500-fold dilution of 25% metalaxyl-M emulsifiable concentrate (EC), an 800-fold dilution of 50% metalaxyl-M WP, a 1600-fold dilution of 25% prochloraz EC, or a 1600-fold dilution of 30% prochloraz EC. Alternate these treatments and apply every 7–10 days based on disease progression, for a total of three to four applications.

1.3 Lily Black Spot Disease

Lily black spot is a common fungal disease on lilies, widely distributed and causing severe outbreaks where the wilting rate of plants can reach 50% to 70%. This significantly impairs the normal development, yield, and quality of lilies.

Symptoms
Lily black spot primarily affects the leaves of the plant. Initial symptoms appear as water-soaked spots that later develop into nearly circular, oval, or irregularly shaped lesions. These lesions are brown, gray-brown, or dark brown, often surrounded by a yellow halo. Black, mold-like substances with concentric rings appear on the surface of the lesions. As the disease progresses, lesions can spread across the entire leaf, leading to severe cases where the leaves drop off and the entire plant turns yellow and withers (Fig. 1.7).

Pathogen
The causative agent of lily black spot is the asexual stage of the ascomycete Alternaria alternata (Fr.) Keissl. When cultured for 3 to 5 days, the underside of the colony appears white to brown with a light brown halo, presenting a velvety or powdery texture. After 7 days, the mycelium turns from gray to black and grows rapidly. Conidiophores are either solitary or clustered, brown, erect or curved, and septate. Conidia are produced in chains or singly, egg-shaped, inverted baton-shaped, or nearly oval, brown, with three to six transverse and one to four oblique or longitudinal septa, slightly constricted at the septa. Conidia measure (22.5–40) × (8–13.5) μm, have a short columnar or conical beak, are pale brown, uniseptate or with one septum, with most capable of transforming into a pseudobeacon and producing secondary conidia (Fig. 1.8).

Occurrence Pattern
The pathogen survives winter as mycelium and conidia in the soil or on diseased plant debris, with infected bulbs also serving as significant sources of infection. In the spring, under favorable environmental conditions, the mycelium and sclerotia produce conidia, which are spread by wind and rain, splashing onto lower leaves. The pathogen has a very short incubation period, manifesting symptoms 2 to 3 days after infection and quickly spreading upward. New conidia produced on diseased parts spread the infection further. The disease thrives in rainy seasons or when air humidity is high and plants are under-fertilized.

Control Measures
1. Field Management: Practice crop rotation where feasible. Plant densely and promptly remove and destroy diseased leaves.
2. Chemical Control: At the initial stage of infection, spray with a 400–500 times dilution of 40% difenoconazole suspension concentrate (SC), a 1000 times dilution of 50% difenoconazole WP, or a 1500 times dilution of 80% allicin EC. Apply approximately every 10 days, repeating three to four times.

Fig. 1.7 Symptoms of Lily Leaf Spot Disease

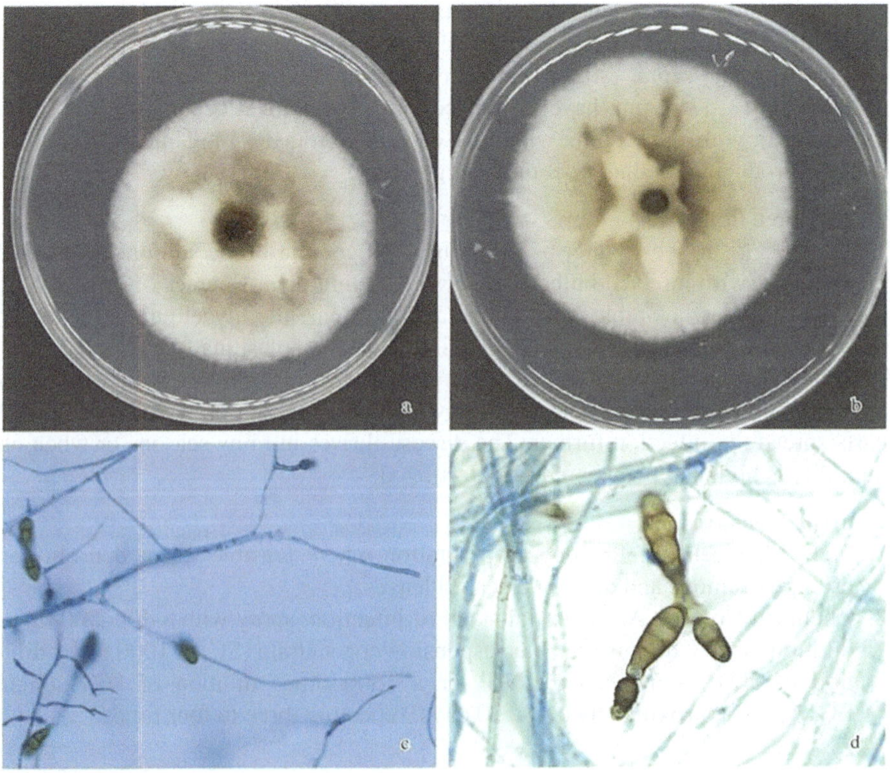

Fig. 1.8 Morphology of Lily Leaf Spot Pathogen. (**a, b**) Culture characteristics. (**c, d**) Conidia and mycelium morphology

1.4 Lily Gray Mold Disease

Lily gray mold is one of the significant fungal diseases in lily production. It is widespread, with a general incidence rate of 5% to 20%, and in severely affected areas, it can reach over 40%.

Symptoms

Lily gray mold primarily affects the stems and leaves of lilies. Affected stems display oval lesions, with the impacted parts turning brown and constricting, which may lead to breaking. The growth points soften and decay, and in severe cases, the entire plant dies. The leaves show symptoms initially at the tips and edges, forming yellow or yellow-brown spots. These lesions are circular to oval, surrounded by water-soaked edges, and expand rapidly under suitable climatic conditions, displaying concentric rings. The centers of the lesions start to whiten and thin but do not split. In humid weather, a gray mold layer forms on the affected parts. During hot and dry conditions, the lesions dry out and thin, turning light brown and gradually expand, causing the entire leaf to die. Some flowers and individual bulbs also become infected, leading to rot. In the later stages, black, granular sclerotia are visible on the lesions (Fig. 1.9).

Pathogen

The pathogen of lily gray mold is the asexual form of the ascomycete Botrytis elliptica (Berk.) Cooke. Initially, colonies are white to gray-white with rings and the mycelium is cottony. Later, small black sclerotia are produced, and the growth rate is fast. After sclerotia germinate, they produce conidiophores; conidiophores are erect, simple-branched at the top, gray-white, with conidiogenous cells enlarging.

Fig. 1.9 Symptoms of Lily Gray Mold

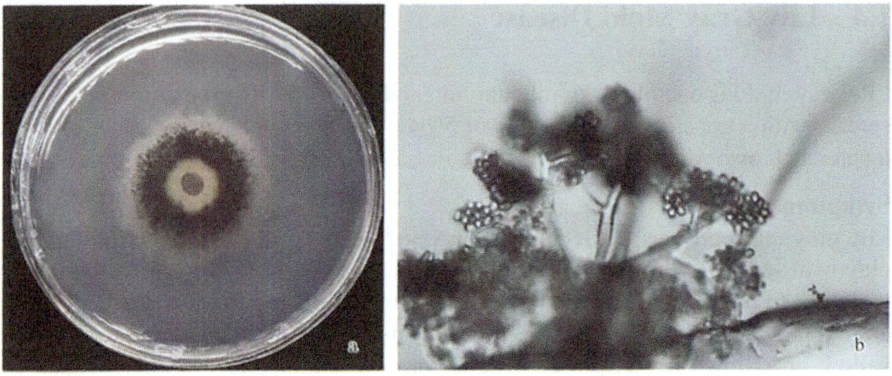

Fig. 1.10 Morphology of Lily Gray Mold Pathogen. (**a**) Culture characteristics. (**b**) Conidia morphology

Conidia are unicellular, transparent, oval or inverted-oval, with a wrinkled surface, measuring (18.1–38.4) × (16.4–27.2) μm, with a length to width ratio of 1.31 (Fig. 1.10).

Occurrence Pattern

The pathogen primarily overwinters as mycelium and sclerotia on diseased plants and plant debris. Bulbs used for planting can also carry the disease. In the following spring, when environmental conditions are favorable, the mycelium and sclerotia produce conidia. Conidia germinate fastest at about 16 °C and over 95% humidity, spread by wind and rain, forming the initial infection. The disease spreads rapidly when temperatures reach around 22 °C and humidity exceeds 90%. Cool, moist environments favor conidia production, with outbreaks beginning around April and becoming widespread by mid to late May.

Control Measures

1. Site Selection: Choose sites where previous crops were not lilies, tomatoes, onions, leeks, garlic, sweet potatoes, or other crops susceptible to gray mold, and ensure good drainage. Avoid planting in soils high in nitrogen, phosphorus, or potassium. Opt for fertile, well-drained, and well-aerated soils.
2. Cultural Control: Timely deep plowing and sunning of ridges to a depth of over 25 cm, followed by crushing and leveling after drying.
3. Chemical Control: Options for chemical treatment include a 600-fold dilution of 500 g/L iprodione SC, a 1000-fold dilution of 10% benomyl water-dispersible granules (WG), a 1500-fold dilution of 50% procymidone WP, a 1500-fold dilution of 43% propiconazole SC, and a 500-fold dilution of 64% metalaxyl-mancozeb WP. Rotate these sprays, applying every 7–10 days, for two to three cycles.

1.5 Lily Virus Diseases

As the cultivation area for lilies continues to expand, the prevalence of viral diseases has increasingly become a concern, with natural infection rates typically ranging from 20% to 30%. In severely affected areas, rates can reach 80% to 90%, causing significant economic losses to the lily cultivation industry.

Symptoms
Symptoms of lily virus diseases include yellowing of leaves, loss of green pigmentation, mottling, color breaking, variegation, prominent veining, necrotic spots, deformities, stunted growth, deformed flowers, or color breaking in flowers. Some buds may fail to open at all (Fig. 1.11).

Pathogen
Currently, more than 20 viruses are known to infect lilies. The most widespread include Lily Symptomless Virus (LSV), Cucumber Mosaic Virus (CMV), and Lily Mosaic Virus (LMoV). LSV, part of the *Carlavirus* genus, specifically parasitizes lilies. It has elongated, flexuous rod-shaped particles approximately 640 nm long and 17–18 nm in diameter, with a distinctive grooved structure along the axis. CMV belongs to the *Cucumovirus* genus within the Bromoviridae family. Its particles are isometric icosahedral, non-enveloped, about 29 nm in diameter, and consist of single-stranded positive-sense RNA. LMoV, a member of the Potyvirus genus within the Potyviridae family, has flexuous filamentous particles, measuring about 740 nm × 14 nm (Fig. 1.12).

Occurrence Pattern
In nature, LSV can be transmitted nonpersistently by peach and lily aphids and persistently by whiteflies. It can also be spread through leaf contact and mechanical damage. This virus spreads very rapidly and is prevalent in commercial lily bulbs. LSV primarily spreads during June and July, with slower transmission rates in May and from August to September. CMV can be spread by aphids and mechanical damage, with over 60 aphid species capable of transmitting this virus. After spring vegetable germination, aphids begin to become active and migrate, serving as the main vectors for CMV. The optimal temperature for disease outbreak is around 20 °C, with infections often remaining latent above 25 °C. The virus typically overwinters in perennial vegetables, field weeds, and trees. LMoV is mainly transmitted through plant sap and aphids, with peach aphids being the primary vectors. Bulbs can also spread the virus.

Control Measures

1. Strict Quarantine: Implement stringent tracking and monitoring of imported lily bulbs and enhance postentry quarantine supervision to prevent the entry of infected bulbs from abroad. Strengthen transport and origin quarantine in both affected and non-affected areas domestically. Strictly control and isolate affected areas to prevent the spread through bulbs.

Fig. 1.11 Symptoms of Lily Virus Disease

2. Production of Virus-Free Seedlings: Establish a virus-free seedling production line using healthy lily bulbs for propagation. This addresses the issue of virus accumulation due to self-propagation by farmers. Expose lily seedlings to 38 °C with 16 h of light and 25 °C with 8 h of darkness for 15 days; use meristem tips

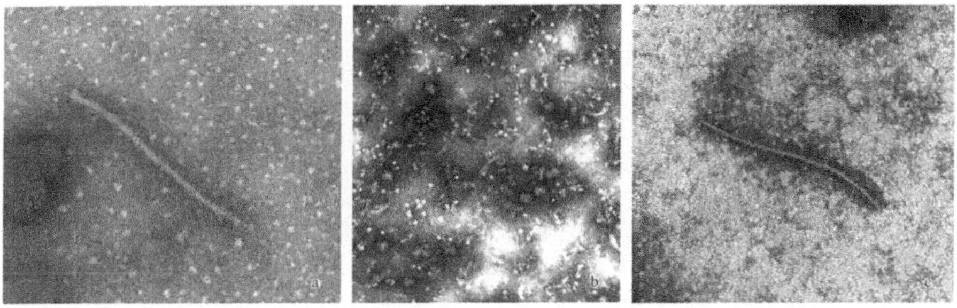

Fig. 1.12 Morphology of Lily Virus Pathogens. (**a**) LSV (Lily Symptomless Virus) particles. (**b**) CMV (Cucumber Mosaic Virus) particles. (**c**) LMoV (Lily Mottle Virus) particles

measuring 0.8–1.2 mm, cultured in a medium with a virus concentration of 6 mg/mL for virus elimination.

3. Agricultural Control: Where feasible, use insect-proof nets to block the spread of virus-carrying insects and avoid mixed planting with vegetables and fruit trees that are susceptible to viruses. Promptly remove infected weeds from the surroundings. Ensure that agricultural tools are disinfected to prevent cross-contamination. Avoid agricultural practices that may cause wounds to plants.
4. Chemical Control: Chemical treatments primarily include controlling vectors and preventing viral diseases. Before or at the initial outbreak of the virus, apply a 700-fold dilution of 5% amino oligosaccharide aqueous solution (AS) mixed with a 4000-fold dilution of 50% dinotefuran water-soluble powder (SP), an 800-fold dilution of 2% ningnanmycin AS mixed with a 2000-fold dilution of 50% picoxystrobin water-dispersible granule (WG), or a 1000-fold dilution of 6% oligosaccharide-peptide WP. Depending on the severity of the outbreak, spray every 7 to 10 days, for a total of two to three applications.

Chapter 2
The Diseases of Fritillaria

Fritillaria, belonging to the lily family, has been a valued medicinal herb in traditional Chinese medicine for centuries. The "Pharmacopoeia of the People's Republic of China" includes several species such as *Fritillaria cirrhosa*, *F. maximowiczii*, *F. pallidiflora*, *F. walujewii*, *F. thunbergii*, and *F. hupehensis*.

Fritillaria is a perennial herbaceous plant, prized medicinally for its dried underground bulbs. It predominantly grows in the middle and lower Yangtze River regions, the Qinghai-Tibet Plateau and adjacent Hengduan Mountains, Xinjiang, and the Northeast of China. Fritillaria thrives in riverbeds, grassy slopes, gravel cracks, or under forests, and is renowned for its properties to cool, moisten the lungs, suppress coughs, and resolve phlegm.

The primary diseases affecting Fritillaria are rust, root rot, and gray mold, which have significant impacts on both yield and quality.

2.1 *Fritillaria maximowiczii* Rust Disease

Known also as yellow rust, this disease is one of the most severe afflictions of *F. maximowiczii*, with a widespread occurrence. Annually, the infection rate stands at 40% to 70%, but in severely affected areas, it can exceed 90%. This disease mainly infects the stems and leaves, causing premature withering and death of the above-ground parts, which severely impacts the yield and quality of *F. maximowiczii*.

Symptoms

The disease primarily targets the stems and leaves, leading to early plant wilting. Initially, yellow, elongated spots appear on the underside of the leaves and the lower parts of the stems. Soon after, golden yellow rust sori emerge on these spots. Upon bursting, they release yellow, powdery rust spores that disperse with the wind. The affected areas develop holes, and the stems and leaves turn yellow and wither. In the

© China Agriculture Press Co., Ltd. 2026
T. Wang et al., *Atlas of Chinese Medicinal Plant Diseases*,
https://doi.org/10.1007/978-981-95-6326-5_2

later stages, the stems and leaves are covered with dark brown pustules, which are the pathogen's winter sori, leading to gradual wilting and premature death of the plant (Fig. 2.1).

The causative agent of *F. maximowiczii* rust disease is *Uromyces aecidiiformis*[*Uromyces aecidiiformis* (*F. Strauss*) Rees].

Pathogen
[(Synonym: U. lili (Link.) Fuck.], a basidiomycete fungus. The aecia are spherical, yellow-brown, located between rust cavities. Rust cavities predominantly form on the underside of leaves, petioles, and stems, turning yellow. When mature, they split open. The rust spores are nearly spherical, yellow, and warty, measuring 24–32 μm in diameter. The winter spores are oval or elongated oval, single-celled, brown, with a nipple-like protrusion at the tip and small warts, measuring (24–45) μm × (19–28) μm. The stalk of the winter spore is colorless and easily detached (Fig. 2.2).

Occurrence Pattern
The pathogen overwinters in the form of winter spores on plant debris, serving as the primary source of infection for the following year. After germination, these spores infect *F. maximowiczii*, producing aecia and rust cavities and forming rust spores that continue to infect the plant. Summer spores do not appear on *F.*

Fig. 2.1 Symptoms of Rust Disease in *Fritillaria ussuriensis*

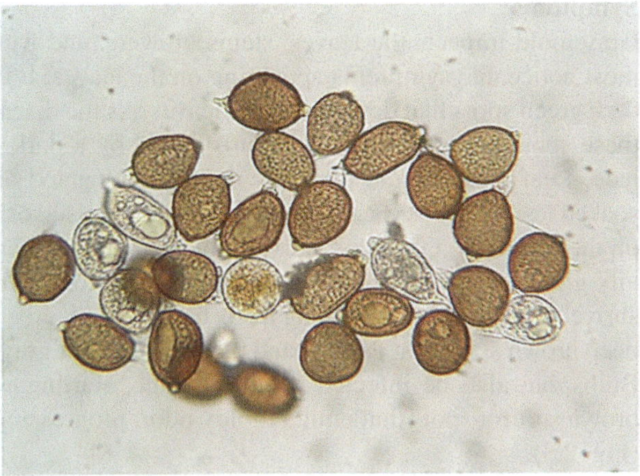

Fig. 2.2 Morphology of *Fritillaria ussuriensis* Rust Pathogen Teliospores

maximowiczii. The disease outbreak is correlated with temperature, humidity, rainfall, soil texture, planting density, and the duration of cultivation. Typically, in the Northeast region, the disease starts around early May when temperatures reach 15–18 °C, with peak infection occurring in mid to late May.

Control Measures

1. Field Hygiene: Reduce the base number of overwintering pathogens by destroying plant debris in autumn and winter to decrease the number of overwintering winter spores.
2. Cultural Management: Maintain an appropriate planting density and avoid overly dense cultivation. Ensure timely drainage after rain to reduce field humidity. Apply phosphorus and potassium fertilizers judiciously, control nitrogen application, and enhance plant disease resistance. Timely weed control can also reduce disease incidence.
3. Chemical Control: At the onset of the disease, apply a 300–500-fold dilution of 15% triadimefon WP, an 800 fold dilution of 70% thiophanate-methyl WP, or a 400-600 fold dilution of 20% myclobutanil EC. Spray once every 7–10 days, for a total of three applications.

2.2 Fritillaria Gray Mold Disease

Gray mold disease, also known as "early blight," "ring spot," or "green collapse," is a significant affliction on Zhe Bei Mu. This disease has a broad occurrence, severely affecting leaves, stems, flowers, and fruits, and often causing substantial economic losses to growers, with yield reductions ranging from 10% to 30%.

Symptoms

Gray mold impacts the leaves, stems, flowers, and fruits of Zhe Bei Mu, with the most noticeable symptoms appearing on the leaves. Initial symptoms include small dark green spots that develop on the leaves. As the disease progresses, the centers of these spots turn yellow-brown, surrounded by a dark green border with a yellow halo. Leaf margins and tips may develop elongated or irregularly shaped water--soaked lesions that expand continuously. In later stages, gray moldy growth appears on these lesions. On stems, early symptoms start as dark green lesions that eventually girdle the stem, leading to plant death. Affected flowers turn yellow-gray and shrivel without opening. Young fruits become dark green, while larger fruits develop deep brown spots that expand and may also exhibit gray mold during high humidity. Bulbs can also be infected during storage, starting with water-soaked, white to brown soft rot spots that emit a moldy odor, progressing to a gray-white mold layer (Fig. 2.3).

Pathogen

The causal agent of gray mold in Zhe Bei Mu is Botrytis elliptica (Berk.) Cooke. This fungus produces erect, light brown to brown conidiophores with several septa, terminating in branches that cluster many conidia. Conidia are single-celled, colorless to pale brown, predominantly oval, occasionally spherical, with one pointed end, measuring 16–35 microns by 10–24 microns. Sclerotia are small, about 0.5–1 mm in diameter (Fig. 2.4).

Occurrence Pattern

The pathogen overwinters as sclerotia in plant debris or soil, germinating in the spring to produce conidia, which serve as the primary infection source. As temperatures reach 19–23 °C in early April and coincide with rainy conditions, the conidia germinate and spread rapidly. By late April, as temperatures stabilize around 20 °C and if consecutive rainy, muggy days follow, the infection rate can reach 100% by early May, severely impacting Zhe Bei Mu yields.

Control Measures

1. Cultural Practices: Implement deep plowing and crop rotation, use disease-free bulbs for planting, and establish disease-free seed fields. During hot summers, cover the soil with plastic film to utilize solar energy for soil sterilization.
2. Chemical Control: At the onset of symptoms, apply 40% pyrimethanil SC at an 800–1,000-fold dilution, 50% iprodione WP at a 1,000-fold dilution, or 50% carbendazim WP at a 1,500-fold dilution. Spray every 7–10 days, repeating two to three times to effectively control the disease.

2.3 Fritillaria Root Rot

Root rot disease, also known as dry rot, is a soil-borne disease transmitted through soil moisture, subterranean insects, and nematodes. It occurs in the main production areas of Zhe Bei Mu, with an incidence rate generally between 10% and 30%. In

Fig. 2.3 Symptoms of Gray Mold Disease in *Fritillaria thunbergii*

severe cases, especially from September to November following autumn planting, the bulbs rot in the soil, leading to a high rate of missing seedlings reaching 40% to 60%. From May to August the following year, root rot during the summer months can prevent seed preservation, resulting in catastrophic losses.

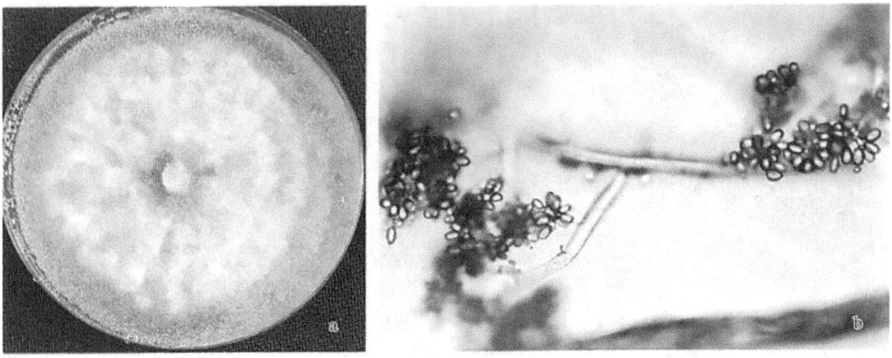

Fig. 2.4 Morphology of *Fritillaria thunbergii* Gray Mold Pathogen. (**a**) Culture characteristics. (**b**) Conidia

Symptoms

Root rot in Zhe Bei Mu primarily affects the roots, causing partial or complete decay; the stems and leaves wilt and droop due to the inability to absorb water, eventually leading to the death of the entire plant. The disease mainly occurs during the seedling phase, but mature plants can also be affected. Early symptoms include individual feeder roots and root hairs showing signs of disease, gradually spreading to the main roots. Initially, there are no visible symptoms on the plant, but as the disease progresses, wilting appears during strong midday sunlight, recovering overnight initially. In the later stages, this wilting does not recover overnight. At this point, the affected bulbs appear honeycombed, with the base or sides showing dark black or blue-black spots, or decaying into irregularly sized cavities of black-brown to blue, ultimately resulting in plant death (Fig. 2.5).

Pathogen

The causal agent of Zhe Bei Mu root rot is the blue variant of the *Fusarium* species, *Fusarium solani* (Mart.) Appel et Wollenw. ex Snyder et Hansen var. coeruleum Sacc. Cultured on PDA medium for 2–3 days, the fungus exhibits vigorous growth with gray-white, fluffy mycelium; the colony's reverse appears felt-like and deep purple-blue, generally producing both large and small conidia. The small conidia are ovate, zero to one septate, borne on elongated conidiophores; the large conidia are slightly curved, blunt, three to five septate, measuring 25–30 × 3.8–4.9 μm; the chlamydospores are nearly spherical, pale yellow (Fig. 2.6).

Occurrence Pattern

The pathogen overwinters in the soil and plant debris as mycelium, conidia, or chlamydospores, generally beginning to infect in late March to early April, with May seeing peak infection. The occurrence of the disease is highly dependent on environmental and climatic conditions, most commonly in low-lying, moist areas. Low temperatures, high humidity in seedbeds, and insufficient light are the primary triggers of this disease. Sticky, compacted soils that impede root development are also

Fig. 2.5 Symptoms of Root Rot Disease in *Fritillaria thunbergii*. (**a**, **b**) Field symptoms. (**c**) Leaf symptoms. (**d**) Bulb symptoms. (**e** ~ **h**) Affected bulbs from different angles

Fig. 2.6 Morphology of *Fritillaria thunbergii* Root Rot Pathogen. (**a**, **b**) Culture characteristics. (**c**) Conidia

prone to disease. Furthermore, injuries caused by subterranean pests and nematodes provide entry points for the pathogen.

Control Measures

1. Crop Rotation Is Critical: Avoid continuous cropping and rotate with other crops after a minimum of 3 years to improve the terrain and drainage, and deep plow the soil to reduce disease incidence.
2. Enhanced Cultivation Practices and Plant Resistance: Postharvest in autumn and winter, ensure thorough field sanitation by deeply burying or destroying infected plant debris to reduce infection sources for the following year. Use fertile, well--drained, and aerated soils, preferably sandy loam for seedbeds. Raise the beds to prevent waterlogging and ensure good drainage during the rainy season. Soil disinfection can be done using 21% peracetic acid AS diluted 500-fold. Increase phosphorus and potassium fertilization, and apply dipotassium phosphate foliar sprays two to three times to enhance plant disease resistance.
3. Chemical Control: At the onset of the disease, apply 50% carbendazim WP at a 500-fold dilution, 2.5% prochloraz-mancozeb AS at a 300-fold dilution, or 50% thiram WP at a 500-fold dilution to irrigate the roots, or use a 30% prochloraz AS at a 1,500–3,000-fold dilution to irrigate the infected soil areas.

Chapter 3
The Diseases of Solomon's Seal

The genus *Polygonatum* (Solomon's Seal), belonging to the Liliaceae family, encompasses over 60 species globally. China as the primary center of diversity for this genus, is home to around 40 species. The 2020 edition of the Pharmacopoeia of the People's Republic of China designates three species as the source plants for herbal materials: *Polygonatum kingianum* Collett & Hemsl., *P. sibiricum* Red., and *P. cyrtonema* Hua. Polygonati Rhizoma is derived from the rhizomes of these plants and is regarded as both food and medicine. It has a sweet and neutral property. Its functions include nourishing qi, enriching yin, strengthening the spleen, moisturizing the lungs, benefiting the kidneys. It is used for treating conditions such as qi deficiency of spleen and stomach, fatigue, insufficient stomach yin, dry mouth with poor appetite, dry cough due to lung deficiency, hemoptysis from taxation, deficiency of essence and blood, soreness and weakness in the waist and knees, premature graying of the hair, and internal heat with thirst.

Polygonatum kingianum, commonly referred to as "large Huang Jing" due to its large tubes, is predominantly found in Yunnan, Sichuan, and Guizhou provinces. *P. sibiricum*, known for its slender rhizomes, is called "chicken-head Huang Jing" and is mainly cultivated in the Northeast, North China, as well as Shaanxi, Ningxia, Gansu, Henan, Shandong, Jiangsu, Anhui, Zhejiang, Fujian, Guangdong, Guangxi, Hunan, Jiangxi, and Guizhou. *P. cyrtonema*, often called "ginger-shaped Huang Jing," is primarily produced in Hunan, Hubei, Anhui, Jiangsu, Zhejiang, Jiangxi, Fujian, Sichuan, and Guizhou.

The primary diseases of *Polygonatum* include rhizome rot, anthracnose, gray spot, and southern blight disease of which rhizome rot causes the most severe damage.

© China Agriculture Press Co., Ltd. 2026

T. Wang et al., *Atlas of Chinese Medicinal Plant Diseases*,
https://doi.org/10.1007/978-981-95-6326-5_3

3.1 Rhizome Rot of *Polygonatum*

Rhizome rot is the most severe disease of *Polygonatum*, occurring commonly across its growing regions. Field surveys in Yunnan and Guizhou provinces reveal that the severity of rhizome rot in *Polygonatum* exacerbates with increasing cultivation years, a phenomenon attributed to its perennial growth habit. Fields propagated through mature rhizome division are particularly susceptible compared to those employing seed-derived seedlings. Disease prevalence also correlates with soil texture, climatic conditions, and field management practices. Due to dispersed cultivation patterns, disease incidence varies significantly among sites. Generally, fields with seed-derived seedlings under 3 years show disease incidence below 5%, whereas some areas with clay-heavy soil using rhizome cuttings report incidence exceeding 75% after 5 years of cultivation.

Symptoms

Rhizome rot of *Polygonatum* is a soil-borne disease. Initially, brown water-soaked lesions appear on the rhizomes while the aerial parts remain asymptomatic. As the disease progresses, the rhizomes develop depressions, cankers, cracks, and rots, changing color from yellow-white to reddish-brown. The diseased areas on the rhizome surface develop a black mold layer consisting of the pathogen's conidia. When the disease severely affects the rhizome nodes connected to the stems, it disrupts water transport, leading to wilting, yellowing, and eventual withering of the aerial parts of the plants (Fig. 3.1).

Fig. 3.1 Symptoms of Rhizome Rot Disease in *Polygonatum sibiricum*. (**a**) Massive plant mortality caused by rhizome rot. (**b**, **c**) Rotted rhizomes

Pathogen

The pathogen causing *Polygonatum* rhizome rot belongs to the genus *Colletotrichum* (anamorphic Ascomycota). Ma Weisi et al. (2021) named it *Colletotrichum kingianum*. This fungus grows rapidly on the potato dextrose agar (PDA) medium, reaching a colony diameter of about 8.5 cm after 10 days at 25 °C. The marginal hyphae are initially white, and then gradually darken to grayish-brown. The colony produces black setae with two to four septa, measuring 142.1-243.1 μm in length. The fungus readily produces a large quantity of conidia on PDA medium. Germinating conidia form 1–several-celled colonies visible as orange, viscid conidial masses, typically covering the colony center. Conidiophores are densely packed, unbranched, and terminate in cylindrical to elongate-ampulliform phialides, each producing one conidium apically. The conidia are slightly curved, crescent-shaped, acutely pointed at both ends, aseptate, and contain a single central oil droplet. The size of the conidia is (8.3 - 31.1) μm × (5.2 - 7.0) μm, with a length-to-width ratio of 3.9. The appressoria derived from conidial germination are dark brown, spherical, elliptical, or ovoid, with a size of (8.5 - 20.2) μm × (6.6 - 12) μm. The appressoria formed by the hyphae are often irregular in shape with fissured margins (Fig. 3.2).

Occurrence Pattern

The rhizome rot of *Polygonatum* can occur throughout the year, with infections initiating even before seedlings emergence. Symptoms are typically not apparent on aerial plant parts of the plant until after emergence. The disease is most severe during the rainy season from June to September, with disease incidence exceeding 70% in plots with long-term cultivation or severe disease pressure, leading to extensive mortality.

The primary mode of disease transmission to new areas is through infected planting materials, with plants propagated from rhizome cuts more prone to disease than seed-derived seedlings. The infected tissues produce abundant conidia that are dispersed distances via rainwater and irrigation. Soils that are heavy, poorly drained, and poorly aerated are more susceptible to disease outbreaks. The pathogen overwinters on Polygonatum rhizomes or in the soil.

Control Measures

As an underground disease, rhizome rot of Polygonatum is difficult to detect in its early stages. Symptoms such as wilting and yellowing in the aerial parts of the plant typically indicate that the underground tissues have already sustained significant damage, rendering treatment at this stage generally ineffective. Consequently, prevention should be the primary strategy.

1. Scientific Selection of Planting Sites—It is advisable to choose planting sites with a certain slope, good drainage, and loose soil texture to facilitate effective water management during the rainy season. For flat areas, it is crucial to construct raised beds to prevent waterlogging. Fields previously planted with Polygonatum should undergo at least one year of rotation, while fields with a

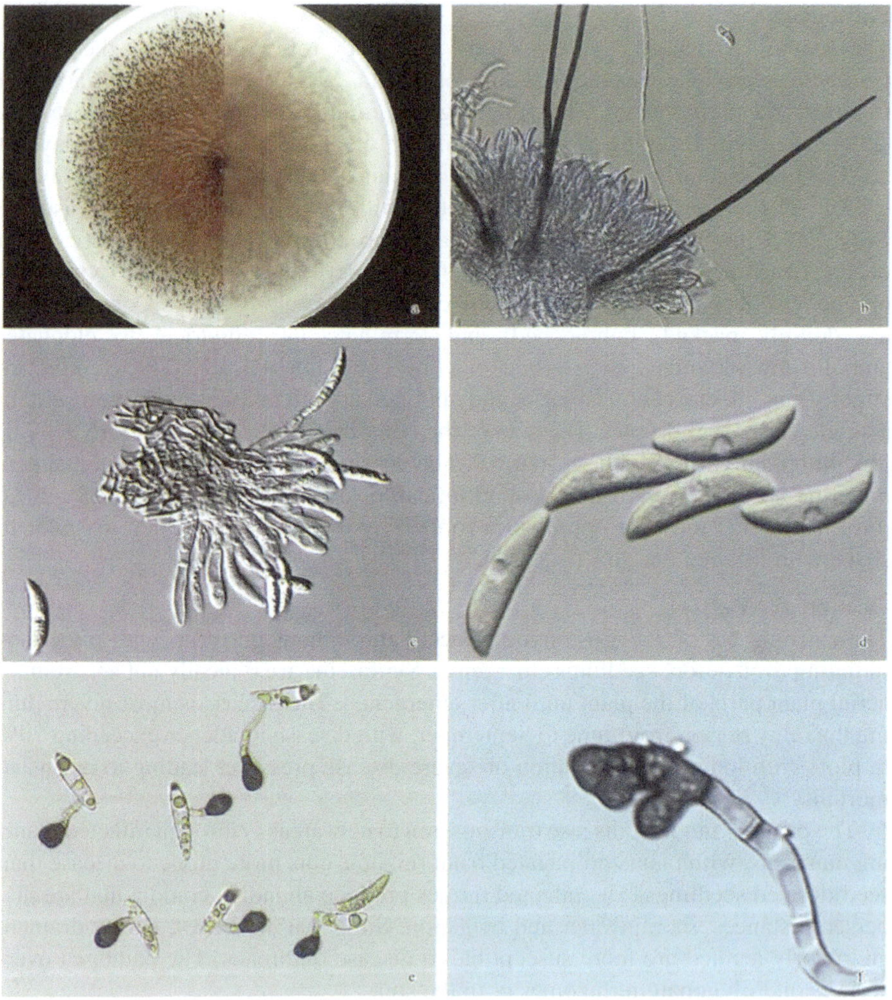

Fig. 3.2 Morphology of *Polygonatum sibiricum* Rhizome Rot Pathogen. (**a**) Culture characteristics. (**b**) Setula and Conidiophores. (**c**) Conidiophores and Conidia. (**d**) Conidia. (**e**) Appressoria derived from conidial germination. (**f**) Appressorium derived from a hypha

history of rhizome rot should be left fallow or planted with non-host crops for at least two years before replanting with *Polygonatum*.

2. Use of Healthy Plant Material—*Polygonatum* can be propagated through rhizome cuttings or seeds. Seed-derived seedlings are generally preferred as planting material. When sourcing planting materials, it is essential to inspect the source of seeds or seedlings to ensure they are from areas free of rhizome rot. Care should be taken during the transportation and storage of seeds and seedlings to avoid conditions that promote disease. Before transplantation, sort the seeds and seedlings, remove the diseased and weak plants, and carry out disinfection.

3. Biological Control—Before sowing *Polygonatum* seeds, treat them with a Bacillus subtilis wettable powder formulation (1 billion CFU/g) at a 1: (10–15) ratio (powder:seed). For seedlings, apply a 600–800x dilution solution of the same formulation evenly to the rhizomes before transplanting, ensuring thorough drying of surface moisture. Upon detecting initial disease symptoms in the field, immediately remove infected plants apply the B. subtilis solution (600–800x) to soil and roots in affected areas repeating every 10–15 days for 2–3 consecutive applications. Avoid mixing B. subtilis with copper-based fungicides or ethyllicin.
4. Chemical Control—Prior to transplanting, select one type of highly effective and low-toxicity protective fungicide and one type of systemic fungicide, and mix them. Disinfect the seedlings through methods such as spraying, seed soaking, or seed coating. Then, place the seedlings in a shady and ventilated place, and wait until the pesticide on the surface of the seedlings dries before planting. Suitable protective fungicides and their application concentrations can be selected as follows: 50% iprodione wettable powder (diluted 600–800 times), 75% chlorothalonil wettable powder (diluted 600-800 times), 80% mancozeb wettable powder (diluted 1200 times), and 50% captan wettable powder (diluted 600-800 times), etc. Systemic fungicides and their application concentrations can be selected as follow: 15% metalaxyl-M wettable powder (diluted 1000–1500 times), 45% sodium ferbam wettable powder (diluted 200 - 500 times), 50% carbendazim wettable powder (diluted 800 - 1000 times), 37% thiophanate - methyl + cyproconazole water - dispersible granules (diluted 3500 - 4500 times), 25% azoxystrobin emulsifiable concentrate (diluted 2000 - 2500 times), etc.

When the disease occurs sporadically in the field, promptly remove the rhizomes of diseased plants. Bury them deeply in a place far away from the field or destroy them with quicklime. Never discard the diseased plants randomly in the field, as this may cause the spread of the disease. Use an 800-fold diluted solution of 15% metalaxyl-M WP, a 200-fold diluted solution of 45% sodium ferbam WP, or a 500-fold diluted solution of 50% iprodione WP. For the diseased plant pits and the rhizosphere of surrounding plants, conduct root irrigation treatment 3 times, with an interval of 7 days. When the disease spreads over a relatively large area in the field, use metalaxyl - M, sodium ferbam, iprodione alone or their compound preparations to conduct root irrigation over the entire field 3 times with an interval of 7 days. If the disease still cannot be effectively controlled, timely excavate Polygonatum sibiricum to minimize losses. The excavated rhizomes cannot be used as propagation materials anymore.

3.2 Anthracnose of *Polygonatum sibiricum*

Anthracnose is one of the primary diseases of *Polygonatum sibiricum*. It commonly occurs in the main production provinces of *P. sibiricum*, such as Yunnan, Sichuan, Guizhou, Hunan, Hubei, and Anhui. The incidence of this disease is influenced by climatic conditions, planting years, and the level of field disease management, resulting in significant variability in disease incidence across different cultivation sites. Typically, the disease occurs in autumn. In some plots, the diseased plant rate exceeds 70%, and the leaf incidence rate of some individual plants reaches 100%. This disease reduces the photosynthetic capacity of the leaves, cause leaf abscission, and makes the plants lodge prematurely, thus adversely affecting the production of *P. sibiricum* as a medicinal herb.

Symptoms

The disease occurs on the leaves of *P. sibiricum*, forming semicircular, circular, or oval lesions on the leaf tips, leaf edges, or the middle of the leaves. These lesions are sunken, reddish-brown with distinct edges, and surrounded by a yellow-green halo. As the disease progresses, the leaf tissue in the center of the lesion dies, fades in color, and is prone to forming perforations. Black dots appear on the lesions, representing clusters of setae and conidia. In fields with mild infection, the plant infection rate is below 10%, but in severely affected fields, the infection rate can exceed 50% during the later growth stage of the plants, with leaf infection rates surpassing 70%. Severe disease causes leaves to wilt and fall off, and the plants to lodge (Fig. 3.3).

Fig. 3.3 Symptoms of Anthracnose Disease in *Polygonatum sibiricum*. (**a** ~ **c**) Red-brown lesions on leaves. (**d**) Clusters of Setula and Conidia on infected area

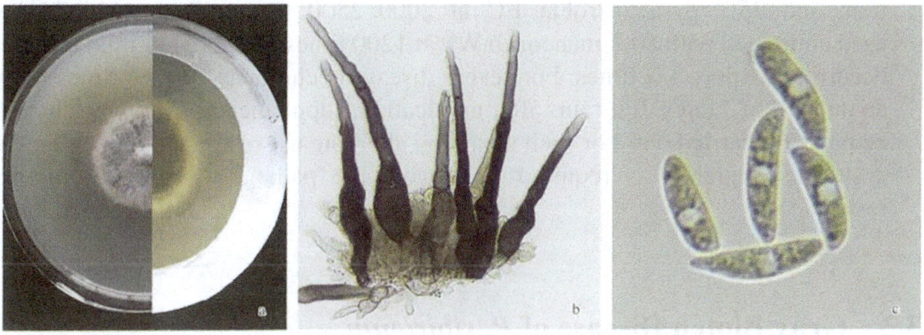

Fig. 3.4 Morphology of *P. sibiricum* Anthracnose pathogen. (**a**) Colony. (**b**) Setula. (**c**) Conidia

Pathogen

The causal agent of *P. sibiricum* anthracnose is the asexual form of the ascomycete *Colletotrichum circinans*. On the leaves, the conidial discs are dark brown; after breaking through the leaf epidermis, they produce abundant dark brown setae. These setae are septate, smooth, with a swollen base and a pointed tip, measuring 106 - 257 micrometers in length, and arranged radially from the conidial discs. The conidiophores are rod-shaped or cylindrical, densely growing on the conidial discs. The conidia are crescent-shaped, pale, unicellular, with a size of (20.4 - 28.5) μm x (6 - 8.5) μm, containing one oil globule. On PDA medium, the colony edges are neat, with relatively well-developed, villous mycelia, which are initially white and later turn gray-brown. Dark brown conidial discs are produced close to the surface of the medium, bearing setae, conidiophores, and conidia; there are no conidia on the aerial mycelia (Fig. 3.4).

Occurrence Pattern

The anthracnose of *P. sibiricum* begins to appear in July and persists until the plants lodge, with August to September being the peak period of the disease. Conidia formed on the lesions are dispersed by air currents, rain, and agricultural activities. The high temperature and high humidity conditions during the rainy season are conducive to the outbreak of this disease.

Control Measures

1. Strengthen Cultivation Management: Plant at a reasonable density, and adopt measures such as erecting trellises, pulling lines, tying branches, and supporting seedlings to promote upright growth and prevent lodging, and at the same time facilitate field ventilation and moisture removal. When the diseases occurs sporadically, remove and destroy the diseased leaves. Clean the field in winter. Clear away the residual branches and fallen leaves and destroy them centrally to reduce the overwintering pathogen.
2. Chemical Control: Chemical control is a crucial method for managing this disease. After the onset of infection, alternate applications of systemic fungicides such as 50% carbendazim WP at 800–1000 times dilution, 25% triadimefon WP at 1500–2000 times dilution, 37% difenoconazole WG at 3500–4500 times dilu-

tion, and 25% pyraclostrobin EC at 2000–2500 times dilution should be used combined with 80% mancozeb WP at 1200 times dilution. For mild disease occurrence, spray 1-2 times. For severe disease occurrence, spray 3 times with an interval of 7 days. If it rains after application, supplementary spraying should be properly carried out. For each spraying, alternate use of the above-mentioned systemic fungicides is required to prevent the pathogens from developing resistance.

3.3 Gray Blotch Disease of *P. sibiricum*

Gray blotch disease is one of the primary foliar diseases during the growth period of *P. sibiricum*. It occurs in southwestern provinces of China, such as Guizhou, Sichuan, and Guangxi, with Yunnan as the core area, mainly affecting *P. kingianum*. In some seriously affected plots, the disease incidence rate of plants reaches 100%. The pathogen competes with the plant for nutrients, reducing the photosynthetic capacity of the leaves, causing leaf abscission, and making the plant senesce prematurely, thus affecting the production of medicinal materials.

Symptoms

The pathogen typically infects the leaf tips and edges to form dark brown lesions. The lesions are semicircular or nearly circular, with a diameter of 11 - 23 mm, showing concentric rings patterns. The edges are distinct and accompanied by a yellow halo. A gray-brown mold layer can be seen on both the upper and lower surfaces of the leaf lesion (more on the lower surface). Multiple scattered lesions can form on a single leaf. As the disease worsens, the lesions may coalesce into patches. The leaf tissue outside the lesions turns yellow, and the leaves gradually wither and fall off. When leaf wilting is severe, it can cause premature lodging (see Fig. 3.5).

Pathogen

The causative agent of *P. sibiricum* gray blotch disease is a fungus of the asexual form of *Cercospora* (*Cercospora* sp.) in the class Ascomycetes. On PDA culture media, it grows slowly; after 30 days of cultivation at 25°C, the colony reaches about 5 cm in diameter, green, with mycelium embedded in the medium forming a dense, nearly circular colony with radiating grooves and a dark brown underside with 7-8 fissures. The pathogen does not easily produce conidia on PDA culture media. For newly isolated pathogens, a small number of conidia can be seen on the colonies of the initial transfer, and after multiple transfers, the colonies no longer produce conidia.

On the underside of infected leaves, the pathogen forms dense and black conidiomata. On these conidiomata, 30-50 conidiophores are clustered and borne. The size of the conidiophores is (42.2-183.6) μm x (3.1-9.6) μm, brown, unbranched, with 1 to 3 septa and 0 to 6 knee-like bends. The hila are significantly thickened, with a diameter of 1.1 to 3.2 μm. Conidia are solitary, with a size of (26.3-86.3) μm x (4.4-8.0) μm, colorless or light brown, often curved, cylindrical to obclavate, with

Fig. 3.5 Symptoms of Gray Spot Disease in *Polygonatum sibiricum*

3-9 (mostly 6-8) septa, blunt at the tip. The hila are significantly thickened, with a diameter of 1.2–2.9 μm, and the top is slightly narrowed to a rounded tip (see Fig. 3.6).

Occurrence Pattern

Gray blotch disease of *Polygonatum sibiricum* tends to occur in the rainy season from June to October. High temperature and high humidity are conducive to the outbreak. The pathogen produces a large number of conidia on the surface of *Polygonatum sibiricum* leaves. These conidia are spread by media such as air currents, rain, irrigation water, humans, and insects. If fungicidal spraying is not carried out in a timely manner after the disease occurs, it can cause large-area incidence in a short period, with plant infection rates in some areas reaching 100% and leaf infection rates exceeding 90%.

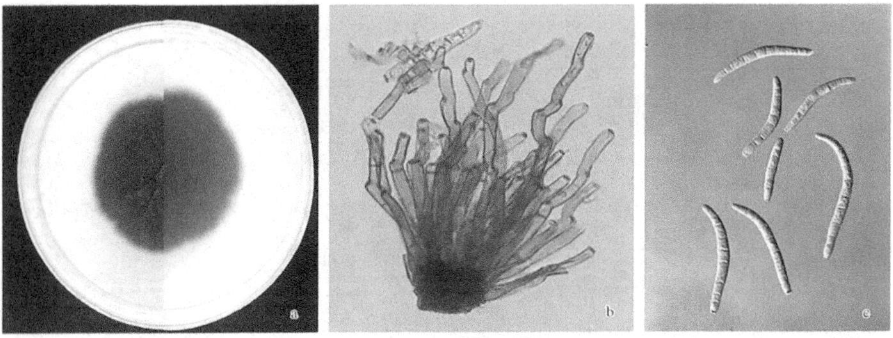

Fig. 3.6 Morphology of *Polygonatum sibiricum* Gray Spot Pathogen. (**a**) Culture characteristics. (**b**) Conidiophores. (**c**) Conidia

Control Measures

1. Strengthen Cultural Practices: Practice rational close planting and remove weeds in a timely manner. During the rainy season, enhance field ventilation to reduce air humidity. Infected plants and leaves should be promptly destroyed or buried deeply in a timely manner. Winter field sanitation, including the removal of all plant debris, is crucial to reduce overwintering pathogens.
2. Chemical Control: Chemical control is a main and effective measure for managing this disease. Once the disease occurs, alternately select fungicides such as 50% carbendazim WP (800-1000 times dilution), 25% triadimefon WP (1500-2000 times dilution), 37% difenoconazole AS (3500-4500 times dilution), and 25% pyraclostrobin EC (2000-2500 times dilution). Then, tank-mix them with 80% mancozeb WP (1200 times dilution), and spray 2-3 times. When spraying, ensure that the undersides of the leaves are covered with the fungicide.

3.4 Sclerotinia Disease of *Polygonatum*

Sclerotinia blight of *Polygonatum* is commonly found in southern provinces of China, with a wide host range, over 60 plant families and more than 200 plant species are known to be susceptible. In recent years, during the hot and rainy seasons from June to October, in producing areas such as Hunan, severe outbreaks frequently occur in *Polygonatum* planting bases, with an infection rate of about 30%. Once the disease becomes epidemic, it can lead to large-scale withering and death of *Polygonatum* plants, severely impacting both the yield and quality of the medicinal material.

Symptoms

In the field, the symptoms of infected *Polygonatum* plants mainly manifest as water-soaked, brown lesions at the stem base. Dense white mycelium and numerous white-to-brown sclerotia are scattered around the lesions, and the mycelium can

Fig. 3.7 Symptoms of Southern Blight Disease in *Polygonatum sibiricum*

spread to the soil surface. The aerial parts of the plant wilt, leaves fall off, and eventually, the entire plant dies. The underground parts rot and the cortex easily peels off (Fig. 3.7).

Pathogen

The causative agent of *Polygonatum* southern blight is Sclerotium delphinii Welch, an asexual form of basidiomycete in the genus Sclerotium. The mycelium is white, aerial, growing radially outward, with an average growth rate of (20.54 ± 0.52) mm/day. White small sclerotia can be observed 5 days after inoculation in the culture medium, and brown mature sclerotia can be observed after 14 days. The number of mature sclerotia produced per petri dish is 8 to 23, and the size of the sclerotia is (5.95 ± 2.34) mm x (7.51 ± 2.88) mm. The hyphae have septa and clamp connections can be observed, while no spores are found. The optimal growth temperature is 25–30°C and the most suitable growth pH is 5–9 (Fig. 3.8).

Occurrence Pattern

The pathogen overwinters in the soil or on infected roots as mycelium or sclerotia. When the temperature is suitable the following year, new mycelium is produced. The pathogen can spread in the soil with surface water flow, and the mycelium spreads in the soil by growth to infect rhizomes of *Polygonatum*. The disease mostly occurs in hot and rainy seasons. High temperature and high humidity are important conditions for the disease. In addition, it is prone to occur in acidic to neutral, sandy soils. High soil moisture is conducive to the occurrence of the disease. Especially

Fig. 3.8 Morphology of *Polygonatum sibiricum* Southern Blight Pathogen. (**a**) Characteristics after 3 days of culture. (**b**) Characteristics after 7 days of culture. (**c**) Mycelium

after continuous drought followed by rain, it can promote the germination of sclerotia and increase the chance of infecting hosts. In continuous cropping fields, due to the accumulation of a large number of pathogens in the soil, the disease is more serious. In clay soils, in plots with poor drainage, insufficient fertility, and high planting density, the disease is serious.

Control Measures

1. Crop Rotation: Fields with infection rates reaching 10% should implement crop rotations. Generally, a 2-3-year crop rotation is carried out, and severely diseased fields should undergo crop rotations for more than three years.
2. Strengthen Cultural Management: It is crucial to promptly remove diseased plant debris. After harvest, deeply plow fields to expose it to freezing, reducing the overwintering pathogen sources in the field. Properly adjust the planting density to avoid overly high density that cause poor ventilation and leads to high temperature and humidity in the planting base. During the rainy season, promptly clear drainage channels after rainfall to ensure smooth drainage.
3. Chemical Control: Options include using 30% prochloraz aqueous solution (500-fold dilution), 29% lime sulfur mixture (400-fold dilution), 10% silicon - fluoride aqueous emulsion (8000-fold dilution), or biological agents such as *Bacillus subtilis*.

3.5 Miscellaneous Diseases of *Polygonatum*

Polygonatum occasionally suffers from diseases such as root rot caused by *Fusarium* spp., leaf spot diseases by fungi including *Phyllosticta* sp., *Alternaria* sp., *Phomopsis* sp., and *Pestalotiopsis trachicarpicola*, as well as gray mold disease caused by *Botrytis* sp. (Table 3.1).

Table 3.1 Symptoms and occurrence patterns of miscellaneous diseases affecting *Polygonatum*

Disease (Pathogen)	Symptoms	Occurrence pattern
Root Rot [*Fusarium solani*, *Fusarium oxysporum*]	Initially, irregular black lesions appear on the rhizome. These lesions gradually expand and merge. As the infection progresses, the rhizome darkens to a blackish-brown color and decays, eventually leading to wilting of the aerial stems	The pathogen is a soil-dwelling inhabitant or introduced through other means. Sticky and easily waterlogged fields are prone to the disease. Aging or damaged *Polygonatum* rhizomes are susceptible to pathogen infection. The disease can occur throughout the year, with the most serious occurrence from June to September. The pathogen produces spores on the diseased parts, spreads in the soil through rainwater or irrigation water, and overwinters in *Polygonatum* roots or in the soil as a parasite or saprophyte
Leaf Spot Diseases [*Phyllosticta* sp.]	It starts from the leaf tips with elliptical or irregular spots. The outer edge is brownish-green, and the middle is off-white. As the disease spreads, the spots expand into nearly circular or irregular, water-soaked, yellowish-brown spots. Later, the center of the spots becomes off-white, and the edge is brown. When humidity is high, a gray mold layer, which are pycnidia, grows on the lesions. In severe cases, the entire leaf withers and dies	High temperatures and rainy conditions promote disease spread. Conidia are mainly dispersed by air currents and rain. The pathogen overwinters on infected plant tissue in the form of mycelium and other forms, and produces new conidia to infect when the growth conditions are suitable the next year
Leaf Spot Disease *Alternaria* sp.]	After infection, leaves develop circular, nearly circular, or irregularly dark brown lesions, which are initially small. As the disease progresses, the lesions expand. When multiple lesions coalesce, the leaves wither or dry out	High temperatures and rainy seasons facilitate disease occurrence. Conidia produced in diseased parts are primarily dispersed by air currents or rain. The pathogen overwinters on host residues in the form of hyphae, and when the growth conditions are suitable the next year, it produces new conidia, to cause infection
Leaf Spot Disease [*Phomopsis* sp.]	Leaf lesions appear as circular, yellow-brown spots with purplish-red edges. Multiple lesions may coalesce into larger patches, with numerous small black dots scattered in the center, which are the pycnidia of the pathogen. The center of these lesions is prone to rupture, forming perforations; in severe cases, the entire leaf withers and dies	High temperatures and rainy seasons facilitate disease occurrence. Conidia produced in infected areas are primarily dispersed by air currents or rain. The pathogen overwinters on host residues in the form of mycelium, and under favorable growth conditions in the following year, it germinates to produce new conidia, initiating further infection

(continued)

Table 3.1 (continued)

Disease (Pathogen)	Symptoms	Occurrence pattern
Leaf Spot Disease [*Pestalotiopsis trachicarpicola*]	Lesions are circular or irregularly shaped, with purple-brown to dark brown edges and light brown centers that have small black dots (fruiting bodies) and yellow halos. In severe infections, the tissue in the center of the lesions becomes necrosis, turna thin and membranous or forms perforations. When multiple lesions merge, large-area necrosis is formed, eventually causing the aboveground parts of the plant to wither and die	
Gray Mold [*Botrytis* sp.]	The infected parts wilt and droop, and a large amount of grayish-white mold layer forms on the surface, consisting of hyphae and conidia of the pathogen	Cool, humid conditions favor the development of gray mold, which tends to be more severe during the rainy season. Conidia produced in infected areas are dispersed mainly spread by air currents. The pathogen has a broad host range and can survive as either a saprophyte or a parasite. It overwinters on *Polygonatum* or other plant debris in the form of mycelium or sclerotia. When environmental conditions are suitable, it produces conidia to start a new infection cycle

Chapter 4
Diseases of *Ophiopogon japonicus*

Ophiopogon japonicus (Thunb.) Ker Gawl., which belongs to the genus Ophiopogon in the Liliaceae family, is a perennial herb that primarily utilizes its dried tuberous roots for medicinal purposes. Its earliest mention can be found in the "Shan Hai Jing - Zhong Shan Jing," noting, "The region is abundant with locust and paulownia trees, and among its herbs are peonies and xun dong," where "xun dong" refers to *Ophiopogon japonicus*. Documented as "Maidong" in the "Shennong Bencao Jing" during the Eastern Han Dynasty, the name was consistently used since the Ming Dynasty, as noted by Du Wenxian in "Yao Jian." The 2015 edition of the "Pharmacopoeia of the People's Republic of China" lists "Maidong" as its official designation. Authentic Maidong medicinal materials are mainly from Santai, Sichuan (known as Sichuan Maidong), and Hangzhou, Zhejiang (known as Zhejiang Maidong); the latter is recognized for its higher flavonoid content, while the former is noted for its higher diosgenin content.

Ophiopogon japonicus is an evergreen perennial herb with thick roots often swelling in the middle or ends into small ovoid or spindle-shaped tubers. The plant features very short stems, with leaves basal, tufted, grass-like, and lanceolate bracts tapering at the tips. The seeds are spherical. Harvesting occurs in summer when the roots are washed, repeatedly sun-dried, piled until approximately 70-80% dry, fibrous roots removed, and then fully dried. Maidong is widely distributed across China, with commercial medicinal materials primarily coming from cultivated sources. Suitable for cultivation across most of the Yangtze River basin, it is predominantly produced in places like Santai, Hangzhou, Yuyao, Wuxi and Zhenjiang. Maidong possesses properties such as nourishing yin and generating fluids, moisturizing the lungs, and clearing the heart. It is indicated for treating conditions like dry cough due to lung dryness, consumptive cough due to yin deficiency, throat paralysis, sore throat, fluid depletion causing thirst, internal heat causing thirst, restlessness, insomnia, and constipation due to intestinal dryness.

Main diseases of Maidong is susceptible to both leaf and root diseases, with leaf diseases including leaf blight, black spot disease, and anthracnose, and root diseases

© China Agriculture Press Co., Ltd. 2026

T. Wang et al., *Atlas of Chinese Medicinal Plant Diseases*,
https://doi.org/10.1007/978-981-95-6326-5_4

primarily involving root-knot nematodes. Typically, large-scale outbreaks of leaf diseases are rare, but root-knot nematode infestation can significantly reduce Maidong yields.

4.1 Maidong Leaf Blight

Symptoms

Maidong leaf blight is a common fungal leaf disease with complex causative factors, potentially affecting 25–40% of leaves during severe outbreaks. It is prevalent in all regions where Ophiopogon species are cultivated, including Fujian, Zhejiang, Hubei and Sichuan. The disease first manifests with leaf tips turning yellow and wilting, and then it progressively spreads to the leaf base, forming brown or yellow spots until the leaves become dry and yellow, accompanied by wilting. Outer leaves are more susceptible, curling and wilting progressively, impacting growth.

Pathogen

The disease is predominantly caused by the asexual stage of *Fusarium* spp. fungi, including *Fusarium oxysporum, F. equiseti, F. concentricum, F. proliferatum, F. graminearum, F. fujikuroi, F. tricinctum, F. chlamydosporum, F. verticillioides*, and *F. incarnatum*. Surveys in Sichuan have identified *F. oxysporum, F. equiseti*, and *F. proliferatum* as the dominant strains, representing 37%, 23%, and 17% of cases, respectively. *F. oxysporum* colonies exhibit cottony mycelia that are dense centrally and sparse peripherally, turning red or light yellow in later stages. Conidia vary significantly in shape and size among strains, from sickle-shaped to oval or ovoid. *F. equiseti* colonies are circular, off-white to pale yellow, with short, cottony aerial hyphae, and large crescent-shaped conidia with three to six clear segments and a conical apical cell. *F. proliferatum* exhibits abundant wool-like aerial hyphae, initially white then turning pale yellow, with large sickle-shaped conidia curved at the tip, three to five segmented, measuring (2.4–4.2) μm × (32–52) μm, and conidiophores are either branched or unbranched, borne in clusters or singly, with small conidia attached in chains or in a false head arrangement (Figs. 4.1 and 4.2).

Occurrence Pattern

Pathogens overwinter in diseased leaves left in the soil, serving as the primary source of infection for the following year. The disease begins in mid-April, with its progression closely linked to rainfall, intensifying during the rainy season. Disease centers are easily identifiable in fields and spread rapidly under favorable temperature and humidity conditions.

Control Measures

1. Select disease-free seedlings, choosing healthy, green-leaved plants for propagation.

Fig. 4.1 Field symptoms of Maidong Leaf Blight

2. Enhance field management by applying 100 kg of wood ash per mu (1 mu = 666.7 square meters) before the morning dew dries during the early stages of an outbreak. Ensure timely drainage after rainfall to reduce field humidity.
3. When the disease severity is high, initiate by pruning-affected Maidong leaves, then apply treatments such as 12% Metalaxyl-M EC at a 1,000-fold dilution, 30% Prochloraz EC at a 1,000-fold dilution, or 70% Propineb WP at 500–700-fold dilution. Integrate the use of biological fungicides like *Bacillus subtilis* to promote plant recovery. Chemical control options include 80% mancozeb WP at an 800-fold dilution, spraying once every 10–14 days, for three to four consecutive applications. Alternatively, bordeaux mixture (copper sulfate, lime and water at a 5:5:2 ratio) at 400–500-fold dilution can be sprayed every 15 days, up to four times, with a safety interval of 30 days. Another option is to use 3% metalaxyl-M + mancozeb WP at 500–800-fold dilution, limited to one application per season.

4.2 Maidong Black Spot Disease

Maidong black spot disease is a prevalent foliar disease that severely impacts Maidong, affecting its growth and secondary metabolites. This disease leads to leaf yellowing and defoliation, reducing photosynthesis, which significantly affects both the yield and quality of Maidong, potentially causing substantial economic losses. It is primarily observed in the provinces of Sichuan, Hubei, and Fujian.

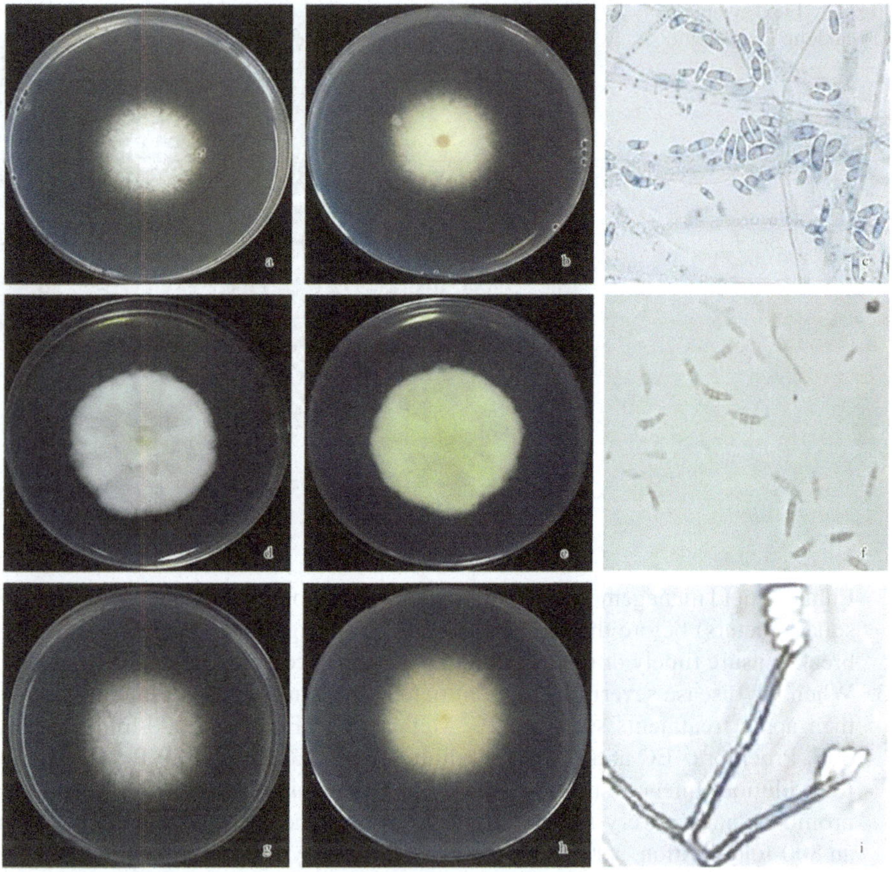

Fig. 4.2 Morphology of the pathogens causing Maidong Leaf Blight. (**a**, **b**) Cultural characteristics of *Fusarium oxysporum*. (**c**) Conidia of *Fusarium oxysporum*. (**d**, **e**) Cultural characteristics of *Fusarium equiseti*. (**f**) Conidia of *Fusarium equiseti*. (**g**, **h**) Cultural characteristics of *Fusarium proliferatum*. (**i**) Conidia of *Fusarium proliferatum*

Symptoms

Initially, the disease presents with the yellowing and browning of leaf tips, eventually leading to complete desiccation. The infection progressively spreads toward the leaf base. The interface between healthy and infected tissues generally appears darker. Occasionally, leaves develop water-soaked, brownish-yellow spots. As the disease progresses, the spots will deepen to a dark brown color. In severe cases, the entire leaf turns yellow and dies, which can lead to extensive plant death across the field (Fig. 4.3).

Pathogen

The causal agent of Maidong black spot disease is the asexual stage of the fungus *Alternaria alternata*. The fungus produces conidiophores that are either solitary or

Fig. 4.3 Symptoms of Maidong Black Spot Disease

in bundles of 2 to 30, which are dark brown with a lighter tip and slightly enlarged basal cells. These conidiophores are straight or slightly curved, non-branching, and lack knee-like joints. The conidia are club-shaped, yellowish-brown to brown, measuring (17.5–40) μm × (7.5–15) μm. The beaks of the conidia are columnar or conical, measuring (6–20) μm × (2.5–5) μm, with two to six transverse septa and zero to five longitudinal (oblique) septa. Conidia frequently form chains, typically in long chains of more than ten (Fig. 4.4).

Occurrence Pattern

The pathogen overwinters on host leaves, fruits, and seeds, germinating under high humidity conditions the following year. It enters through the host's stomata or epidermis, establishing the initial infection, which accumulates and leads to disease outbreaks. Infection begins in April, with outbreaks typically starting from May to June, becoming common from August to September, and peaking between October and November, especially during rainy and overcast seasons. The disease's development is closely associated with rainfall, with severe outbreaks during rainy seasons. Infected plant centers are clearly visible in the field and spread rapidly under favorable temperature and humidity conditions, leading to widespread death of plants.

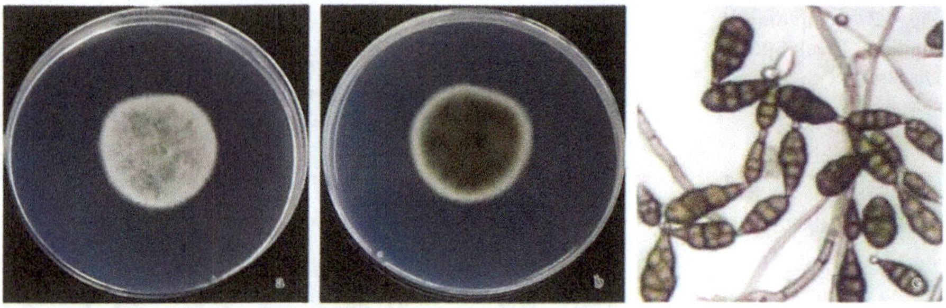

Fig. 4.4 Morphology of the Maidong Black Spot Disease Pathogen. (**a**, **b**) Cultural characteristics. (**c**) Conidia

The pathogen is highly resistant to adverse conditions, capable of overwintering and surviving for several years.

Control Measures

1. Soil and Seed Disinfection: Disinfect soil and seeds before planting to eliminate the pathogen source, minimizing the risk of initial infection.
2. Cultural Management: Apply wood ash early in the morning before dew dries and ensure timely drainage following rain to reduce field humidity. Prompt removal and destruction of diseased leaves are crucial upon detection of the disease.
3. Chemical Control: Use 75% chlorothalonil WP diluted 500 times, spraying once every 7–10 days for a total of three to four times; 75% thiram WP diluted 1,000-fold or 66% mancozeb + copper oxide WP diluted between 1000 to 1500 times, spraying once every 10–15 days for a total of two to three times, in conjunction with 30% propiconazole SC diluted between 1200 to 2400 times, spraying once every 15 days, with no more than two applications.

4.3 Maidong Anthracnose Disease

Maidong anthracnose disease is a common ailment in Maidong, widely distributed across major cultivation areas such as Sichuan, Zhejiang, Hubei, and other regions including Henan and Anhui.

Symptoms

In the early stages, the pathogen affects the tips and edges of leaves, and as the disease progresses, it may spread to the middle parts of the leaves. Lesions are circular, oval, or irregular in shape, initially appearing as small water-soaked chlorotic spots that gradually turn red or yellow-brown. In later stages, the centers of these lesions become gray-black, with irregular small black dots surrounding them. The lesion edges are distinctly chlorotic, with a yellow halo around them. Lesions along the leaf margins appear semicircular or elongated-oval. Affected leaf tips turn from brown to gray-black and eventually dry out, progressing downward. At the boundary between healthy and infected tissues, the coloration is often red or yellow--brown. Lesions may expand vertically or merge with one another, leading to leaf breakage, clustering, or entire patches of foliage drying out. In the advanced stages, numerous small black dots and blackish, hairy structures appear on the lesions, which are the acervuli and setae of the pathogen (Fig. 4.5).

Pathogen

The causal agents of Maidong anthracnose disease are fungi from the genus *Colletotrichum*, including *Colletotrichum liriopes* (American anthracnose), *C. dematium* (black-line anthracnose), and *C. gloeosporioides* (glue-bearing anthracnose). Among these, *C. liriopes* and *C. dematium* are the dominant strains, representing 58% and 31% of cases, respectively. *C. liriopes* colonies grow rapidly on PDA medium, with round shapes and smooth edges that transition from gray-white to gray, dark green, or black, with well-developed aerial mycelia. *C. dematium* colonies are initially white, later becoming gray-brown to dark brown, with velvety aerial mycelia and gray-white conidial masses. The average daily growth rate is 1.2 cm. Its conidiophores are branched, light brown, nearly flask-shaped, colorless at the apex, where conidia are produced. The conidia are sickle-shaped, unicellular, colorless, with a slightly blunt tip, measuring (22.7–27.5) μm × (3.6–4.8) μm, with an average size of 25 μm × 3.9 μm (Fig. 4.6).

Occurrence Pattern

The pathogen overwinters in soil on infected plant residues and tissue, with conidia dispersed by wind and water droplets, entering the plant through wounds. High planting density, high rainfall, or months and years with concentrated precipitation increase infection rates. The pathogen's optimal growth temperature is around 25 °C. The disease is most likely to occur in July and August when temperatures are high, and rainfall is abundant.

Fig. 4.5 Field symptoms of Maidong Anthracnose Disease

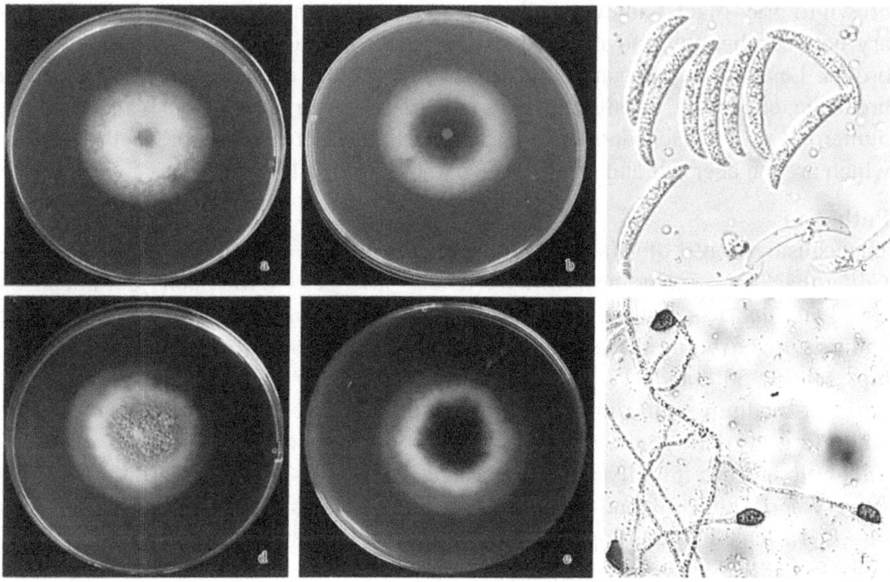

Fig. 4.6 Morphology of the pathogens of Maidong Anthracnose Disease. (**a**, **b**) Cultural characteristics of *Colletotrichum liriopes*. (**c**) Conidia of *Colletotrichum liriopes*. (**d**, **e**) Cultural characteristics of *Colletotrichum dematium*. (**f**) Conidia of *Colletotrichum dematium*

Control Measures

1. Cultural Management: In winter and early spring, promptly remove infected plant residues and trim diseased leaves. Disinfect the soil surface using a 2,000-fold solution of 45% lime sulfur or 50% dimethomorph WG. Avoid dense planting to allow airflow and light penetration.
2. Chemical Control: During the rainy season, spray 75% mancozeb WP at a 400-fold dilution, 50% carbendazim WP at an 800-fold dilution, 70% chlorothalonil WP at an 800-fold dilution, or 80% anthracnose + thiram WP at an 800-fold dilution on leaves every 7–10 days, for two to three consecutive applications, with a 10–15 day interval between applications.

Chapter 5
Diseases of *Anemarrhena asphodeloides*

Anemarrhena asphodeloides Bunge, a perennial herbaceous plant belonging to the genus *Anemarrhena* in the Liliaceae family, is used medicinally for its rhizomes. It is cultivated nationwide in China, primarily in regions such as Hebei, Shanxi, Shandong, Shaanxi, and Gansu. The rhizome is in the shape of long strip, slightly curved, and somewhat flattened, with a yellow-brown to brown surface. It is hard, easy to break, with a yellow-white cross-section. It has a slightly sweet smell and somewhat bitter taste, and a sticky texture when chewed. It has the effects of clearing heat, purging fire, nourishing yin, and moisturizing dryness, and can be used for exogenous febrile diseases, high fever with polydipsia, intestinal dryness and constipation, etc.

Anemarrhena asphodeloides has strong resistance to adverse conditions, with relatively few types of diseases. Currently, the diseases reported in studies include anthracnose and gray mold.

5.1 *Anemarrhena asphodeloides* Anthracnose

Anemarrhena asphodeloides anthracnose, also known as *Anemarrhena asphodeloides* leaf spot disease, is a major disease affecting *Anemarrhena asphodeloides*, primarily damaging the leaves. The disease becomes particularly severe during the later growth stages, with an incidence rate of over 80%. It is widely distributed in provinces such as Hebei, Anhui, Shanxi, and Jilin.

Symptoms

Anemarrhena asphodeloides anthracnose primarily affects the stems and leaves. In the early stages, brown spots of 3–6 mm appear on leaves or stems, ranging from oval to irregular in shape. As the disease progresses, the lesions become slightly sunken with a distinct red-black edge, and black small granules (i.e., acervuli and conidia) appear on the lesion surface, ultimately hindering the plant's normal growth (Fig. 5.1).

© China Agriculture Press Co., Ltd. 2026

T. Wang et al., *Atlas of Chinese Medicinal Plant Diseases*,
https://doi.org/10.1007/978-981-95-6326-5_5

Fig. 5.1 Symptoms of Anthracnose Disease in *Anemarrhena asphodeloides*

Pathogen

The causative agent of anthracnose in *Anemarrhena asphodeloides* is *Colletotrichum spaethianum*, an asexual morph of ascomycetes belonging to the genus *Colletotrichum*. On PDA medium, colonies initially appear gray-white, and gradually turn mouse-gray; on SNA medium, colonies are flat, lacking aerial hyphae, and produce abundant orange-red to flesh-pink conidial masses. The conidia are hyaline, smooth, unicellular, sickle-shaped, or slightly curved, measuring (17.3 - 21.4) μm x (3.1 - 4.1) μm, and each contains one oil globule. Appressoria are either solitary or loosely clustered, dark brown, irregularly shaped, with deep lobes, measuring (6.7 - 16.7) μm x (6.3 - 13.3) μm (Fig. 5.2).

Occurrence Pattern

The pathogen overwinters primarily as conidia but can also overwinter as acervuli and mycelia on infected plant debris in the soil, becoming the initial infection source for the following year. Under favorable conditions, conidia produced by overwintering mycelia or overwintering conidia spread through air currents or rainwater to initiate primary infection. After the disease occurs, new conidia form on lesions, leading to repeated cycles of infection. Conidia typically enter through wounds, but they can also penetrate directly through the host's epidermis. High temperatures and heavy rainfall promote disease spread, and poor ventilation, inadequate drainage, and high planting density also favor disease development. The disease becomes particularly severe in the later growth stages of *Anemarrhena asphodeloides*.

Control Measures

1. Scientific Site Selection: Practice crop rotation and choose well-drained, loose soil for planting.
2. Cultural Management: Strictly control planting density to improve ventilation and light transmittance within the field. Ensure timely drainage after rainfall to reduce humidity. Apply phosphorus and potassium fertilizers as needed to enhance plant resistance to disease. Maintain field sanitation by promptly removing diseased plant debris. Early removal of fallen leaves and other plant residue, followed by centralized destruction or deep burial, can reduce the pathogen source.

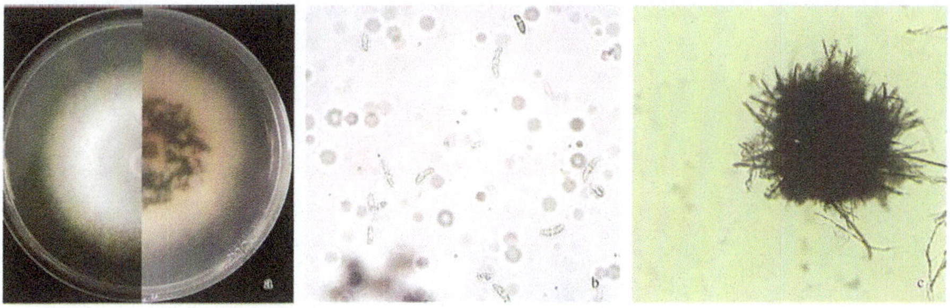

Fig. 5.2 Morphology of the Anthracnose Pathogen in *Anemarrhena asphodeloides*. (**a**) Cultural characteristics. (**b**) Conidia. (**c**) Acervul

3. Seedling Selection: Choose healthy, disease-free rhizomes for propagation, and disinfect rhizomes before planting.
4. Chemical Control: Before and after the onset of the disease, apply a 1:1 Bordeaux mixture or 65% Mancozeb settable powder at a 500-fold dilution, once every 7 days, and spray 3-4 times continuously.

5.2 Occasional Diseases of *Anemarrhena asphodeloides* (Table 5.1)

Table 5.1 Characteristics of occasional diseases on *Anemarrhena asphodeloides*

Disease (Pathogen)	Symptoms	Occurrence pattern
Gray Mold (*Botrytis cinerea*)	The disease affects the lower parts of the stems and the more mature leaves. Initially, brown or reddish-brown cloud-patterned lesions appear, and then gradually expand into oval brown to dark brown lesions. Lesions on the stems can reach up to 10 cm, with a layer of grayish mold (i.e., mycelium) forming on the surface of infected leaves	The pathogen overwinters as mycelium on infected plant residues. In spring, overwintered fungal bodies release spores, acting as the primary source of infection. The disease is severe in rainy years in spring. High planting density and high humidity favor disease development are conducive to the occurrence of the disease

Chapter 6
Diseases of *Paris polyphylla*

Paris polyphylla var. *yunnanensis* (Franch.) Hand.-Mzt. and *P. polyphylla* var. *chinensis* (Franch.) Hara are plants in the Liliaceae family, valued for their dried rhizomes. *P. polyphylla* var. *yunnanensis* is primarily distributed in Yunnan, Guizhou, and Sichuan, while P. polyphylla var. *chinensis* is mainly found in Guangdong, Guangxi, Fujian, Hubei, and Sichuan. *Paris polyphylla* has the effects of clearing heat and detoxifying, reducing swelling and alleviating pain, cooling the liver and calming convulsions. It is widely used to treat conditions such as furuncles and carbuncles, sore throat, venomous snake bites, traumatic injuries, and convulsions.

The main diseases affecting *Paris polyphylla* include leaf spot disease, anthracnose, soft rot, and gray mold, which affect large areas and cause significant damage. Root rot, damping-off, white mold, and bacterial shot hole disease are observed occasionally.

6.1 Leaf Spot Disease of *Paris polyphylla*

Leaf spot disease is a common fungal disease of *P. polyphylla*, occurring in all major cultivation regions, particularly during the rainy season, with an incidence rate ranging from 7% to 50%. This disease presents various symptoms and involves complex pathogenic factors, significantly impacting the production of *P. polyphylla*.

Symptoms

Leaf spot disease primarily affects the leaves but can also occur on petioles, young branches, flower stalks, and immature fruits. In the early stages, small yellow spots appear on the leaf tips and edges, gradually enlarging into irregular brown lesions. During periods of abundant rainfall, lesions may coalesce, leading to leaf wilting and, in severe cases, plant death. There are multiple types of leaf spot disease, with diverse symptoms observed in the field (Fig. 6.1).

© China Agriculture Press Co., Ltd. 2026

T. Wang et al., *Atlas of Chinese Medicinal Plant Diseases*,
https://doi.org/10.1007/978-981-95-6326-5_6

Fig. 6.1 Symptoms of Leaf Spot Disease in *Paris polyphylla*

Pathogen
Leaf spot disease in *P. polyphylla* is caused by infections with single or multiple fungi, with a complex array of pathogens involved. The dominant pathogenic fungi belong to the genus *Alternaria*, with *Alternaria tenuis* Nees being the primary pathogenic species. *A. tenuis* grows prostrate on PDA medium, producing dense, velvety mycelia. Colonies have a regular shape, with white mycelia at the center and gray-green or light green on the periphery. The colony edge is white, with a brown center on the reverse side and lighter outer areas arranged in concentric rings. Conidiophores are yellow-brown or light brown, solitary or clustered, branched or unbranched, erect or curved, with swollen tips. Conidia are club-shaped, ovoid, or nearly oval, light brown to brown, with a smooth or warty surface. They may have transverse or longitudinal septa, with constrictions at the septa where the color deepens. Some spores have short beaks, measuring 5.9–26.3 μm in width and 14.6–69.3 μm in length (Fig. 6.2).

Occurrence Pattern
Paris polyphylla includes two distinct varieties, with variations in disease onset and severity across regions. The disease generally begins in April to May with the start of the rainy season, though in some areas, it starts as early as March. Rainfall is the primary factor for disease occurrence and spread, as the warm, humid conditions during the rainy season favor pathogen proliferation. Conidia can spread through wind and rain. The peak infection period is from July to September, with disease severity increasing with rainfall, which may lead to plant death. The pathogen overwinters in plant residues in the soil, re-infecting plants the following year. Conditions such as high summer temperatures, poor drainage, inadequate ventilation, and insufficient sunlight exacerbate the disease, especially in fields with continuous cropping.

Control Measures
1. Agricultural Control: Leaf spot disease in *P. polyphylla* is prone to occur under high-temperature and high-humidity conditions during the rainy season. Proper

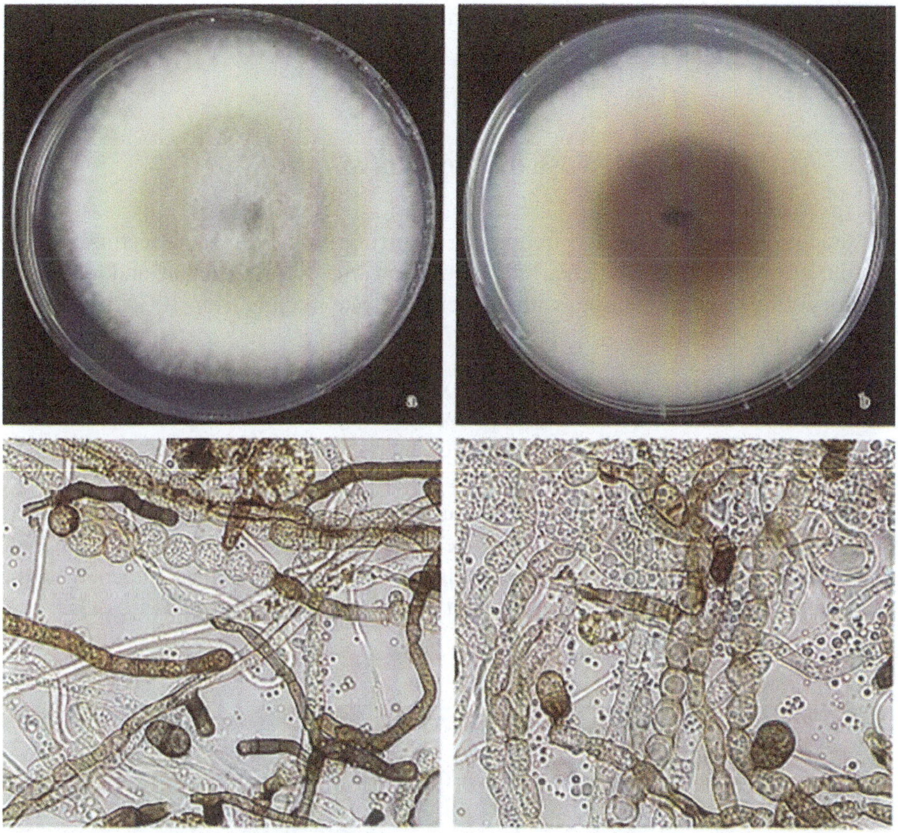

Fig. 6.2 Morphology of *Alternaria tenuis*. (**a**, **b**) Cultural characteristics. (**c**, **d**) Mycelia and conidia

cultivation practices can effectively reduce disease occurrence and spread. As *P. polyphylla* is a rhizome crop that does not tolerate waterlogging, it is necessary to ensure adequate spacing between plants and rows to promote good ventilation in the field. During the rainy season, drain excess water promptly to reduce soil moisture. At the early stage of disease occurrence, remove infected leaves. After plant lodging, remove diseased plants and dead leaves, destroy them to reduce soil-borne inoculum, and reduce infection in the following year. Implement proper crop rotation and avoid continuous cropping.

2. Chemical Control: Fungicides such as Mancozeb, *Bacillus subtilis*, Propiconazole, and allicin are effective against *Alternaria*. At the early stages of disease, apply 40% Mancozeb wettable powder diluted 800–1000 times. Before or at the early onset of the disease, *Bacillus subtilis* can be applied by diluting it 200–500 times with water and spraying it on leaves and rhizomes every 5–7 days. This can be done consecutively, but *Bacillus subtilis* should not be mixed with chemical fun-

gicides, insecticides, herbicides, or sulfur-based fertilizers (e.g., potassium sulfate, straw ash) as they can deactivate the biological agent.

6.2 Anthracnose Disease of *Paris polyphylla*

Anthracnose is a common and frequently occurring fungal disease affecting P. polyphylla. It is widespread, particularly in cultivation areas across Yunnan, and frequently occurs during the rainy season, with an incidence rate ranging from 10% to 30%.

Symptoms

Anthracnose in *Paris polyphylla* typically starts on the lower leaves and gradually spreads upward. Initially, small grayish-white to yellow-brown spots appear on the leaves, which then expand into irregular to nearly circular dark brown lesions. The lesions have a dark brown edge with a grayish-white or brownish center, and are slightly sunken in the middle with a raised edge. Concentric ring patterns may sometimes be visible. In advanced stages, lesions merge to form larger patches, causing the entire leaf to yellow and wither. Dead leaves remain attached to the plant, and in severe cases, the entire plant may turn yellow and die (Fig. 6.3).

Fig. 6.3 Symptoms of Anthracnose Disease in *Paris polyphylla*

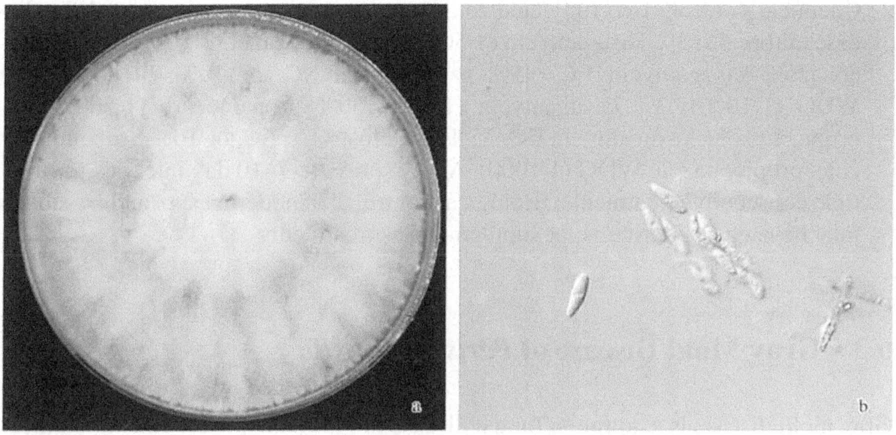

Fig. 6.4 Morphology of the Anthracnose Pathogen in *Paris polyphylla*. (**a**) Cultural characteristics. (**b**) Conidia

Pathogen
The pathogen causing anthracnose in *P. polyphylla* is *Colletotrichum gloeosporioides*, the asexual morph of the Ascomycete fungi. After 7 days of culture on PDA medium, colonies appear white or grayish-white. Conidia are formed after approximately 20 days of culture. Conidiophores are generally colorless, with conidia budding internally. The conidia are short-elliptical, elliptical, or crescent-shaped, colorless, either aseptate or with a single septum, and may contain oil globules (Fig. 6.4).

Occurrence Pattern
The pathogen overwinters as mycelia and acervuli on seedlings or plant debris. In spring, conidia are produced, serving as the primary source of infection. Following initial infection, the pathogen produces large numbers of conidia, leading to secondary infections. Newly emerged leaves during the growing season act as critical reservoirs for repeated infections, facilitating disease spread. Conidia are dispersed by wind, rain, and insects. Upon reaching the leaf surface, conidia germinate, producing germ tubes, appressoria, and infection pegs, which penetrate through stomata, wounds, or directly through the epidermis.

Control Measures
1. Agricultural Control: Implement crop rotation and maintain appropriate plant spacing to improve ventilation and light transmission conditions, and reduce humidity between plants, which can effectively alleviate the occurrence of the disease. Apply an adequate amount of high-quality base fertilizer to promote robust plant growth; however, high concentrations of nitrogen and potassium fertilizers may increase the likelihood of anthracnose outbreaks.

2. Chemical control. Two fungicide rotation programs are recommended for disease outbreaks: 3% kasugamycin (1:500), 30% hymexazol (1:500), 12% difenoconazole–kasugamycin EC, 25% azoxystrobin SC, or 10% difenoconazole WDG (1:1000); 3% kasugamycin (1:500), 30% propamocarb (1:500), 12% propiconazole–kasugamycin EC, 25% pyraclostrobin SC, or 10% benzothiadiazole–propiconazole WDG (1:1000). Apply sprays at 7–10 day intervals for multiple consecutive treatments. Biological control. Trichoderma spp. and B. subtilis may be used as alternative or supplementary treatments.

6.3 Gray Mold Disease of *Paris polyphylla*

Gray mold disease is a common fungal disease in *Paris polyphylla* cultivation areas across Sichuan, Yunnan, Guizhou, and Hubei, with an incidence rate of 40% to 70%. It is one of the major limiting factors for *Paris polyphylla* production.

Symptoms
Gray mold of *P. polyphylla* typically affects the stems, leaves, and flowers. Leaf infections generally begin at the junction of the petiole base and the stem. In the early stages, water-soaked black spots appear on the leaves, gradually expanding as the disease progresses. In later stages, the petioles and nearby stems become soft, causing the leaves to droop, and a gray mold layer appears on the surface. As the disease worsens and field humidity increases, stems collapse, leaves wilt, and the entire plant decays (Fig. 6.5).

Pathogen
The pathogen causing gray mold in *P. polyphylla* is *Botrytis cinerea*, an asexual morph of ascomycetes in the genus *Botrytis*. On PDA medium, the mycelia appear gray-white, and the sclerotia are black, irregularly shaped, and oval, with a diameter of (3.34 ± 0.31) μm. The conidiophores are grape-like in their branching structure, with swollen tips bearing conidia. The conidia are oval, measuring 9.7–13.7 μm in length, with an average length of (11.32 ± 0.82) μm, and 7.05–9.12 μm in width, with an average width of (8.24 ± 0.48) μm (Fig. 6.6).

Occurrence Pattern
The pathogen can overwinter and survive in soil or plant debris, spreading through wind and rain. Initial infection may begin as early as late February. The optimal temperature for pathogen growth is around 20 °C, and cool, rainy weather favors the occurrence of gray mold in *P. polyphylla*. Continuous rain without timely drainage exacerbates the disease. Under conditions of heavy rainfall, large-scale outbreaks can occur in May, with an incidence rate reaching 40% to 70%, leading to extensive leaf decay, stem bending, which seriously affects the normal growth of *P. polyphylla*.

Fig. 6.5 Symptoms of Gray Mold Disease in *Paris polyphylla*. (**a**) Field symptoms. (**b**) Stem symptoms. (**c**) Leaf symptoms

Control Measures

1. Agricultural Control: Maintain appropriate planting density, apply phosphorus and potassium fertilizers to enhance plant resistance, and ensure prompt drainage during the rainy season. Remove infected plants promptly to prevent further spread.
2. Chemical Control: Before transplanting, disinfect the soil with appropriately diluted solutions of fenaminosulf, carbendazim, or hymexazol. The optimal times for application are during the seedling and early flowering stages. Effective treatments include 500 g/L fluazinam suspension concentrate, 8% flusilazole microemulsion, 50% Fludioxonil wettable powder, 430 g/L tebuconazole suspension concentrate, and 500 g/L boscalid suspension concentrate.

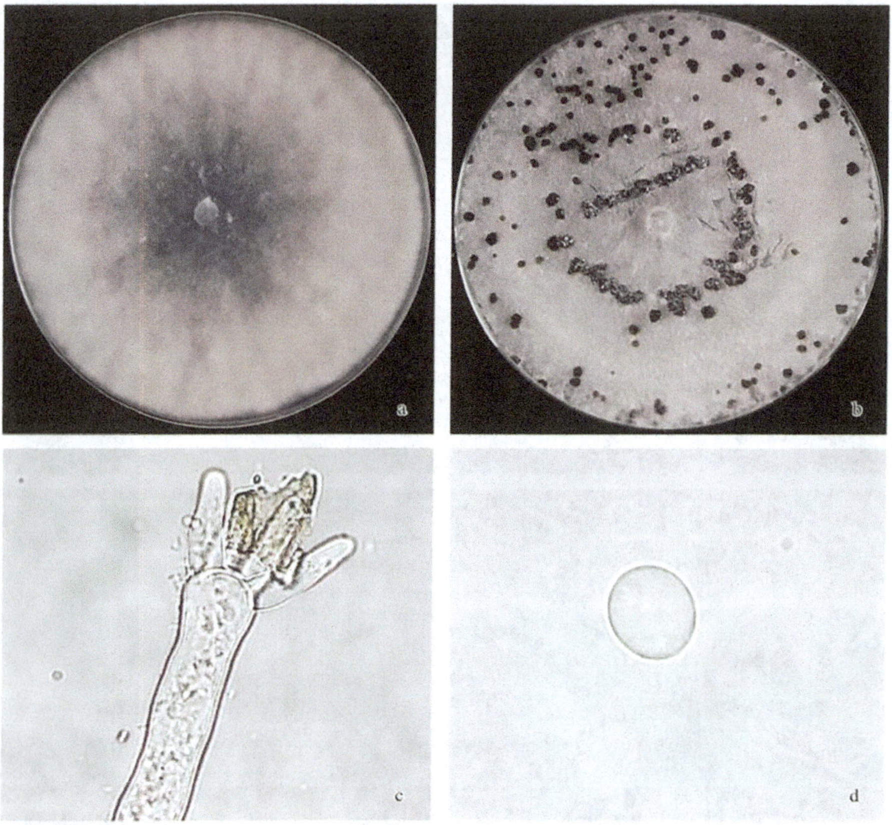

Fig. 6.6 Morphology of the Gray Mold Pathogen in *Paris polyphylla*. (**a**, **b**) Cultural characteristics. (**c**) Conidiophore. (**d**) Conidia

6.4 Soft Rot Disease of *Paris polyphylla*

Soft rot disease is the most common soil-borne bacterial disease affecting *P. polyphylla* cultivation. It spreads rapidly, causes severe damage, and is challenging to control. It occurs in all major *P. polyphylla* production areas, with an incidence rate ranging from 10% to 25%, and in some areas, it can reach as high as 50%.

Symptoms

Soft rot disease infects the leaves, stems, and tubers of *P. polyphylla*, with rapid infection and swift disease progression. The disease initially damges the underground rhizomes, causing partial or complete soft rot. The infected rhizomes' internal tissues become soft, wet, and decomposed with a chalky, granular texture, emitting a foul odor. As the disease advances, stems and leaves exhibit water-soaked, softened decay, and plants may quickly wither and die if exposed to direct sunlight. In the field, the disease usually appears in patches, often spreading outward from infected plants, leading to widespread collapse and death of seedlings (Fig. 6.7).

Fig. 6.7 Symptoms of Soft Rot Disease in *Paris polyphylla.* (**a**) Whole plant symptoms. (**b**) Localized stem symptoms. (**c**) Underground rhizome symptoms

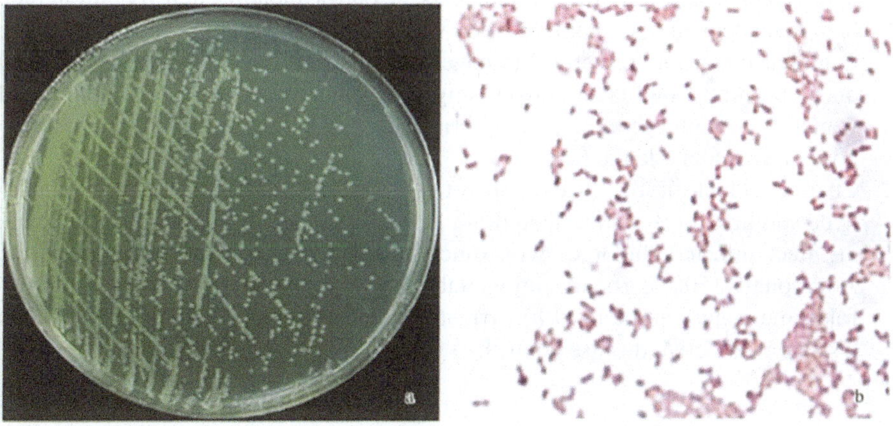

Fig. 6.8 Morphology of the Soft Rot Pathogen in *Paris polyphylla.* (**a**) Cultural characteristics. (**b**) Gram stain results

Pathogen

The causative pathogen of soft rot in *P. polyphylla* is *Pectobacterium carotovorum* subsp. *carotovorum*, a bacterium belonging to the order Enterobacteriales, family Enterobacteriaceae, and genus *Pectobacterium.* On LB medium, colonies appear circular, slightly raised, glossy, transparent, and yellow-white (Fig. 6.8a). Microscopic examination reveals it as a Gram-negative bacterium (Fig. 6.8b) with short, rod-shaped cells, blunt-rounded at both ends, measuring (0.5–0.7) μm × (1.0–2.0) μm, and bearing 4–6 peritrichous flagella. The optimal growth temperature is 25–30 °C, with a maximum growth temperature of 38–39 °C and a minimum growth temperature of 4 °C. The lethal temperature is 48–51 °C.

Occurrence Pattern

The timing of disease onset varies according to regional topography, altitude, and climate. Generally, the disease starts in late March to early April, when temperatures are low, and rainfall is abundant, conditions that favor disease development and spread. As temperatures rise and rainfall continues, the peak infection period occurs from May to July. Poor ventilation, inadequate sunlight, and waterlogging during the rainy season further contribute to disease occurrence. Damaged plants are particularly susceptible under high temperature and humidity. The pathogen spreads through rainwater, intensifying disease severity.

Control Measures

1. Agricultural Control: During the dormancy period of *P. polyphylla* in autumn and winter, plow the field and apply base fertilizer, primarily using decomposed organic fertilizers such as farmyard manure, compost, and oilseed cake. Organic fertilizers can improve soil structure and promote soil microbial diversity, and are beneficial to the healthy growth of *P. polyphylla*. Supplement with phosphorus and potassium fertilizers while reducing nitrogen fertilizer to encourage plant vigor and lower disease incidence. Control planting density, strengthen field management, and maintain good ventilation and light exposure. During the rainy season, ensure effective drainage to reduce soil moisture, remove infected plants promptly, and clear surrounding soil. After plants wither in autumn and winter, thoroughly remove plant debris and dead leaves from the field and dispose of them outside the field.
2. Chemical Control: To prevent soft rot disease, soak seeds in a 0.4% formaldehyde solution for 3–5 min, then drain before planting. Once the disease occurs, disinfect infected rhizomes with quicklime or a 1% ferrous sulfate solution. Additionally, 50% carbendazim wettable powder at a 500-fold dilution or 200-fold lime water can be used for irrigation. Garlic extract (5%) and 80% allicin also show effective disease control.

6.5 Occasional Diseases of *Paris polyphylla* (Table 6.1)

Table 6.1 Characteristics of occasional diseases in *Paris polyphylla*

Disease (Pathogen)	Symptoms	Occurrence pattern
Leaf Spot (*Leptosphaerulina australis*)	Brown spots appear on the leaves, with lesions forming sunken brown patches in the center. In severe cases, lesions merge, resulting in complete leaf death	It starts in June, with peak occurrence from July to August. High temperatures, rainy seasons, and poor drainage significantly exacerbate the outbreak
White Mold (*Ramularia sp.*)	Spots are circular, turning grayish-white or gray	It starts in July, with peak occurrence from August to October. High temperatures, rainy seasons, and poor drainage intensify disease outbreaks
Bacterial Leaf Spot (*Pseudomonas* sp.)	Small, irregular, translucent spots appear, leading to a perforated or "shot-hole" appearance	The peak period is from July to August during hot weather. Poor drainage further exacerbates disease occurrence

Chapter 7
Diseases of *Mentha canadensis*

Mentha canadensis L., a plant in the Lamiaceae family, grows primarily in mountainous areas, wetlands, and along riversides. Its rhizomes spread underground horizontally, and both its stems and leaves can be used medicinally. Mint is a valuable aromatic crop with high economic and medicinal value.

Mint is widely distributed across China, with Jiangsu and Anhui as traditional production areas, although the cultivation area has been decreasing. Mint is known for dispersing wind-heat, clearing the head and eyes, relieving sore throat, promoting rash eruption, and soothing the liver. It is used to treat wind-heat colds, early stages of febrile conditions, headaches, red eyes, sore throat, mouth ulcers, rashes, and chest congestion.

The primary diseases affecting mint include powdery mildew, rust, root rot, stem blight, gray spot, and black stem disease, with powdery mildew and gray mold being particularly severe and frequently occurring.

7.1 Powdery Mildew of Mint

Powdery mildew is a major disease affecting mint and occurs in key mint-producing areas, such as Sichuan, Jiangsu, Anhui, and Shandong provinces. The general incidence rate is 5% to 10%, but it can exceed 30% in severe cases, significantly impacting yields of mint.

Symptoms

Powdery mildew primarily affects the leaves and stems of mint. White powdery spots appear on both sides of the leaves, initially as irregular patches that later coalesce. In the early stages, the powdery mildew spots remain stable, but some may disappear in later stages (Fig. 7.1).

© China Agriculture Press Co., Ltd. 2026

T. Wang et al., *Atlas of Chinese Medicinal Plant Diseases*,
https://doi.org/10.1007/978-981-95-6326-5_7

Fig. 7.1 Symptoms of Powdery Mildew in Mint

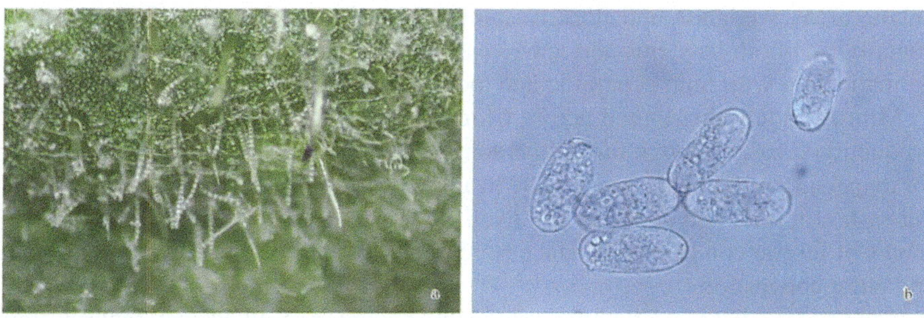

Fig. 7.2 Morphology of the Powdery Mildew Pathogen in Mint. (**a**) Conidiophores and conidia on the surface of mint leaves. (**b**) Conidia

Pathogen

The pathogen responsible for powdery mildew in mint is *Golovinomyces biocellatus* (Ehrenb.) V. P. Gelyuta (synonym: *Erysiphe biocellata* Ehrenb.), a fungus in the phylum Ascomycota. The conidiophores are cylindrical or short club-shaped, unbranched, colorless, and contain two to four septa, bearing chains of conidia that resemble beads. Conidia are colorless, single-celled, cylindrical or elliptical, measuring (22.9–38.1) μm × (13.9–17.8) μm (Fig. 7.2). Chasmothecia (ascocarps) are scattered or nearly clustered, dark brown, and flattened spherical. They possess 12–69 appendages, unbranched or with one to two irregular branches, which may be curved, knee-like, or twisted and intertwined. These appendages measure 45–215 μm in length, are of nearly uniform thickness or irregular, with a width of 4.3–6.3 μm, and have thin, rough walls, though a few may be smooth. The appendages contain one to six septa and are brown to light brown.

Occurrence Pattern

The pathogen overwinters as chasmothecia or mycelia on plant debris. In the following spring (April to May), mature ascospores are released from the chasmothecia, initiating primary infections. Conidia are produced continuously on leaves throughout the mint growing season and are dispersed by wind, causing repeated infections. The disease typically peaks from June to October. In the later stages of growth, chasmothecia are produced again to overwinter. Poor field management and weak plant growth increase susceptibility to the disease. In severe cases, leaves turn yellow, wither, and fall, potentially leading to desiccation and death of the entire plant.

Control Measures

1. Use of Resistant Varieties: Cultivate relatively powdery mildew-resistant mint varieties, such as Fuyou No. 1 and Shanghai 39 (Asia 39).
2. Agricultural Control: After harvest, promptly remove plant debris. Conduct thorough pruning in early spring, removing diseased branches and leaves, and destroy them. Clear the field thoroughly in winter to reduce sources of infection. Strengthen water and fertilizer management by applying more phosphorus and potassium fertilizers while controlling nitrogen application. Use compost with enzyme-treated organic matter or fully decomposed organic fertilizer. Employ formula fertilization techniques, improve management practices, and enhance disease resistance. Ensure adequate drainage during droughts and floods. Prune and thin plants to improve ventilation and increase sunlight exposure.
3. Biological Control: Use a 3% WP of polyoxin at a 150–200-fold dilution for preventive spraying.
4. Chemical Control: At the onset of the disease, spray with a 1000-fold dilution of 10% triadimefon SC, a 2000-fold dilution of 20% diniconazole EC, a 2000-2500-fold dilution of 12.5% flusilazole WP, or a 4000-fold dilution of 25% pyraclostrobin EC. Stop application 7 days before harvest.

7.2 Gray Mold Disease of Mint

Gray mold is a major disease affecting mint, occurring in mint-producing areas such as Sichuan and Shandong. The general incidence rate is 10% to 30%, but severe outbreaks in certain regions can significantly impact both yield and quality.

Symptoms

Gray mold primarily affects the leaves and stems of mint. Irregular or V-shaped brown lesions appear on the leaves, with a gray mold layer visible on the upper leaf surface (Fig. 7.3).

Pathogen

The pathogen causing gray mold in mint is *Botrytis cinerea*, an asexual stage of the Ascomycete fungus in the genus Botrytis. Conidiophores measure (280–550) μm × (12–24) μm, grow in clusters, are initially gray and later turn brown, with swollen or pointed tips bearing small protrusions. Conidia are borne singly on these protrusions, subglobose or ovoid in shape, and measure (9–15) μm × (6.5–10) μm (Fig. 7.4).

Fig. 7.3 Symptoms of Gray Mold Disease in Mint

Occurrence Pattern

The pathogen overwinters as conidia or mycelia on plant debris or alternative hosts in the field. Gray mold is a low-temperature, high-humidity disease. In the following spring (March to April), conidia on plant debris or alternative hosts are dispersed by air currents to young mint tissues. Conidia enter the host through wounds, natural openings, or young tissues, causing primary infections. After infection, the pathogen produces large numbers of conidia, which spread through wind and rain, resulting in multiple secondary infections. When disease occurrence is severe from May to July, leaves may turn yellow, wither, and fall, potentially leading to plant desiccation and death. Poor field management and weak plant growth increase susceptibility to the disease.

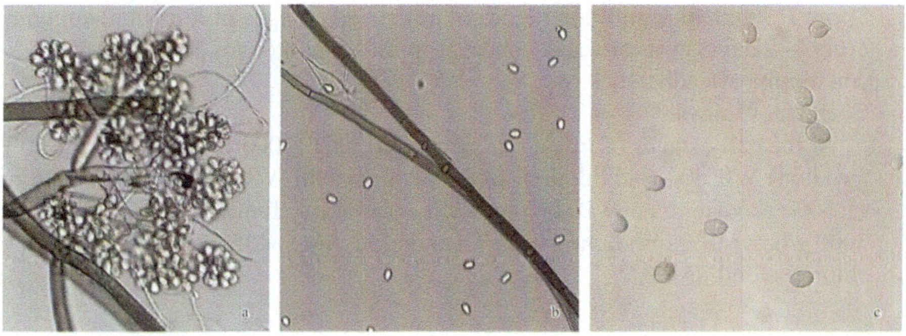

Fig. 7.4 Morphology of the Gray Mold Pathogen in Mint. (**a**) Conidiophores and conidia on the surface of mint leaves. (**b**) Hyphae. (**c**) Conidia

Control Measures

1. Agricultural Control

    (a) In protected cultivation, regulate temperature and humidity carefully. On sunny mornings, open ventilation when the greenhouse temperature reaches 30 °C to release humid air, maintaining a temperature of around 25 °C. Close the vents when the temperature drops to 20 °C in the afternoon, keeping nighttime temperatures between 15 °C and 17 °C. Avoid high nighttime temperatures to reduce plant respiration and conserve photosynthates. In cold weather, protect plants from frost to prevent weakening. Methods for insulation include covering with a small inner greenhouse and using straw mats at night. Irrigate in moderation, preferably with drip irrigation. If topsoil irrigation is used, water in the morning on sunny days, then ventilate promptly to release humid air.

    (b) Remove plant debris to reduce pathogen load. Before transplanting seedlings, thoroughly clear plant debris to minimize sources of primary infection. Deep plow the soil to bury plant residues 20 cm or more below the surface. After infection, promptly remove diseased leaves and bury them deep outside the field to prevent pathogen spread.

2. Physical Control

    Use solarization to kill pathogens with heat. In fallow fields in summer, water the soil thoroughly, cover with plastic film the next day, and maintain for at least 15 days. Before transplanting, close the greenhouse to raise temperatures, using heat to eliminate pathogens in the air, on surfaces, and in the soil. At the early stages of infection, solarize on sunny mornings by closing the greenhouse. Water the day before to enhance effectiveness. After solarization, ventilate to cool and dehumidify, and adjust fertilization to promote plant vigor.

3. Biological Control

    Biocontrol agents like *Trichoderma* spp. and *Ulocladium atrum* have strong antagonistic effects against gray mold pathogens. Antibiotics such as pyrrolnitrin and oxytetracycline produced by *Streptomyces* spp. can also inhibit gray mold.

Spray *T. harzianum* solution before or at the early stages of infection, applying every 5–7 days. In severe cases, shorten application intervals and add organosilicon to improve adherence.

4. Chemical Control

   At the early stages of infection, spray with 1500-2000-fold dilutions of 50% iprodione WP, 600-fold dilutions of 20% vinclozolin WP, or 1500-fold dilutions of 50% procymidone WP. In protected cultivation, fumigate overnight using 3000 g/ha of 15% iprodione smoke formulation (SM) or 3500–4500 g/ha of 20% chlorothalonil SM.

7.3 Occasional Diseases of Mint (Table 7.1)

Table 7.1 Characteristics of occasional diseases in mint

Disease (Pathogen)	Symptoms	Occurrence pattern
Rust Disease (*Puccinia menthae*)	Mint rust disease primarily affects the leaves and stems. In the early stages, yellowish pustules appear on the underside of the leaves, forming raised, orange-brown spots. As the disease progresses, similar spots may also appear on the upper leaf surface. Severely infected leaves turn yellow, wither, and fall off. On the stems, elongated dark brown lesions develop, causing the tissue to dry out and break. Affected leaves turn yellow and wither, leading to plant decline and death	The rust disease pathogen overwinters as spores on infected plant debris, serving as the primary infection source for the following year. Rust spores have limited viability, surviving only 15–30 days, so primary infection largely depends on spores produced during the plant's growing season to initiate secondary infections, which then spread the disease. This disease occurs from May to October, with peak incidence during the rainy season
Root Rot (Pathogen Unknown)	Mint root rot primarily affects the root system. Early symptoms include browning and softening of the roots, which gradually decay. In later stages, infected plants wilt and die. Under humid conditions, a white to pink fungal mycelium may appear on the root surface, causing severe root decay and plant death	The root rot pathogen overwinters in the soil or on infected plant debris. Under favorable conditions, it invades through wounds or natural openings. The disease is primarily transmitted through soil and water
Stem Blight (*Hymenoscyphus repandus*)	Mint stem blight mainly targets stem tissue. In the early stages, light brown to black lesions appear on the stem surface, causing the stem tissue to shrink and wrinkle. In severe cases, the lesions spread extensively, leading to stem collapse and plant death	The stem blight pathogen overwinters in the soil or on infected plant debris. In spring, spores are dispersed over long distances by wind and rain. The disease generally appears from May to October, with peak incidence from July to September. Harvesting from late September to early October can reduce disease occurrence

(continued)

Table 7.1 (continued)

Disease (Pathogen)	Symptoms	Occurrence pattern
Viral Diseases [Reported viruses include cucumber mosaic virus (CMV) and zucchini yellow mosaic virus (ZYMV)]	Mint viral diseases primarily cause systemic infections, resulting in symptoms such as leaf curling, deformation, discoloration, yellowing, and dwarfing. In severe cases, infected plants exhibit stunted, defoliation, and eventually death	Mint viral diseases are highly transmissible and have multiple vectors. Reported pathogens include CMV and ZYMV, which are mainly transmitted by aphids (e.g., *Myzus persicae*, *Aphis gossypii*) and are nonpersistent viruses. They can also spread through mechanical means and by herbivorous insects. CMV is more prevalent in light soils (with an infection rate of 4–8%), while ZYMV is less common. Mint viral diseases generally begin in early April and peak between May and June
Gray Spot Disease (*Diaporthe ganjae*)	Mint gray spot disease primarily affects the leaves. Initially, small black spots appear, which gradually enlarge into irregular brown lesions, eventually spreading across the leaf surface. As the disease progresses, concentric rings develop on the lesions. The infected tissue dries and cracks, and in severe cases, the affected leaves turn yellow and drop, leading to plant decline and increased vulnerability to other infections	The gray spot pathogen overwinters in infected plant debris, soil, or other hosts. Spores are produced from September to November in regions such as Guangdong and Fujian province, causing disease in early spring and spreading during the rainy season

Chapter 8
Diseases of *Salvia miltiorrhiza*

Salvia miltiorrhiza Bunge, commonly known as large-leaved blood-activating danshen, blood ginseng, or red root, is a perennial herbaceous plant in the Lamiaceae family and genus Salvia. It includes the typical variety (*S. miltiorrhiza* var. miltiorrhiza) and a single-leaf variant (*S. miltiorrhiza* var. charbormelii (Levi.) C.Y.Wu).

Danshen is widely cultivated in China, distributed across North China, East China, Central-South China, Northwest China, and Southwest China. The dried root and rhizome are used medicinally, and danshen is one of the most commonly used traditional Chinese medicinal herbs. It was first recorded in the Shennong Bencao Jing, where it was classified as a superior herb. Danshen is effective in dilating coronary arteries, preventing myocardial ischemia and infarction, improving microcirculation, and reducing myocardial oxygen consumption, making it an essential herb in the clinical treatment of various cardiovascular diseases.

In recent years, large-scale continuous cultivation of danshen has led to serious issues related to continuous cropping obstacles, primarily due to soilborne diseases, nematode infestations, and soil acidification and compaction.

8.1 Root Rot Disease of Danshen

Root rot is the most significant disease affecting danshen production, primarily damaging the root system, which leads to reduced yield and quality. The disease can occur throughout the growth period and spreads quickly, causing a substantial reduction in yield and making the roots unsuitable for medicinal use. Root rot is widespread in major danshen-producing areas, with an incidence rate of 10–30%. In some regions, the incidence rate can exceed 80%.

Symptoms
In the early stages, fibrous and lateral roots of danshen develop water-soaked rot, gradually darkening from reddish-brown and eventually spreading to the main root.

© China Agriculture Press Co., Ltd. 2026

T. Wang et al., *Atlas of Chinese Medicinal Plant Diseases*,
https://doi.org/10.1007/978-981-95-6326-5_8

In the later stages, the xylem of the main root completely decays, turning black and fibrous; the aboveground part of the plant withers and can be easily uprooted (Fig. 8.1).

Pathogen

The pathogen causing root rot in danshen is a fungus in the genus *Fusarium*. Research has shown that the pathogen in Sichuan danshen cultivation areas is *Fusarium solani* (Mart.) Appel et Wollenw. ex Snyder et Hansen. This pathogen has also been identified in Shandong's Liaocheng danshen cultivation area. Other studies have reported *Fusarium equiseti* (Corda) Sacc. as a potential pathogen. Since root rot often occurs together with fusarium wilt, for which *Fusarium oxysporum* is the pathogen, further research is needed to determine whether *F. oxysporum* is also a root rot pathogen in danshen. On PDA medium, *F. solani* produces circular colonies with neat edges and well-developed white or pale-yellow woolly mycelia. Under microscopy, two types of conidia are observed: large conidia, which are ovoid or cylindrical (25–26 μm × 4.5–6 μm), and small conidia, which are elongated or sickle-shaped (10–18 μm × 3–5 μm) (Fig. 8.2).

Occurrence Pattern

Root rot disease occurs in regions such as Sichuan, Shaanxi, and Shandong, with an average incidence rate of 10–30%. In core production areas, such as Laiwu, Shandong, the disease peaks from July to August, with incidence rates reaching 60–70%. Root rot often co-occurs with fusarium wilt. As a soilborne disease, the pathogen primarily overwinters as mycelia and chlamydospores in soil, infected danshen roots, and incompletely decomposed manure containing the pathogen, serving as the initial source of infection for the following year. Overwintering pathogens invade the root system through root hairs or wounds. Conidia produced at infection sites spread via soil, irrigation water, rain, cultivation practices, and soilborne pests, causing reinfection. The pathogen can survive in soil for 5–15 years. The incubation period for root rot is generally 10–15 days, with rapid spread at

Fig. 8.1 Symptoms of Root Rot in Danshen

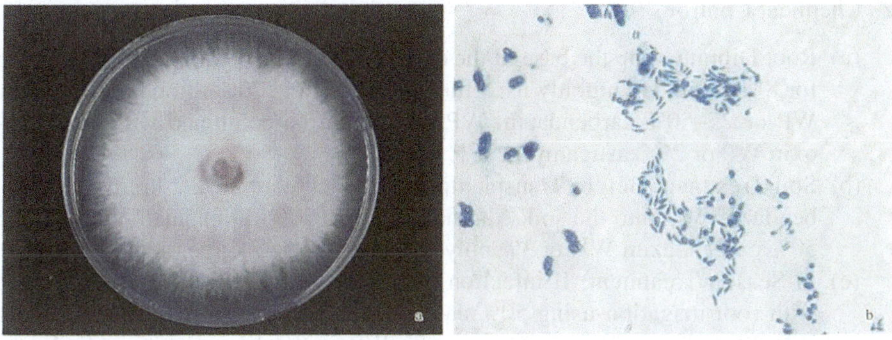

Fig. 8.2 Morphology of *Fusarium solani*, the Pathogen of Danshen Root Rot. (**a**) Colony characteristics on PDA medium. (**b**) Conidia

average daily temperatures of 16–17 °C, and optimal temperatures of 22–28 °C. Disease spread accelerates under conditions of high rainfall, soil moisture, and dense planting. Soil pests and nematodes create wounds on plants, facilitating pathogen invasion and leading to severe infections.

Control Measures

Control of danshen root rot primarily involves a combination of agricultural and chemical management practices.

1. Agricultural Control

   (a) Field Sanitation: After harvesting, promptly remove diseased plant residues to reduce the overwintering pathogen load and sources of primary infection.
   (b) Use of Disease-Free Seedlings: Cultivate seedlings in disease-free areas and test for pathogen presence to ensure that seedlings are free of infection.
   (c) Crop Rotation: Avoid continuous cropping with medicinal plants like *Platycodon grandiflorus* or *Scutellaria baicalensis*, and with vegetables. After cultivating medicinal plants for 1–2 years, switch to crops like corn for 1–2 years.
   (d) Balanced Fertilization: Before transplanting, apply well-decomposed organic fertilizers. Use phosphorus and potassium as base fertilizers, and spray potassium foliar fertilizer during the early flowering stage to promote root growth.
   (e) Appropriate Cultivation Practices: Select well-drained loam or sandy loam with higher elevation. Practice wide-narrow row cultivation with raised beds. Use a wide row spacing of 1.2 m, narrow row spacing of 0.8 m, and bed height of 5–20 cm. Plant four small rows per wide row with 30 cm spacing between small rows, while leaving narrow rows empty to improve ventilation and light exposure.

2. Chemical Control
   (a) Root Dipping: Dip the base of the danshen seedlings in a fungicidal solution for 8–10 min. Commonly used fungicides include 70% thiophanate-methyl WP or 25–50% carbendazim WP, which can be combined with 3% polyoxin WP or 2% kasugamycin WP.
   (b) Soil Treatment Before Transplanting: For each acre, mix 2 kg of 25% carbendazim WP into the soil. Alternatively, use 70% thiophanate-methyl WP, 50% carbendazim WP, or 3% polyoxin WP at 2.5–3 kg per acre.
   (c) In-Season Treatment: If infection occurs during the growing season, treat with root irrigation using 50% carbendazim WP, 70% thiophanate-methyl WP, 3% polyoxin WP, 75% mancozeb WP, or 2% kasugamycin AS. Apply 200–300 mL per plant, with intervals of 1 week, repeating two to three times as needed.

8.2 Southern Blight of Danshen

Southern blight is one of the primary diseases affecting danshen seedlings, commonly occurring in major danshen production areas and posing a serious threat to safe cultivation.

Symptoms
Southern blight primarily affects the root of danshen. In the early stages, white, silk-like mycelia and sclerotia appear around the main root near the soil surface. The root becomes moist, rots, and can be easily pulled out. In later stages, the aboveground parts of the plant wilt and die. Under humid conditions, white mycelia and sclerotia often appear at the base of infected plants (Fig. 8.3).

Pathogen
The pathogen causing southern blight in danshen is *Sclerotium rolfsii* Sacc. After 3 days of growth on PDA medium, the pathogen forms round colonies approximately 70 mm in diameter, with white, silk-like mycelia radiating outward. After 5–7 days, white knots appear on the mycelia, beginning to form sclerotia. Initially white, the sclerotia gradually turn yellow, eventually changing to tea-brown or dark brown. The sclerotia are grayish-white inside, spherical or irregular in shape, with a smooth and glossy surface, measuring 50–120 μm in diameter. They are highly resilient, capable of surviving in soil for 5–6 years or more (Fig. 8.4).

Occurrence Pattern
Southern blight can occur throughout the growing season of danshen, with peak infection from June to August. After the growing season, the pathogen's mycelia and sclerotia can overwinter in plant debris and soil, serving as the primary infection source for the next season. Sclerotia germinate readily under high-temperature and high-humidity conditions, spreading through water in the field and infecting plants upon contact.

Fig. 8.3 Symptoms of Southern Blight in Danshen

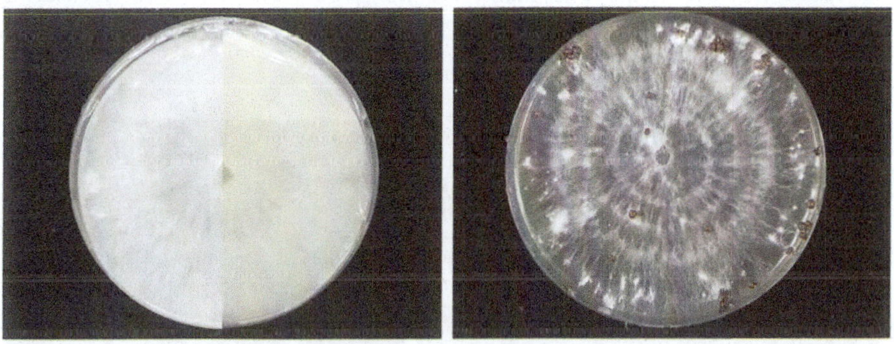

Fig. 8.4 Colony characteristics of Sclerotium rolfsii, the p[athogen of Southern Blight in Danshen

Control Measures

1. Agricultural Control

   Timely removal of infected plants and plant debris from the field to prevent pathogen spread and reinfection. Disinfect soil around diseased plants with quicklime. Improve field management and practice crop rotation with grass crops.

2. Biological Control

   Apply biological control agents such as Trichoderma during the seedling stage and at the early stages of disease onset for effective control.

8.3 Root-Knot Nematode Disease of Danshen

Root-knot nematode disease is a common issue in danshen production, affecting all major production areas. It typically reduces yields by 10–20%, with severe cases leading to reductions exceeding 30%.

Symptoms

After nematodes invade the roots, fibrous and lateral roots of danshen develop nodular swellings of varying sizes. These nodules start off smooth and yellowish-white but eventually turn brown, break down, and rot. When infected by nematodes, the root system's functionality is compromised, affecting nutrient absorption. This leads to nutrient deficiency in the aboveground parts, causing the plant to weaken, yellow, wilt, and eventually die (Fig. 8.5).

Pathogen

The pathogens responsible for root-knot nematode disease in danshen are the southern root-knot nematode (*Meloidogyne incognita*) and the northern root-knot nematode (*M. hapla*), though systematic identification is still lacking. The southern root-knot nematode mainly occurs in Shangluo, Shaanxi. It exhibits sexual dimorphism, with slender, thread-like males and pear-shaped females, which often remain embedded in host tissue. The males are transparent and colorless, measuring 1.0–1.5 mm × 0.03–0.04 mm, while the females measure 0.44–1.59 mm × 0.26–0.81 mm and can lay 300–800 eggs each (Fig. 8.6).

Fig. 8.5 Symptoms of Root-Knot Nematode Disease in Danshen

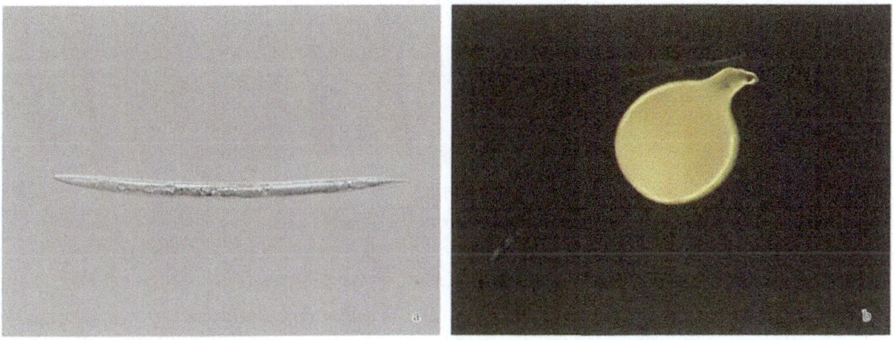

Fig. 8.6 Morphology of Root-Knot Nematodes. (**a**) Second-stage juvenile. (**b**) Adult female

Occurrence Pattern

Root-knot nematodes are mainly found in the 0–5 cm soil layer. They overwinter in soil as eggs or second-stage juveniles within infected plant debris. Infested soil, infected plant debris, contaminated seedlings, and irrigation water are primary transmission pathways. Root-knot nematodes can survive in soil for 1–3 years.

Control Measures

1. Use of Healthy Seedlings

   Establish disease-free seedling bases and use pathogen-free seedlings for planting.
2. Crop Rotation

   Rotate with grass crops, practicing 3–5 years of rotation to disrupt the disease cycle.
3. Field Sanitation

   Ensure manure is fully decomposed to eliminate nematode eggs; promptly remove diseased plant debris to reduce overwintering nematode populations; regularly clean and disinfect farming tools.
4. Soil Disinfection

   Use high-temperature treatments to kill nematodes in the soil. In summer, deeply till the soil, irrigate, and cover with plastic film, allowing sunlight exposure for about 20 days to achieve over 90% control efficacy. Fifteen days before sowing or transplanting, apply 0.2% high-concentration abamectin WP or 10% fosthiazate granules at 30 kg per hectare, mixing thoroughly with 750 kg of soil and incorporating to a depth of 25 cm.
5. Chemical Control

   At the early stage of infection, apply 1.8% abamectin EC diluted 1000–1200-fold to irrigate the roots, using 250–500 mL per plant. Repeat every 7–10 days for two to three treatments.

Chapter 9
Diseases of Patchouli

Pogostemon cablin (Blanco) Benth., commonly known as patchouli, is a member of the Lamiaceae family. The dried aboveground parts are used as medicine, also referred to as pogostemon, huoxiang, or "fragrant herb." Based on origin, Patchouli is classified into various types, including Shipai Patchouli (Shipai Xiang), Zhaoqing Patchouli (Zhao Xiang), Zhanjiang Patchouli (Zhan Xiang), and Hainan Patchouli (Nan Xiang).

Patchouli is a perennial aromatic herb or semi-shrub distributed in Fujian, Taiwan, Guangdong, Hainan, and Guangxi provinces of China, with major production in Guangdong. It is originally native to the Philippines and other tropical Asian regions. The plant has square stems covered with fine hairs, and its distinctive aroma is slightly bitter. It has efficacy of aromatizing turbidity, appetizing and stopping vomiting, releasing and relieving heat.

The main diseases affecting patchouli include bacterial wilt, root rot, angular leaf spot, fusarium wilt, leaf spot and so on. Bacterial wilt and other soil-borne diseases among them cause the most significant damage.

9.1 Bacterial Wilt of Patchouli

Bacterial wilt is one of the most common soil-borne bacterial diseases affecting patchouli cultivation, prevalent in all major production areas. This disease spreads through multiple pathways and can cause extensive damage. It can spread across the entire field once a plant is infected, leading to large-scale mortality and catastrophic losses. In fields, the incidence rate can reach 10–30%. Bacterial wilt is a globally widespread disease and a primary cause of yield reduction in various crops.

Symptoms
In the early stages, the young leaves, tender shoots, and newly unfolded leaves at the top of the plant wilt. The disease progresses rapidly, causing the entire plant to wilt

© China Agriculture Press Co., Ltd. 2026

T. Wang et al., *Atlas of Chinese Medicinal Plant Diseases*,
https://doi.org/10.1007/978-981-95-6326-5_9

and die within 4–6 days, although the stems and leaves remain green. When an infected plant is uprooted and the stem is cut open, the vascular tissue appears brown, and the stem exudes a foul-smelling, sticky substance, indicating acute systemic wilt (Fig. 9.1). The diseased plant's stem, leaves, branches, and other parts show symptoms without visible pathogen signs. Only when the discolored vascular bundles in the stem are cut crosswise and squeezed does a white bacterial ooze appear. When the cut stem is immersed in a saline solution, a milky white, cloudy substance can be seen emerging from the cut, indicating the presence of the pathogenic bacteria.

Fig. 9.1 Symptoms of Bacterial Wilt in Patchouli. (**a**, **b**) Field symptoms. (**c**) Plant symptoms. (**d**, **e**) Root and stem symptoms

Pathogen

The pathogen responsible for bacterial wilt in patchouli is *Ralstonia solanacearum*, a bacterium in the family Pseudomonadaceae within the order Burkholderiales. On LB medium, the pathogen forms round, near round, or spindle-shaped colonies that are milky white, smooth, and raised. On LB medium supplemented with TTC, the colonies appear nearly round or spindle-shaped, raised, with pink centers and milky white edges (Fig. 9.2).

Occurrence Pattern

Bacterial wilt can occur throughout the entire growth period of patchouli, with peak incidence during the hot and rainy summer season. Fields with prolonged continuous cropping are more susceptible. The pathogen has multiple transmission routes, spreading through infected seedlings, soil, irrigation water, farming tools, vehicles, and human or animal activity. It enters through wounds caused by transplanting or damage from wind and rain.

Control Measures

1. Agricultural Control

    (a) Breeding of Resistant Varieties: Patchouli is an introduced species, and traditional vegetative propagation through cuttings results in a narrow genetic base. Developing disease-resistant varieties is essential to enhance patchouli's resilience against bacterial wilt.

    (b) Site Selection and Soil Preparation: Choose sandy loam soil with good permeability, preferably rich in humus, such as brown or black sandy loam. The planting area should be disinfected before cultivation, soil improvement materials like oyster shell powder, plant ash, or biochar should be administered before ridging, and 60 cm wide, 30–40 cm high ridges with 30 cm wide furrows should be prepared for planting.

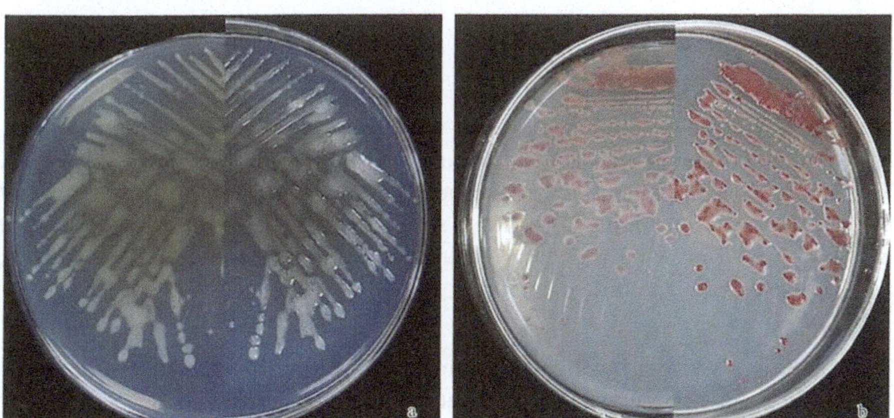

Fig. 9.2 Morphology of *Ralstonia solanacearum*, the pathogen of bacterial wilt in Patchouli. (**a**) Colony morphology on LB medium. (**b**) Colony morphology on LB medium supplemented with TTC

(c) Adjusting Cropping Systems: Practice rotation with rice or intercropping with ginger to effectively reduce soil pathogen levels and mitigate disease occurrence.
(d) Soil Amendment: Pathogens of bacterial wilt prefer acidic soil. 100 kg of quicklime per acre during soil preparation can be used for adjusting the pH to neutral or slightly acidic, effectively suppressing disease.
(e) Enhanced Field Management: Transplant seedlings at the right time to reduce root damage, maintain a clean field, apply fertilizers appropriately, and practice shallow soil hilling and root protection to lower the risk of infection.

2. Chemical Control

Before disease onset, apply 30 mL of metalaxyl – oxadixyl (diluted in 15 kg of water per acre) for root irrigation as a preventive measure, repeating every 7–10 days for two to three times. Immediately remove and destroy infected plants and disinfect the surrounding soil with 2% formalin or 20% lime water. In the early stages of infection, irrigate roots with a 3000–4000–fold diluted nucleoside antibiotic solution or 500–fold diluted 25% cupric ammonium solution. In the mid-stage, spray with 1800 g/ha of 6% validamycin and *Bacillus subtilis* WP every 7–10 days for two to three times. If over 80% of plants become infected in the later stages, terminate production to prevent further spread.

3. Application of Soil Microbial Remediation Agents

Before cultivation, use soil microbial remediation agents combined with plant vaccine agents to treat the patchouli planting area. This method has shown an effectiveness rate of up to 91.87% in preventing bacterial wilt. Alternatively, nonpathogenic strains of the bacterial wilt pathogen were injected into the plant to occupy the ecological niche, thus preventing infection by pathogenic strains.

9.2 Root Rot of Patchouli

Root rot is a common and serious fungal disease affecting patchouli cultivation. This disease has complex pathogenic factors, spreads quickly, causes significant damage, and is challenging to control, severely impacting yield. It is prevalent across patchouli-growing regions, with incidence rates ranging from 5% to 10%, and in severe cases, reaching over 30%.

Symptoms

Patchouli root rot primarily affects the roots, with symptoms beginning in the lower leaves and progressing upward in a sequential wilting pattern. The leaves gradually change color from green to pale yellow, then to dry yellow, and finally to brown. Uprooted roots appear brown, and the vascular bundles in the rotting area near the root collar become brown as well, though the vascular bundles in the stem generally

Fig. 9.3 Symptoms of Root Rot in Patchouli. (**a**) Field symptoms. (**b**, **c**) Plant symptoms. (**d**, **e**) Root symptoms

remain unaffected. The disease causes rot at the junction between the root and stem, eventually spreading to the aboveground parts of the plant. The cortex turns brown and decays, leading to wilting and death of the entire plant (Fig. 9.3).

Pathogen

The pathogen responsible for patchouli root rot belongs to the genus *Fusarium* (Ascomycota, asexual type), and infection is often caused by one or multiple fungal species. Reported species include *Fusarium oxysporum* (Schltdl. ex Snyder et Hansen), *F. graminearum* (Schw.), and *F. solani* [(Mart.) Appel et Wollenw. ex Snyder et Hansen]. *Fusarium oxysporum* produces cotton-like aerial mycelia that are dense and compact, with colors ranging from white, dark red, purple, pink to wax yellow. Some strains may form mycelial balls in the later stages of growth. *F. graminearum* has white, flower-like aerial mycelia and produces pigments in shades of pink, purplish-red, yellow, or orange. *F. solani* appears on PDA medium as having white to pale yellow fluffy mycelia, with small spindle-shaped conidia that are mostly aseptate or occasionally have a single septum (Fig. 9.4).

Recent studies have also identified that *Macrophomina vaccinii*, a fungus from the family Botryosphaeriaceae and genus *Macrophomina*, can cause both root rot and stem rot in patchouli (Fig. 9.5).

Fig. 9.4 Morphology of pathogens causing root rot in Patchouli. (**a**) Culture characteristics of *Fusarium oxysporum*. (**b**, **c**) Culture characteristics of *F. graminearum*. (**d**) Culture characteristics of *F. solani*. (**e**) Small conidia of *F. solani*

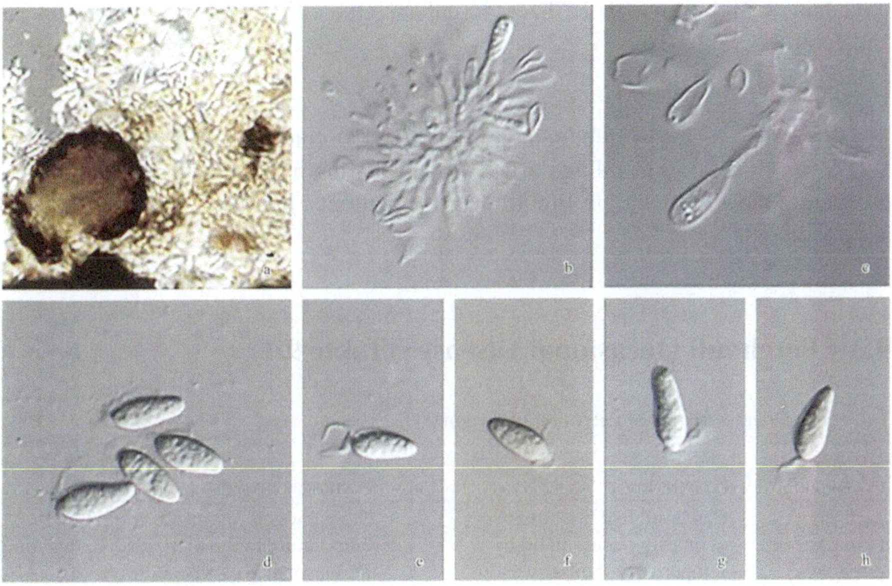

Fig. 9.5 Morphology of Macrophomina vaccinii. (Provided by Zhang Wu) (**a**) Conidiomata. (**b**, **c**) Asci and Ascospores. (**d**–**h**) Spores

Occurrence Pattern

Patchouli root rot commonly occurs after the plants are transplanted. Factors such as pest damage to plant nodes, use of incompletely decomposed manure, continuous cropping, low-lying terrain, high soil clay content, and excessive irrigation increase the likelihood of disease. The peak incidence period is from August to October, particularly following the rainy season when high temperatures, high humidity, and poor soil drainage can lead to large-scale outbreaks.

Control Measures

1. Agricultural Control

Clear Diseased Plants: Remove and destroy diseased plants after harvest to eliminate potential pathogen sources. Apply lime powder in infected areas to eradicate latent pathogens. Maintain proper drainage by clearing ditches during the rainy season to reduce field humidity. Spray potassium dihydrogen phosphate as needed to improve plant resistance. Practice crop rotation and avoid continuous cropping; rotation with rice is especially effective. In summer, intercropping with other crops to provide shade or using mulch can also help.

2. Biological Control

Use nonpathogenic Fusarium strains to suppress root rot. Other biocontrol agents, such as *Bacillus subtilis*, *Streptomyces*, and mycorrhizal fungi, can also be applied to manage patchouli root rot. Whenever possible, combine microbial inoculants with biological fertilizers to enhance the efficacy of root rot prevention and control.

3. Chemical Control

In the early stages of infection, apply 800–1000–fold diluted 50% carbendazim WP or 1000–1500–fold diluted 50% thiophanate-methyl WP to irrigate the roots, preventing the spread and proliferation of the pathogen in every 7 days for a total of 2–3 treatments.

9.3 Patchouli Occasional Diseases (Table 9.1)

Table 9.1 Characteristics of Patchouli Occasional Diseases

Disease (Pathogen)	Symptoms	Occurrence pattern
Leaf Spot (*Corynespora cassiicola*)	Patchouli leaf spot disease primarily occurs in high--temperature, humid seasons, affecting the leaves. The disease manifests as small circular or near-circular spots on the leaf surface. The center of the spots is light brown, bordered by a darker green edge, often with a yellow halo around the lesions	The pathogen can grow within a temperature range of 10–35 °C, matching the seasonal growth conditions of patchouli. The disease thrives at an optimal temperature range of 20–28 °C, with 28 °C being the most favorable. During the high-temperature and rainy months of May to June, the disease can spread rapidly, making patchouli particularly susceptible

Chapter 10
Diseases of Schizonepeta

Schizonepeta (*Schizonepeta tenuifolia* (Benth.) Briq.) is an annual herb belonging to the genus Schizonepeta of the Lamiaceae family. It is also known as Xiangjingjie (fragrant schizonepeta), Jiasu (false perilla), and Silengganhao (square-stemmed wormwood). This plant is widely distributed across most regions of China. The whole plant is used as medicine, with an aromatic smell, a slight astringent and pungent-cool taste. It can relieve exterior syndromes and dispel wind, promote eruption, and reduce sores, and is used for colds, headaches, measles, rubella, and initial stages of sores and ulcers.

The primary diseases affecting schizonepeta include stem blight, root rot, black spot, and damping-off. Among these, stem blight occurs most commonly and causes the most serious damage.

10.1 Stem Blight of Schizonepeta

Stem blight, also known as "black shank disease," is the most significant fungal disease affecting schizonepeta, primarily targeting the stems. It spreads rapidly, often forming a disease center that quickly expands outward. The disease incidence typically ranges from 15% to 30%, with some severely affected fields experiencing total crop loss. Stem blight is prevalent. It is currently known to occur in Hebei and Zhejiang provinces.

Symptoms

Stem blight primarily affects the stems of schizonepeta. It usually starts at the stem base and gradually spreads upward, encircling the stem. Occasionally, only the middle part of the stem is affected, with no symptoms at the base. Infected areas appear dark brown, dry, and shrunken, with necrosis around the stem. Early symptoms include numerous brown spots on the stem, and the plant does not wilt. In the

© China Agriculture Press Co., Ltd. 2026

T. Wang et al., *Atlas of Chinese Medicinal Plant Diseases*,
https://doi.org/10.1007/978-981-95-6326-5_10

mid-stage, the spots expand around the stem, causing the affected area to sink and constrict. This leads to wilting of the plant above the infected area, with leaves appearing curled and drooping in a water-loss state. In the later stages, tissues above the infected areas wither and die (Fig. 10.1).

Fig. 10.1 Symptoms of Stem Blight in Schizonepeta

Pathogen

The pathogen causing schizonepeta stem blight is *Phytophthora nicotianae*, a species in the family Peronosporaceae. On V8 agar, colonies of the pathogen appear white, with dense mycelium. Under a microscope, the hyphae show uneven thickness with numerous wart-like projections. Sporangia measure (25–40) μm × (15–20) μm, borne at the tips of hyphal sporangiophores stalks, and are smooth, ellipsoid, or nearly pyriform with a distinct prominent papilla. Mature zoospores measure (2–4) μm × (1–2) μm, released from the orifice of the sporangium, and are single-celled, round, nearly round, or ellipsoid forms (Fig. 10.2).

In addition, the anamorphic *Fusarium* species of Ascomycete, including *Fusarium graminearum*, *F. equiseti*, and *F. semitectum*, are also pathogens of schizonepeta stem blight. The spores of these *Fusarium* species are sickle-shaped. *F. graminearum* conidia have two to seven septa, measuring (3.80 ± 0.02) μm × (48.16 ± 0.46) μm; *F. equiseti* conidia have two to six septa, measuring (5.53 ± 0.05) μm × (31.00 ± 0.32) μm; *F. semitectum* conidia have three to five septa, measuring (4–4.21) μm × (29–34) μm.

Occurrence Pattern

The pathogen overwinters in plant residues or soil. In late May of the following year, under suitable climate and humidity conditions, the pathogen's spores can enter the host plant through root conduits, stomata, or directly through the epidermis. This results in water-soaked lesions on the stem or leaves, which quickly expand up and down the stem, creating a brown necrotic ring around the stem. As disease centers form, the sporangia on infected plants produce abundant zoospores, which are readily spread by rainwater, subsequently infecting nearby healthy plants. Multiple infection cycles can occur within a single growing season.

Control Measures

1. Scientific Site Selection: Choose fertile, loose, well-drained soil on high or dry land for planting.
2. Proper Crop Rotation: Practice 3–5 years of rotation with Poaceae crops.
3. Enhanced Cultivation Management: Sow early, ideally at the beginning of April, so that by the peak disease period in early June, the plants will have developed

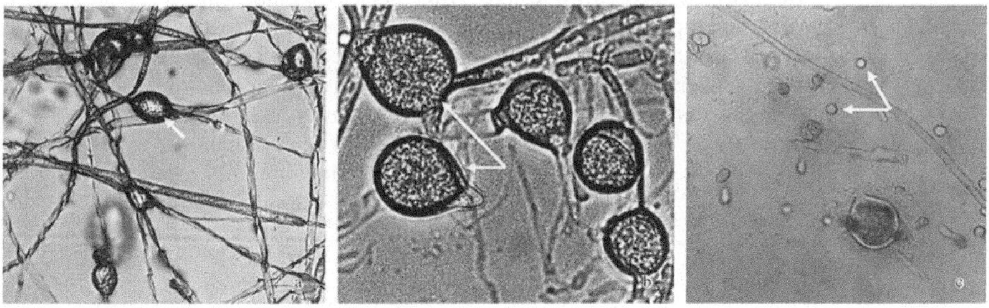

Fig. 10.2 Morphology of *Phytophthora nicotianae*. (**a**) Wart-like projections formed by hyphae. (**b**) Sporangia formed by hyphae. (**c**) Zoospores

some disease resistance. Reduce seeding rates when broadcasting and ensure proper plant spacing. In fields previously planted with wheat, apply sufficient base fertilizer, provide early-stage fertilizer, and balance nitrogen, phosphorus, and potassium to promote growth. Remove diseased plants promptly, carry them off-site for disposal, and apply local treatment to the infected area to eliminate the disease centers and protect surrounding plants.

4. Biological Control: Use biocontrol agents like *Trichoderma viride* and *Trichoderma harzianum*.
5. Chemical Control: At the early stage of disease onset, spray with a 600-fold solution of 25% metalaxyl wettable powder or a 2500-fold solution of 25% dimethomorph suspension every 7–10 days, repeating three to four times.

10.2 Occasional Diseases of Schizonepeta (Table 10.1)

Table 10.1 Characteristics of Occasional Diseases of Schizonepeta

Disease (Pathogen)	Symptoms	Occurrence pattern
Damping-off Disease (Rhizoctonia spp.)	It is a seedling disease. In the early stages of infection, water-soaked, dark brown lesions appear on the base of young stems near the soil surface at the seedling stage. These lesions spread rapidly and encircle the stem. The diseased parts often adhere to small soil-grain-like brown sclerotia. The above-ground parts wilt, and the seedlings lodge and die	The pathogen overwinters as mycelium or sclerotia on residual diseased plants or in the soil, becoming the initial source of infection in the following year. Disease residues, fertilizers, and soil can act as vectors, facilitating spread via water flow, farm workers, livestock, and other carriers. Humid weather conditions favor the outbreak of the disease, and repeated cropping increases disease severity
Black Spot Disease (*Alternaria alternata*)	On the leaves, the disease typically starts from the tips or along the edges, initially presenting as small, irregular brown spots. These spots gradually expand into larger semi-circular or irregularly shaped brown lesions. After stems and top shoots affected, they turn brown, and the tops droop or break down. Under high humidity, a black mold layer is produced on the diseased parts	The pathogen overwinters in the soil with diseased plant residues, and in the following year, it produces conidia that serve as the primary source of infection. The conidia produced on diseased plants can be spread by wind and rain, causing multiple rounds of reinfection. The disease typically begins in June, and the disease gradually lessens after mid-July. Rainy seasons can lead to severe seedling mortality

Chapter 11
The Diseases of Licorice

Licorice (*Glycyrrhiza uralensis* Fisch.), also known as Guolao, Sweet Grass, and Red Licorice, is a perennial herbaceous plant in the Fabaceae family. It typically includes Swollen Licorice and Smooth Licorice. The roots and rhizomes are harvested in spring and autumn, with fibrous roots removed and the remaining parts dried for medicinal use. Wild licorice is distributed across regions from Northeast China (Heilongjiang, Liaoning, Jilin), North China (Hebei, Shanxi, Inner Mongolia), to Northwest China (Shaanxi, Gansu, Ningxia, Qinghai, and Xinjiang). Cultivated licorice is found in provinces including Heilongjiang, Jilin, Liaoning, Inner Mongolia, Shanxi, Hebei, Ningxia, Gansu, Qinghai, Shaanxi, and Xinjiang, where it forms a long east-west distribution and narrow north-south distribution. Licorice has various medicinal properties, including tonifying the spleen and qi, clearing heat and detoxifying, resolving phlegm and stopping cough, relieving pain, and harmonizing other medicines. It is used for treating spleen and stomach weakness, fatigue, palpitations, shortness of breath, cough with copious phlegm, abdominal pain, limb spasms, carbuncles, boils, and for mitigating drug toxicity. Known as the "Nine Herbs in Ten Formulas," licorice is susceptible to diseases such as root rot, leaf spot, rust, brown spot, and powdery mildew, with root rot being particularly severe.

11.1 Licorice Root Rot

Licorice root rot is a common and severe soil-borne disease in licorice plantations in Xinjiang and Gansu. The disease usually appears in mid-May, with some areas reporting cases as early as April, and peaks from June to September. Field infection rates typically range from 5% to 10%, with severe cases reaching up to 80%. Licorice root rot significantly limits licorice production, causing substantial economic losses due to reduced yield and quality.

© China Agriculture Press Co., Ltd. 2026

T. Wang et al., *Atlas of Chinese Medicinal Plant Diseases*,
https://doi.org/10.1007/978-981-95-6326-5_11

Symptoms

In the early stages, leaves turn yellow and wilt, with the main and fibrous roots showing a yellow-brown color, eventually darkening to deep brown. The disease spreads upward from the root to the stem. In the middle stage, the aboveground part of the plant becomes stunted, with leaves wilting, yellowing, and shriveling. In the advanced stage, leaves wither and fall off, stems decay, and the epidermis separates from the woody parts, leaving woody fibers and debris behind, or roots appear water-soaked and rot, sometimes leading to complete plant collapse and death (Fig. 11.1).

Pathogen

The causative agents of licorice root rot are fungi from the genus Fusarium, which may infect plants either singly or in combinations. These pathogens include *Fusarium solani* (Mart.) Appel et Wollenw. ex Snyder et Hansen, *F. oxysporum* Schltdl. ex Snyder et Hansen, and *F. verticillioides* (Sacc.) Nirenberg (Fig. 11.2). Among these, *F. oxysporum* and *F. solani* are the dominant pathogens in various licorice-growing regions. *F. oxysporum* produces dense, cottony aerial mycelium in white, dark red, purple, pink, and waxy yellow colors, with some strains forming mycelial balls in the later stages of growth. Large conidia are spindle-shaped with pointed apical cells and foot cells, mostly having three to five septa, measuring (29.59–52.62) μm × (6.35–8.44) μm. Small conidia appear kidney-shaped or ovoid,

Fig. 11.1 Symptoms of Licorice root rot. (**a**, **b**) Stem symptoms. (**c**, **d**) Root symptoms. (**e**) Whole plant symptoms

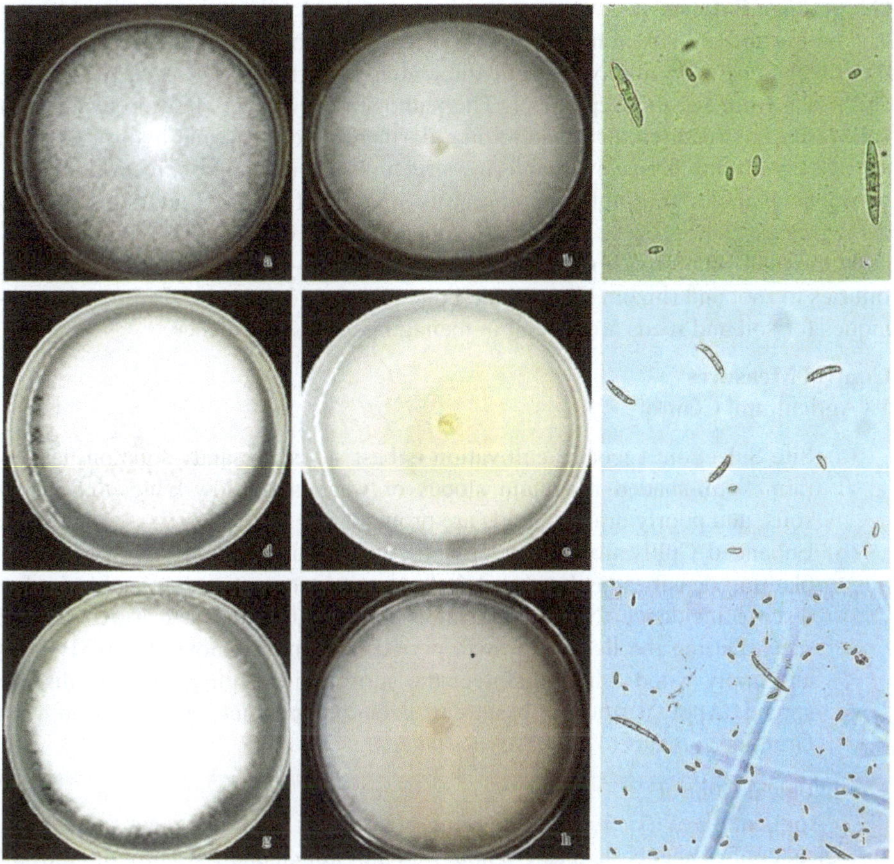

Fig. 11.2 Pathogen morphology of Licorice root rot. (**a**, **b**) *Fusarium oxysporum* culture characteristics. (**c**) *Fusarium oxysporum* Conidia. (**d**, **e**) *Fusarium solani* culture characteristics. (**f**) *Fusarium solani* Conidia. (**g**, **h**) *Fusarium verticillioides* culture characteristics. (**i**) *Fusarium verticillioides* Conidia

measuring (7.87–14.12) μm × (3.92–6.72) μm. *F. solani* produces sparse, yellow-white to light gray aerial mycelium with concentric ring patterns and a yellow-brown central clump. Conidia vary significantly in shape and size, including sickle, oval, and ovoid forms, often with basal foot cells. Small conidia have zero to one septum, measuring (4.24–8.46) μm × (1.64–3.63) μm, and large conidia have three to five septa, measuring (13.46–19.21) μm × (2.59–3.98) μm. Colonies grow vigorously with curled, woolly mycelium forming a felt-like layer, initially white but darkening over time, with orange pigments on the reverse side of the culture medium. *F. verticillioides* produces long, slender large conidia with pointed apical cells and slight curvature, usually having three to five septa, measuring (11.84–29.36) μm × (1.07–2.81) μm, and small conidia are mostly rod-shaped without septa, measuring (6.69–16.10) μm × (2.87–4.84) μm.

Occurrence Pattern

The timing and severity of licorice root rot vary by region and plantation. Generally, the disease begins in mid-May, with some areas reporting cases as early as April, and peaks from June to September. The pathogen overwinters in crop residues or soil, forming structures such as sclerotia, dormant mycelium, and hyphae to resist adverse conditions. These overwintering structures act as primary infection sources, germinating under favorable conditions to infect root cortexes and subsequently damaging vascular tissues. The disease spreads extensively in areas with high temperatures and humidity, poor soil drainage, and long-term continuous cropping. Injuries to root and rhizome tissues facilitate pathogen invasion. Disinfection techniques for soil and seeds are crucial in managing disease incidence.

Control Measures

1. Agricultural Control
   (a) Site Selection: Licorice cultivation is best suited to sandy soils on flat terrain. Semi-shaded mountain slopes or wastelands, low-lying areas, clay soils, and poorly drained fields are prone to disease.
   (b) Enhanced Cultivation Management: Avoid continuous cropping and high planting densities, as longer periods of continuous cropping lead to higher disease incidence. Newly cultivated or rotated fields have lower disease rates. During the licorice growth period, promptly remove diseased plants and apply wood ash to disinfect the surrounding soil, preventing disease spread. Apply sufficient basal fertilizer and practice crop rotation with Gramineae plants to increase resistance.
2. Biological Control
   Application of *Trichoderma*: During planting or seedling stages, apply biological control agents such as *Trichoderma harzianum* and *T. koningii* to the soil for disease prevention. Follow product instructions based on soil and environmental conditions, with the option of adding rice bran or soybean meal to enhance *Trichoderma* proliferation in the field. Use a mixture of *Trichoderma* and organic materials as bio-fertilizer, integrating 1–1.5 kg of *Trichoderma* per ton of organic fertilizer or adding it to crop waste as bio-fertilizer.
3. Chemical Control
   (a) Soil Disinfection: Treat soil with 7–8 g/m^2 of 50% carbendazim WP or a disinfectant powder mix (50:50:2 of wood ash, lime, and sulfur powder) at 7.5 kg/m^2.
   (b) Seedling Drenching: Soak seedlings for 5 min in a 1:1 mixture of carbendazim and methyl thiophanate-methyl diluted 200 times, then air-dry for 1–2 h before transplanting.
   (c) Spray or Root Irrigation: At the onset of infection, apply a foliar or root drench of 50% carbendazim or thiophanate-methyl WP at 800–1000-fold dilution, or 50% benomyl WP at 1500-fold dilution.

11.2 Licorice Leaf Spot Disease

Licorice leaf spot disease is a significant foliar disease affecting licorice plantations in Xinjiang and Gansu. When severe, the infection rate can reach up to 80%, posing a major limitation to licorice production.

Symptoms
The disease primarily affects leaves but can also infect stems. It usually starts on the older, lower leaves and gradually spreads upward. In the early stages, lesions appear as small, brown spots, the size of pinheads, round or oval, occasionally irregular in shape. The center of the lesion is darker, with lighter edges and a faint yellow halo. The underside of the leaf exhibits yellow-brown spots. As the disease progresses, lesions expand into round, oval, or irregular shapes, dark brown in color. In severe cases, lesions merge, leading to leaf necrosis (Fig. 11.3). When the stem is infected, similar symptoms appear, often originating at the leaf axils and spreading to new shoots and the main stem. This results in the affected area becoming brown, sunken, and cylindrical or elongated in shape. In severe cases, lesions can cover the entire stem surface, forming continuous patches that darken the stem's epidermis to a black-brown color, significantly disrupting nutrient transport, weakening the plant, and in some cases, leading to the death of the infected plants.

Pathogen
The pathogens of licorice leaf spot disease belong to the genus *Alternaria*, including *Alternaria alternata* (Fr.) Keissl., *A. tenuissima* (Fr.) Wiltshire, and *A. rosae* E.G. Simmons et C.F. Hill. Based on samples collected from major licorice--producing areas across China, *A. alternata* and *A. tenuissima* are the dominant species.

Fig. 11.3 Symptoms of Licorice leaf spot. (**a**) Field symptoms. (**b**) Leaf symptoms

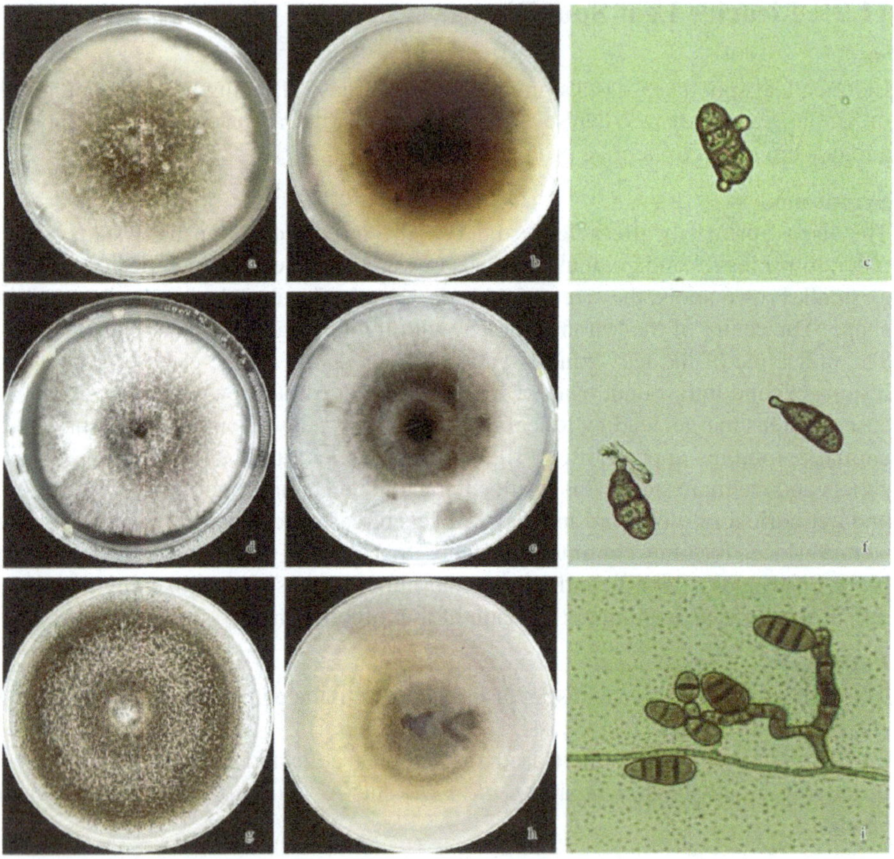

Fig. 11.4 Pathogen morphology of Licorice leaf spot. (**a**, **b**) *Alternaria alternata* culture characteristics. (**c**) *A. alternata* Conidia. (**d**, **e**) *A. tenuissima* culture characteristics. (**f**) *A. tenuissima* Conidia. (**g**, **h**) *A. rosae* culture characteristics. (**i**) *A. rosae* Conidia

In culture, *A. alternata* initially produces white, radiating mycelium, which gradually turns gray-white, gray, or black. The colonies are typically velvety with neat edges, producing dark green or black pigments on the reverse side. Microscopically, the hyphae are colorless with septa, light brown or brown. Conidiophores are upright or knee-bent, with septa, arising from the ends or sides of hyphae. On PCA medium, conidia form short chains at the tops or sides of conidiophores, with secondary branches forming short, branched chains resembling a tree structure. Conidia are light brown or brown, nearly oval, ovoid, or inverted club-shaped, with one to six transverse septa and zero to four vertical or oblique septa, sometimes possessing a single cell beak or true beak. The septa may be constricted or not, with dimensions of (24.22–44.67) μm × (13.55–23.45) μm, including the beak (Fig. 11.4a-c).

Alternaria tenuissima colonies are circular, with initially white mycelium turning gray or dark green, woolly, with tidy edges and a dark olive green or black reverse side. The aerial mycelium is dense and fast-growing, eventually covering the entire culture plate. On PCA medium, conidia form in chains, with or without branches. Conidia are light brown or brown, long-oval or inverted club-shaped, occasionally pear-shaped, with one to six transverse septa and zero to four vertical or oblique septa. A prominent constriction is present at each septum. Mature conidia typically have a main transverse septum that is darker due to thickening, appearing dark brown, with dimensions of (28.16–62.59) μm × (8.89–20.81) μm, including the beak. Conidiophores are light brown, mostly solitary on the sides of hyphae, with few branches, upright or slightly bent, and have septa. They possess columnar or conical true or false beaks, which are light brown with septa and slightly swollen at the base (Fig. 11.4d-f).

Alternaria rosae colonies initially have a dense dark green felt-like center with white, neat edges and yellow-brown concentric rings on the reverse. The aerial mycelium is sparse and white, with the colony becoming rough as the white turns to dark green from the center to the edge. Colonies grow rapidly, covering the entire plate within 7 days. On PCA medium, conidia form in short chains with secondary branches. Conidia are light brown or brown, oval or inverted pear-shaped, with two to four transverse septa and zero to four vertical or oblique septa, slightly or non--constricted at the septa, with conical true beaks, measuring (18.96–37.11) μm × (11.54–19.11) μm, including the beak. Conidiophores are light brown, mostly solitary on the sides of hyphae, branched, upright or bent, with septa (Fig. 11.4g-i).

Occurrence Pattern

The onset and severity of licorice leaf spot disease vary by region and plantation. Typically, it begins in June, with some areas reporting cases in late May. The disease peaks during the rainy season in August and September, especially in high--temperature, high-humidity fields. The pathogen overwinters in crop residues in the field, which serves as the primary infection source, with seed-borne infections as a secondary source. The disease incidence and spread are closely related to temperature and humidity, with high humidity being the primary factor in disease occurrence and spread. Higher, drier fields have lighter infections, while crop rotation effectively reduces incidence.

Control Measures

1. Agricultural Control
   (a) Crop Rotation: Implement a rotation of at least 2 years to allow complete degradation of infected residues.
   (b) Improved Cultivation Management: Promptly remove and destroy infected plants to reduce the inoculum in the field. Overwintering infected plants are the primary initial infection sources. When licorice stems and leaves wither, cut and collect them for disposal to eliminate field residues. After licorice harvesting, clear the field of residues and nearby weeds, deep burying them to reduce the next year's initial infection sources. Control planting density appropriately by selecting suitable plant spacing and pruning interior branches for better ventilation and light exposure.

(c) Fertilization Management: Apply balanced fertilizers, combining nitrogen, phosphorus, and potassium, to avoid overuse of nitrogen and increase potassium application to promote healthy plant growth and enhance disease resistance.

2. Biological Control

The biocontrol agent *Trichoderma viride* has demonstrated an inhibitory effect on the growth of *Alternaria* species, with an inhibition rate of 73.9%. Studies show that its efficacy remains significant even at low spore concentrations, meaning *Trichoderma viride* effectively suppresses *Alternaria* at minimal doses. Additional research indicates that *Bacillus subtilis* also provides a substantial control effect against *Alternaria*. Since different biocontrol agents have distinct active strains, application rates vary accordingly. The following application methods are provided for *Bacillus subtilis* (40 billion CFU/g):

Seedling Nursery: Mix 2–5 grams of *Bacillus subtilis* per square meter with soil and apply to the seedbed, ideally in combination with a suitable amount of fertilizer.

Drip Irrigation: Apply through drip irrigation two to three times during the growing season, adding *Bacillus subtilis* to the nutrient solution.

Spraying: Before disease onset or at an early stage, dilute at a 200–500 ratio and spray on foliage and root collars, ensuring even coverage. Spray once every 5–7 days and continue as needed. To minimize UV damage to *Bacillus subtilis*, it is best to spray in the evening. If rain occurs within 3 h of spraying, reapply. However, if it rains more than 3 h after spraying, additional spraying is not necessary. Drenching: After preparing a mother solution, apply via irrigation, using 200–400 grams per mu (approximately 0.0667 hectares). Broadcast Application: Apply alone or mixed with fine soil, manure, or fertilizer, then plow the field to incorporate it into the soil, using 200–400 grams per mu. Root Drenching: After transplanting, dilute the agent at a 500–1000 ratio and drench the roots. Seed Treatment: Use 50–200 grams per mu. For seed quantities exceeding 10 kg per mu, double the dosage. For treating tubers or rootstocks, the dosage generally does not need to be increased. Alternatively, a broad-spectrum pyrimidine nucleoside antibiotic can be applied at a dilution of 1200–1500 times, spraying every 7–10 days for 2–3 consecutive applications.

3. Chemical Control

Several fungicides effectively inhibit the growth of *Alternaria*, though efficacy may vary among different products. Based on comprehensive assessments of fungicidal efficacy, metalaxyl-mancozeb, Bordeaux mixture, thiophanate-methyl, tebuconazole, pyraclostrobin, and boscalid are all suitable for controlling licorice leaf spot disease. Apply Bordeaux mixture (copper sulfate: lime: water = 1:1:100–160), or 70% thiophanate-methyl WP at a 1,500–2,000-fold dilution. Alternatively, spray 80% tebuconazole WG at 5,000–7,000-fold dilution, 50% pyraclostrobin WG at 2,000-fold dilution, or 10% boscalid WG at 750-fold dilution. Spray once every 10 days, with two consecutive applications, for effective control of licorice leaf spot disease.

11.3 Licorice Rust Disease

Licorice rust disease is one of the most severe diseases affecting licorice, damaging the entire above-ground portion of the plant. It severely impacts the growth and photosynthesis of licorice, and is prevalent across the entire growing season and in licorice plants of various ages. The disease is widespread in licorice-growing regions such as Gansu, Xinjiang, Inner Mongolia, and Ningxia. In some planting areas, the infection rate can reach up to 100%, significantly affecting licorice quality and yield.

Symptoms

Licorice rust disease mainly affects leaves, petioles, and young stems. Wild licorice often exhibits systemic symptoms, causing stunted growth, chlorosis, and deformities, eventually leading to plant death. Lesions primarily appear on the underside of leaves and young stems, sometimes producing yellowish-brown spore pustules on the leaf surface. These pustules are oblong, releasing a large amount of brown powder (summer spores) when ruptured. In August and September, blackish-brown, small blister-like structures called teliospores form (Fig. 11.5).

Pathogen

The pathogen responsible for licorice rust disease is *Uromyces glycyrrhizae* (Rabenh.) Magnus, a basidiomycete rust fungus. Under microscopic observation, teliospores are oval or obovate, with short stalks or no stalks, measuring (22.9–30.9) μm × (19.1–23.1) μm, with an apical wall thickness of 1.5–3.8 μm and a lateral wall thickness of 1.2–2.4 μm. Uredospores are nearly spherical with spiny surfaces, measuring (22.9–27.3) μm × (17.7–26.1) μm, with a lateral wall thickness of 1.3 μm. Slight morphological differences may occur depending on the collection area (Fig. 11.6).

Fig. 11.5 Licorice Rust Disease symptoms. (**a**, **b**) Field symptoms. (**c**) Leaf symptoms. (**d**) Teliospore pustules

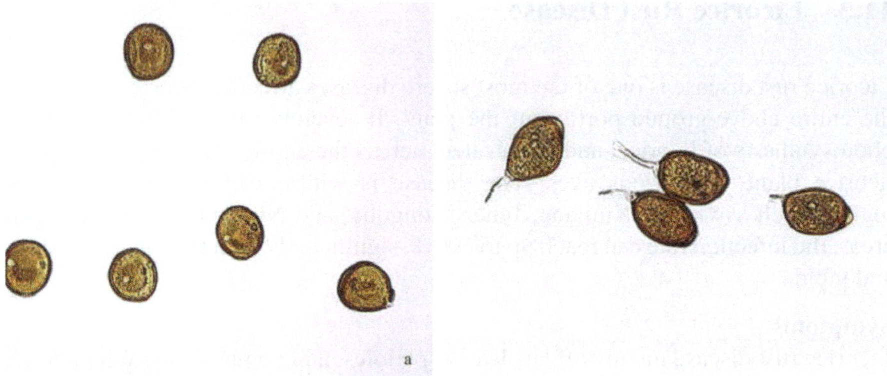

Fig. 11.6 Pathogen morphology of Licorice Rust Disease. (**a**) Uredospore morphology. (**b**) Teliospore morphology

Occurrence Pattern

Licorice rust disease affects both wild and cultivated licorice, making it one of the most common diseases in licorice-growing areas. The pathogen *Uromyces glycyrrhizae* is a full-cycled rust fungus, producing all five spore types (spermatia, aeciospores, uredospores, teliospores, and basidiospores). As an obligate parasite, *U. glycyrrhizae* overwinters on licorice without needing an alternate host. The initial inoculum includes three types of spores: uredospore-type, mixed-type, and aeciospore-type. Uredospore-type infections originate from overwintered uredospores, while aeciospore-type infections stem from overwintered teliospores. Mixed-type infections involve a combination of both. Both uredospores and teliospores overwinter by infecting licorice root buds or axillary buds, providing resistance to cold environments.

In favorable conditions, overwintered uredospores germinate and directly infect licorice plants. Similarly, overwintered teliospores germinate to infect licorice plants, with primary hyphae from teliospores gradually forming pycnidia, which then produce numerous aecia on the leaves, eventually leading to the production of uredospores that cause field infections. Uredospore infections are more infectious and are the main cause of rust disease outbreaks, unlike aeciospores, which are less infectious and serve as an intermediate stage. Rust disease has two peak infection periods: the first peak occurs in June, caused by uredospores, leading to leaf necrosis. This is also the disease's most active phase. The second peak occurs in September, primarily caused by aeciospores, which damage leaves, marking another high infection period. The earlier rust disease appears, the more severe the damage, potentially leading to mass plant death. There is a noticeable difference in disease resistance among plants, with some severely infected plants dying while neighboring plants remain unaffected, warranting further research into its mechanism.

Control Measures

1. Agricultural Control

   (a) Enhanced Cultivation Management: Thoroughly remove crop residues after harvest. Select healthy, rust-free plants for seed propagation. Irrigate during winter and spring, and remove above-ground stems and leaves in autumn to reduce disease occurrence.
   (b) Selection and Use of Resistant Varieties: Since the rust fungus is an obligate parasite, using resistant varieties is the most effective control method.
   (c) Proper Sowing: Avoid monocultures of a single licorice variety within the same area. Instead, mix several licorice varieties to prevent the spread of rust and reduce losses. Increase row spacing in seed-production fields, and harvest rust-infected plants early to lower disease incidence and transmission.
   (d) Balanced Fertilization: Avoid over-application of nitrogen; apply phosphorus, potassium, and calcium fertilizers to enhance resistance to rust.
   (e) Irrigation Management: Avoid waterlogging, ensure proper drainage, and prevent excessive humidity in the canopy to mitigate disease.

2. Biological Control

Research on biological control for licorice rust is limited domestically and internationally. However, studies have shown that *Trichoderma* spp. and *Bacillus* spp. effectively control rust diseases in various crops. *Trichoderma harzianum* and *Bacillus amyloliquefaciens* have demonstrated antagonistic effects against rust mycelia. Additionally, pyrimidine nucleoside antibiotics have shown efficacy in rust disease control. Use a 4% pyrimidine nucleoside antibiotic solution at 600–800× dilution or a 2% solution at 300–500× dilution, spraying at the early stage of infection. Repeat every 7–10 days for two to three applications.

3. Chemical Control

At the early infection stage, apply a 1,000-fold dilution of 25% triadimefon WP. Alternatively, spray with 0.3° Baumé lime sulfur, 65% zineb WP at 500-fold dilution, or 67% selenium compound WP at 200-fold dilution. For enhanced efficacy, apply with preader–sticker adjuvant to improve the protective effect, spraying once every 7–10 days for 2–3 consecutive applications.

11.4 Licorice Powdery Mildew

Licorice powdery mildew is a common foliar disease found in licorice fields, primarily in regions such as Gansu and Xinjiang. Severe outbreaks have been reported in Minqin, Jingning, and Yuzhong in Gansu, with Jingning showing the highest incidence rate of 38.6% and the most severe cases. Infected leaves in the field often also exhibit symptoms of other foliar diseases. This disease significantly impairs photosynthesis and metabolism in licorice; in severe cases, affected leaf tissues turn yellow and dry out, eventually leading to leaf drop and a substantial impact on licorice production.

Symptoms

Licorice powdery mildew manifests primarily on leaves. In the early stages, small white powdery spots appear on the leaves, initially the size of a pinhead. These spots gradually expand to the size of a mung bean and eventually spread across the entire leaf, flower stalks, and young stems, ultimately covering them with a white powdery layer. In later stages, the white layer falls off, leaving behind numerous small black dots on the lesions (Fig. 11.7).

Pathogen

The causative agent of licorice powdery mildew is *Erysiphe polygoni* DC., a fungus in the family Erysiphaceae under the phylum Ascomycota. The pathogen's chasmothecia are nearly spherical, brown to dark brown, with filamentous appendages. Each chasmothecium contains four to six asci, each bearing two to six oval ascospores. Conidia are oval and occur singly on conidiophores (Fig. 11.8).

Occurrence Pattern

Licorice powdery mildew typically peaks in spring (May–June) and autumn (September–October). In northern China, it primarily occurs in spring, with limited incidence in autumn, although some areas experience year-round infections. The pathogen spreads through airborne spores, causing secondary infections in the field. The optimal temperature for spore germination is 25–28 °C, with an ideal relative humidity of 50–80%; strong sunlight inhibits the pathogen's growth. Consequently, conditions such as a sunny spell after rain or humid weather with significant day-night temperature fluctuations are particularly conducive to the occurrence and spread of powdery mildew. As the pathogen invades host epidermal tissues directly or through stomata and lenticels, the characteristics of host epidermal tissues

Fig. 11.7 Field symptoms of Licorice powdery mildew

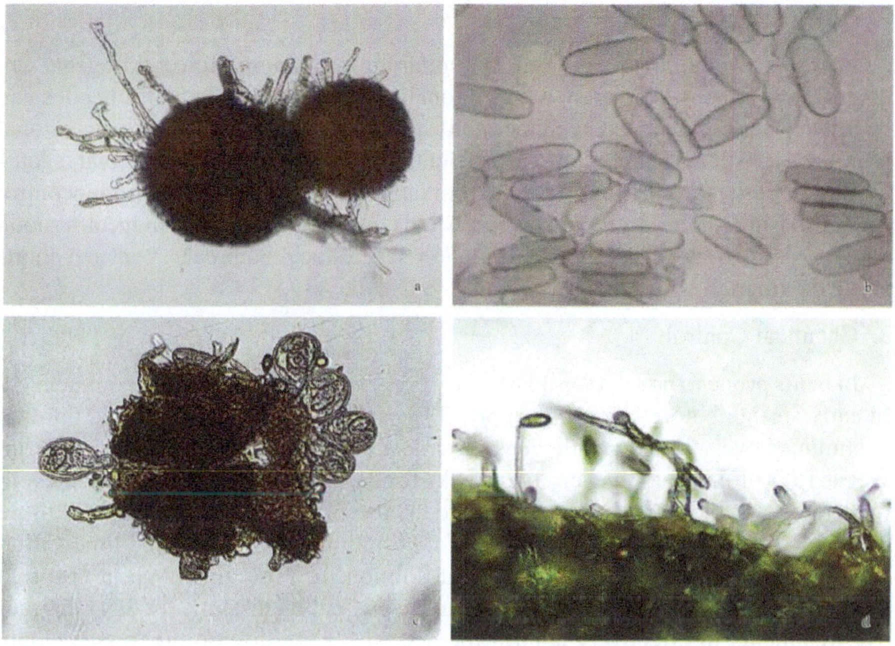

Fig. 11.8 Pathogen morphology of Licorice powdery mildew. (**a**) Chasmothecia. (**b**) Conidia. (**c**) Ascospores. (**d**) Conidiophore structure

significantly influence susceptibility. The pathogen overwinters in soil with plant residues as chasmothecia. The disease incidence and severity are higher in plants with thin epidermal tissues, often resulting from insufficient light, excessive nitrogen fertilization, or excessive growth.

Control Measures

1. Agricultural Control

   (a) Enhance Field Management: Remove and destroy diseased leaves promptly to reduce sources of infection. Improve ventilation and light conditions; avoid excessive planting density. Before planting, optimize light exposure and ventilation.

   (b) Selection of Resistant Varieties: Significant differences in powdery mildew resistance exist among licorice varieties, making the choice of resistant varieties essential.

   (c) Balanced Fertilization: Supplement calcium and silicon fertilizers. Licorice with thin epidermal tissues is more susceptible to powdery mildew infection. In early growth and rapid growth stages, particularly in densely planted areas, maintaining balanced soil fertility and supplementing calcium and silicon foliar fertilizers significantly enhance disease resistance.

2. Biological Control

Bacillus amyloliquefaciens and *Trichoderma harzianum* biological agents can control licorice powdery mildew. Additionally, mycophagous ladybird beetles can be utilized for biocontrol. Given the rapid spread of powdery mildew, conventional biological pesticides are often insufficient, making combined microbial agents (*T. harzianum* + *B. amyloliquefaciens*) a viable option. When applying biocontrol agents, ensure thorough coverage, especially after morning warming, afternoon cooling, or clear weather following rain. For severe local outbreaks, consider shortening the interval between applications.

3. Chemical Control

In fields prone to powdery mildew, early preventive treatments are recommended. In early spring, before sprouting, apply 50% carbendazim WP at 600-fold dilution to eliminate overwintering pathogens. After leaf expansion, apply 50% carbendazim WP at 1,000-fold dilution every two weeks for 3–4 applications to solidify preventive effects. At the onset of powdery mildew, use 50% benomyl WP at 1,000-fold dilution, 15% triadimefon WP at 1,000-fold dilution, 0.2–0.3° Baumé lime sulfur, 50% thiophanate-methyl WP at 800-fold dilution, or 50% zineb WP at 600-fold dilution. Apply every 7–10 days, spraying leaves first, then stems, for 3–4 consecutive treatments to effectively control the disease. If the disease progresses, apply 40% carbendazim WP at 1,500-fold dilution or 25% triadimefon WP at 1,500-fold dilution, spraying every 10–15 days for 2–3 treatments.

11.5 Licorice Viral Diseases

The primary viral diseases affecting licorice include licorice mosaic virus, licorice clover yellow vein virus, licorice tobacco streak virus, licorice phytoplasma disease, licorice wilt disease, licorice leaf spot virus, and licorice stem and leaf spot virus. These diseases can occur under favorable environmental conditions and may have a significant impact on the yield and quality of field-cultivated licorice (see Table 11.1).

Table 11.1 Characteristics of Licorice Viral Diseases

Disease (Pathogen)	Symptoms	Occurrence pattern
Cercospora Leaf Spot Disease (*Cercospora glycyrrhizae*)	Primarily affects leaves, with lesions appearing as irregular or triangular spots, usually located in the central area of the leaf, along the main or side veins. These spots are often elongated and strip--shaped, yellowish to reddish-brown in color. As the disease progresses, the lesions develop a dense layer of black mold-like substance on the surface	During the warm and humid months of July and August, with frequent rainfall, heavy dew, and high humidity, the disease incidence is high. Cultivated licorice tends to suffer more than wild licorice (with the exception of licorice mosaic virus, which affects both wild and cultivated varieties). Disease severity increases notably in the second year of cultivation
Clover Yellow Vein Virus (*Ascochyta onobrychidis*)	Lesions on leaves may also appear as round or nearly round spots, brown in color, some surrounded by faint yellow halos. In later stages, the spots may crack, producing numerous spore masses on the lesions	The pathogen overwinters in infected plant residues or soil debris and also on licorice seeds. Cool and moist conditions favor the disease's onset and spread. The severity of the disease generally intensifies with the age of the plant
Gray Mold Disease (*Botrytis cinerea*)	The disease primarily damages leaves, with lesions ranging from small to large spots located at the center or edges of the leaves. These spots are generally round, reddish-brown, or dark brown, often with a lighter center. The diseased tissues do not display a clear boundary but are covered with a layer of dark moldy hyphae	Warm and humid environmental conditions encourage the disease's development, especially if plant management is lax or if plants show weak growth
Alternaria Black Spot Disease (*Alternaria azukiae*)	Lesions frequently appear along leaf margins and extend inward in a semicircular or irregular shape. The spots are reddish-brown, shallow, and pale. In later stages, the affected areas darken, turning a deep brown and cracking, eventually producing spore masses	The disease generally begins spreading at the end of June, peaks from July through August, and may continue through September. The amount of rainfall, dew, and ambient humidity levels are closely related to the disease's spread and intensity
Virus Disease (Pathogen Undetermined)	The disease can cause flower structures to become pale green, ultimately leading to wilting of the leaves	Disease severity tends to lessen from June to August

(continued)

Table 11.1 (continued)

Disease (Pathogen)	Symptoms	Occurrence pattern
Damping-Off Disease (*Rhizoctonia solani*)	Primarily infects young seedlings at the base of stems or roots. Initial symptoms are irregularly shaped, water-soaked spots that turn brown. During rainy weather, symptoms intensify, leading to rapid spread. Infected seedlings exhibit wilting and eventually collapse	
Stem and Leaf Spot Disease (*Phoma herbarum*)	Mainly damages stems and branches, causing lesions that appear as circular to irregular, reddish-brown to dark brown spots. As the disease progresses, the spots enlarge and darken, merging to form sunken areas that affect internal tissues, eventually leading to stem rot and breakage at the base, resulting in plant wilting and death	Licorice damping-off often occurs during the seedling stage in cultivated nursery beds. Inadequate seed treatment and poorly drained soil significantly increase the incidence rate, leading to high seedling mortality

Chapter 12
The Diseases of Kudzu

Kudzu [*Pueraria montana* var. lobata (Willdenow) Maesen & S. M. Almeida ex Sanjappa & Predeep], belongs to the Fabaceae family, also known as kudzu vine or wild kudzu. Its dried roots, known as kudzu root, are used medicinally. Kudzu is distributed across China except in Xinjiang, Qinghai, and Tibet. Kudzu root is one of the most commonly used medicinal materials in China, first recorded in "Shen Nong Ben Cao Jing," known for its properties to relieve muscle tension, reduce fever, quench thirst, clear rashes, stop diarrhea, facilitate blood circulation, and detoxify from alcohol.

Kudzu is primarily affected by diseases such as pseudo-rust, rust, and root rot.

12.1 Kudzu Pseudo-Rust Disease

Kudzu pseudo-rust is one of the most severe diseases in kudzu cultivation. It mildly affects the photosynthesis in leaves leading to a decline in both yield and quality of kudzu roots; in severe cases, it causes leaf shrinkage, curling, vine twisting, and yellowing and dropping of leaves. It can destroy plantations within about 3 days of onset, causing significant financial losses to farmers and serious impediments to the industry. The disease is common in large-leaf kudzu, fine-leaf kudzu, and wild kudzu, with typical infestation rates in fields ranging from 5% to 10%, and up to 90% in severe cases.

Symptoms

Kudzu pseudo-rust affects the leaves, petioles, and stems of the plant, with most lesions occurring along the veins (Fig. 12.1a, b). Early symptoms visible to the naked eye include chlorotic spots around the veins and petioles. As the disease progresses, these spots become yellowish and blister-like. When these blisters are manually ruptured, they exude a yellow pus-like substance (comprising the pathogen's sporangia and motile spores). In later stages, these blisters burst, releasing an

© China Agriculture Press Co., Ltd. 2026

T. Wang et al., *Atlas of Chinese Medicinal Plant Diseases*,
https://doi.org/10.1007/978-981-95-6326-5_12

Fig. 12.1 Symptoms of Kudzu Pseudo-Rust Disease (**a**, **b**). Symptoms on leaves (**c**). Symptoms on stems

orange-yellow powder similar to rust, hence the name "pseudo-rust." Due to the swelling of affected cells, the lesions eventually appear wart-like and lumpy on stems (Fig. 12.1c), with a rough surface. Cutting through the epidermis reveals a mass of orange-yellow sporangia inside. Severely affected leaves and petioles become deformed, turn yellow, wilt, and eventually die.

Pathogen

The pathogen responsible for kudzu pseudo-rust is *Synchytrium puerariae* Miy, also known as *Synchytrium minutum* (Pat) Gaum. It is an endophytic, obligate parasitic fungus that specifically infects kudzu species. The sporangia are spherical or nearly so, granular, containing orange material, measuring between 20.00–26.25 μm × 18.75–25.00 μm. The sporangia can germinate between 10 and 35 °C, with the optimal germination temperature at 35 °C and lethal at 46 °C; they can germinate across a pH range of 3–11, with an optimal pH of 6–8. Mature sporangia burst upon contact with water, releasing single-flagellated motile spores.

Occurrence Pattern

The pathogen's sporangia and motile spores are spread through wind and rainwater, causing repeated infections and transmission.

Poor ventilation, low-lying moist areas, and shaded parts of the plant tend to suffer more severely, but during rainy years or seasons, the upper leaves and young stems can also become infected. Well-ventilated, dry areas, and those with structures facing the sun have fewer incidents of the disease, indicating that humidity is a key factor in disease occurrence. Sporadic cases begin appearing in May and continue through to the harvesting period in January of the following year. To date, no alternate hosts for this pathogen have been identified.

Control Measures

1. Agricultural Management: Enhance cultivation practices; use high, wide planting beds to improve air circulation among plants and reduce field humidity.
2. Chemical Control: Treat with 58% metalaxyl-M WP at a 1500-fold dilution, 64% mancozeb-metalaxyl-M WP at a 1000-fold dilution, or 72% mancozeb-metalaxyl WP at a 1500-fold dilution early in the disease cycle. Rotate fungicides to avoid resistance development.

12.2 Kudzu Rust Disease

Kudzu rust is one of the primary diseases affecting kudzu, prevalent across all major kudzu cultivation areas in the country, with an average infection rate of 5–30%.

Symptoms

The disease primarily targets kudzu leaves. Initially, pinhead-sized pustules, yellowish-white to light brown, appear on the undersides of the leaves. As the disease progresses, these pustules rupture, releasing a yellowish-brown powdery substance—the summer spore masses of the pathogen. In severe cases, the undersides of leaves are densely covered with rust-colored powder, causing leaf deformation, impaired photosynthesis, increased transpiration, and ultimately, gradual leaf desiccation. This significantly affects the growth of underground tubers and reduces yields (see Fig. 12.2).

Pathogen

The causative agent of kudzu rust is *Phakopsora pachyrhizi* Syd. et P. Syd., a basidiomycete fungus from the rust family. The summer spore masses are brownish, located beneath the host's epidermis, and are pear-shaped or oval with lateral filaments and a basal connection with an apical pore. Upon maturity, the epidermal layer of these masses ruptures, releasing the summer spores (see Fig. 12.3). These spores are nearly spherical to oval, single-celled, yellowish-brown, densely covered with spines, featuring four to five subtle germination pores, and measure (22.4–35.2) μm × (14.4–25.6) μm. The winter spore masses are embedded within the host tissue, pale yellowish-brown to light brown, elongated cylindrical or irregularly shaped, smooth on the surface, and measure (13–25) μm × (8–12) μm. The outermost wall of the spore often thickens. Each basidium produces one to three stalks, each bearing terminal basidiospores (see Fig. 12.4).

Fig. 12.2 Symptoms of Kudzu Rust Disease

Fig. 12.3 Summer spore masses of Kudzu rust pathogen

Occurrence Pattern

The rate of leaf infection by kudzu rust ranges from 10% to 50%, with significant variation depending on the kudzu's growth stage and plot location. In northern Guangxi, the pathogen overwinters as winter spores, which germinate the following year to produce basidia and basidiospores, serving as the initial sources of infection. The disease is further propagated through air currents carrying the summer spores, which act as secondary infection sources. In southern Guangxi, summer spores function as both primary and secondary infection sources, completing the disease's annual cycle. Here, winter spores are either not produced or play a negligible role. Besides kudzu, the pathogen also affects other legumes, such as soybeans. Warm, rainy weather, particularly high humidity, significantly facilitates the disease's spread. No notable differences in resistance among varieties have been identified thus far.

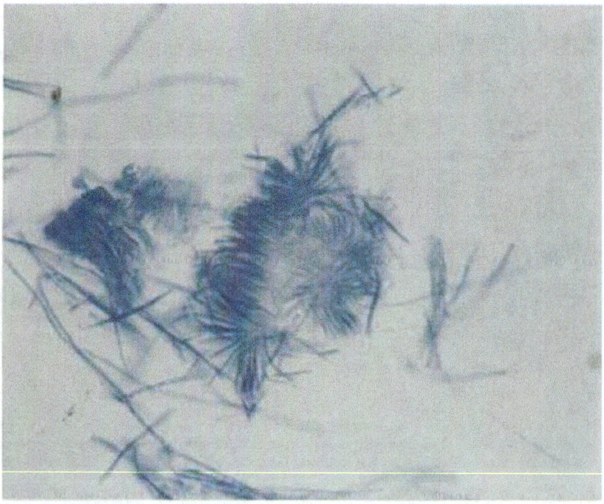

Fig. 12.4 Winter spores of Kudzu rust pathogen

Control Measures

1. Agricultural Control: Managing the prevalence of kudzu rust involves primarily regulating humidity. It is crucial to enhance plant spacing to improve air circulation and reduce moisture levels in the fields.
2. Chemical Control: In the early stages of the disease, treat with a 15% triadimefon WP or EC diluted at 1500–2000-fold. Ensure thorough and uniform application to both the upper and lower surfaces of the leaves.

12.3 Kudzu Root Rot Disease

Kudzu root rot is a common soil-borne disease that poses a significant threat to kudzu's growth and development, ultimately reducing both yield and quality.

Symptoms

Kudzu root rot primarily affects the tuberous roots and basal stems, with infection occurring during both the seedling stage and the tuber formation period. When seedlings are infected, water-soaked brown lesions appear on the epidermis of the root tips or mid-root sections. In severe cases, the roots decay and turn brown, resulting in stunted growth, slow development, and premature yellowing and shedding of the basal leaves. The upper parts of the plants exhibit wilting symptoms. When infection occurs during tuber formation, the initial symptoms are reddish--brown, nearly circular to irregular, slightly sunken spots on the tuber surface. As the disease advances, these spots enlarge and merge into extensive dark-brown patches with cracked surfaces and discolored, dry, rotting subdermal tissue. Cross-sections of the tubers reveal reddish-brown vascular bundles, and in later stages, the tubers develop a dry rot with a dark-brown, pithy center (see Fig. 12.5).

Fig. 12.5 Symptoms of Kudzu Root Rot Disease. (**a**) Symptoms on the plant. (**b**) Symptoms on the stem. (**c**) Symptoms on the roots

Pathogen

The pathogen responsible for kudzu root rot is *Fusarium solani* (Mart.) Appel & Wollenw. ex Snyder & Hansen, an asexual fungus from the genus Fusarium within the Ascomycota phylum. On PDA medium, colonies of *Fusarium solani* appear white, nearly circular, and velvety, with septate hyphae dividing into segments, each containing cytoplasm and one or more nuclei. The pathogen produces two types of conidia: large and small. The large conidia are fusiform to crescent-shaped, colorless, transparent, with blunt ends, and contain two to four septa (typically 3), measuring (22.5–37.5) μm × (3–4) μm. The small conidia are spindle-shaped to ovoid, with 0 to 1 septum, and measure (4.5–24) μm × (2.5–4) μm (see Fig. 12.6).

Occurrence Pattern

Kudzu root rot starts appearing sporadically from April to May and can persist until the harvest period in January of the following year. Peak disease incidence occurs from May to September, closely correlated with the rainy season during this period. In mildly affected fields, the disease incidence rate ranges from 10% to 20%, whereas in severely affected fields, it can reach 50% to 80%. The primary transmission pathways include infected soil, contaminated seedlings, and waterborne spread.

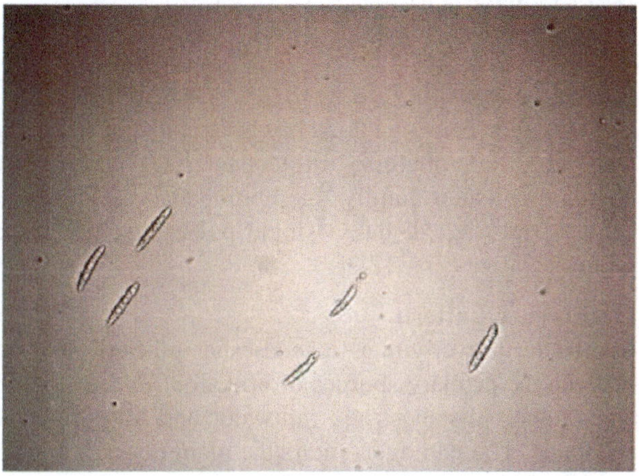

Fig. 12.6 Morphology of Kudzu root rot pathogen

Control Measures

An integrated control strategy, centered on agricultural practices to enhance plant resistance and supplemented with appropriate chemical control, is recommended.

1. Field Sanitation: After harvest, thoroughly remove all plant residues to reduce infection sources.
2. Soil Disinfection: Combine soil preparation with lime application before kudzu transplanting. Apply 50–70 kg of quicklime per mu to adjust soil pH and reduce pathogen populations.
3. Cultivation of Disease-Free Seedlings: Select healthy, disease-free kudzu vines or tubers as propagation material. Before planting, disinfect seed tubers by soaking them in hot water at 50–55 °C for 15–20 min.
4. Chemical Control: In fields with a disease incidence rate above 10%, chemical control can complement agricultural practices. Mix 50% carbendazim WP (800-fold dilution) with a suitable amount of fine soil to create treated soil for application in planting holes, avoiding direct contact with kudzu roots to prevent phytotoxicity. Avoid chemical applications within 80 days before harvest.

12.4 Kudzu Bacterial Leaf Spot Disease

Kudzu bacterial leaf spot disease is a major affliction of kudzu foliage, widespread across primary kudzu-growing regions nationwide, with an average infection rate of 10–30%.

Symptoms

Affected leaves initially show water-soaked, light green lesions that appear semi-transparent when viewed against the light. As the disease progresses, the lesions turn light brown or brown. The spots are polygonal, constrained by the leaf veins.

Under high humidity, a thin bacterial ooze can be seen on the leaf's underside, drying to form a glossy, egg-white-like film (Fig. 12.7).

Pathogen

The pathogen causing kudzu bacterial leaf spot is Xanthomonas campestris, a member of the phylum Proteobacteria, class Gammaproteobacteria, order Xanthomonadales, family Xanthomonadaceae, and genus *Xanthomonas*. The bacterium is rod-shaped, has a single polar flagellum, is chemoheterotrophic, and is Gram-negative (Fig. 12.8).

Occurrence Pattern

The bacteria overwinter in seeds and infected plant debris within the soil, which become the primary sources of infection for the next growing season. The bacteria spread short distances via rainwater and long distances through the transport of seedlings. The pathogen enters the plant through wounds or natural openings, with repeated infections occurring during warm, rainy weather, leading to increased disease spread.

Control Measures

1. Use Disease-Resistant Varieties: Select kudzu varieties that are resistant to bacterial leaf spot disease.

Fig. 12.7 Symptoms of Kudzu Bacterial Leaf Spot Disease

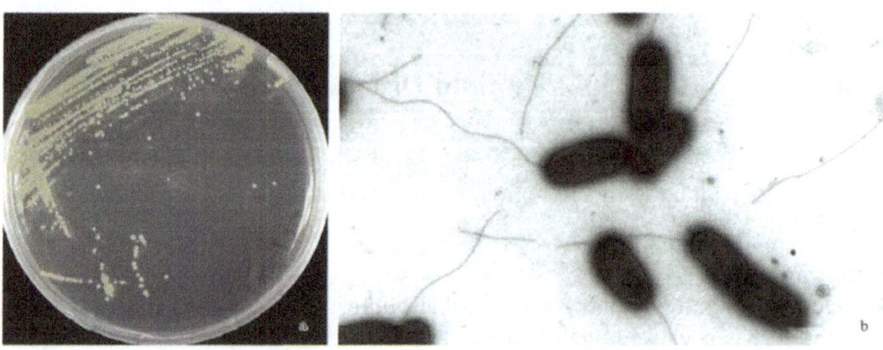

Fig. 12.8 Morphology of Xanthomonas campestris. (**a**) Cultural characteristics. (**b**) Bacterial cell structure

2. Nutrient Management: Apply phosphorus and potassium fertilizers while avoiding excessive nitrogen to prevent vulnerability.
3. Chemical Control: At the onset of disease symptoms, apply a 500–600-fold dilution of 77% copper hydroxide WP or a 500–600-fold dilution of 25% copper ammonium solution as a foliar spray.

Chapter 13
The Diseases of Astragalus (Huangqi)

Astragalus, a perennial herbaceous plant in the legume family, includes two main species: Mongolian Astragalus (*Astragalus membranaceus* var. mongholicus (Bunge) P. K. Hsiao) and membranous Astragalus (*A. membranaceus* (Fisch.) Bunge). It is primarily cultivated in Inner Mongolia, Shanxi, and Gansu provinces, with extensive plantations in Longxi County, Ding'an District, and Weiyuan County in the Gansu Province. Longxi County is known as the "Hometown of Chinese Astragalus." Astragalus roots are used medicinally for their sweet, warm nature, acting on the lungs and spleen to replenish energy and strengthen the middle. They are commonly used in treating symptoms such as fatigue, reduced appetite, spontaneous sweating, difficult wound healing, organ prolapse, and edema.

The main diseases affecting Astragalus include root rot, powdery mildew, and downy mildew.

13.1 Astragalus Root Rot Disease

Astragalus root rot is one of the major diseases affecting Astragalus and is a common root disease. In recent years, improper crop rotation and continuous cropping have led to an increased prevalence of Astragalus root rot. Currently, it has been reported in provinces such as Beijing, Inner Mongolia, Gansu, Shanxi, Jilin, and Hubei.

Symptoms

Infected plants exhibit weak growth above ground, stunted size, and pale to gray--green leaves; in severe cases, the entire plant's leaves turn yellow and fall off (see Fig. 13.1a~c). The epidermis of the root and stem base appears rough and slightly brown, with many horizontal wrinkles, eventually developing longitudinal cracks and fissures (see Fig. 13.1d). The brown phloem in the root cross-section shows numerous cavities, resembling a foam-like structure with small purple spots, and

© China Agriculture Press Co., Ltd. 2026

T. Wang et al., *Atlas of Chinese Medicinal Plant Diseases*,
https://doi.org/10.1007/978-981-95-6326-5_13

Fig. 13.1 Symptoms of Astragalus Root Rot Disease (**a**, **b**) Field symptoms. (**c**) Above-ground symptoms. (**d**) Cracked epidermis of the root. (**e**) Browning of the root pith. (**f**) White mycelium on the root surface

decays to a brown color with an easily peeled epidermis. The pith in the xylem initially forms pale yellow circular rings, which later expand to pale purple-brown or light yellow-brown, spreading down to the root tip (see Fig. 13.1e).

The above-ground parts wilt and lose color, with the plant dying from the base upward. The root tips soften, producing dense white mycelium (see Fig. 13.1f) that encircles the root tips, while the roots above the infected area remain normal. Some stem bases turn grayish-white or light brown, forming white streaks of feather-like fungal cords covered with dense white mycelium. Some roots show browning in the middle or lower parts, with white mycelium on the surface; in some cases, the entire lower portion of the root becomes brown and decays. This disease often co-occurs with black root rot.

Pathogen

The pathogen causing Astragalus root rot includes asexual fungi from the genus Fusarium (*Fusarium* spp.) within the Ascomycota phylum, with infections caused by single or multiple strains. Pathogenic species include *Fusarium oxysporum* Schltdl. ex Snyder & Hansen, *F. solani* (Mart.) Appel & Wollenw. ex Snyder & Hansen, as well as a smaller presence of *F. equiseti* (Corda) Sacc. and *F. acuminatum* Ellis & Everh.

F. oxysporum: Colonies are white, woolly, dense, and distinctly raised, with an off-white color on the underside and a central region varying from gray-orange to pale gray-black or pale purple-gray. The large conidia are crescent-shaped with one to five septa (typically three to four), (20–38.8) μm × (2.9–4.7) μm (average 29.9 × 3.8 μm), with a length-to-width ratio of 7.9. The small conidia are mostly single--celled, occasionally double-celled, and elliptical with tapered ends, (7.1–12.9) μm × (2.4–3.3) μm (average 9.7 μm × 2.5 μm). Conidiophores are single-bottle-shaped and very short. Chlamydospores are produced within the hyphae, either in chains or singly (see Fig. 13.2).

F. solani: Colonies are earth-gray to light yellow, thin, spread flat, with a powdery-gray appearance, and colorless hyphae. Large conidia are sickle-shaped, widest at the upper two-thirds, slightly curved, with three to four septa, measuring (23.5–36.5) μm × (3.5–4.7) μm (average 30.1 μm × 4.1 μm). Small conidia are oval

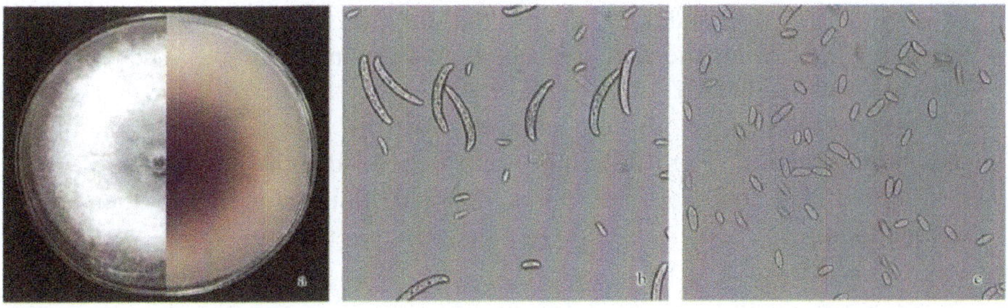

Fig. 13.2 Morphology of *Fusarium oxysporum*. (**a**) Colony characteristics. (**b**) Large conidia. (**c**) Small conidia

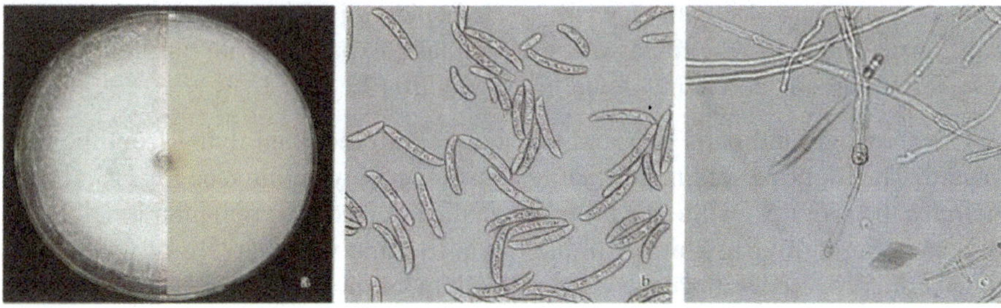

Fig. 13.3 Morphology of *Fusarium solani*. (**a**) Colony characteristics. (**b**) Large conidia. (**c**) Chlamydospores

to long-oval, kidney-shaped, colorless, single- or double-celled, measuring (5.9–11.8) μm × (2.4–4.1) μm (average 7.6 μm × 2.7 μm). Conidiophores are single--bottle-shaped, longer, measuring 16.8–28.2 μm (average 21.1 μm), with some conidiophores being very long (see Fig. 13.3).

Occurrence Pattern

The pathogen can survive in soil for over 5 years. It invades through root wounds, and various mechanical injuries from soil pests, nematodes, and tillage practices facilitate pathogen entry. The pathogen spreads through water flow, soil tillage, and farming tools. Conditions such as water accumulation in low-lying areas, dense weed growth, poor ventilation, sudden temperature increases after rain, and continuous cropping contribute to severe outbreaks. In Longxi County and Weiyuan County in Gansu Province, the disease incidence exceeds 35%, with severity levels ranging from grade 2 to 4, and seedling infection rates between 18.5% and 26%.

Control Measures

1. Cultivation Measures: Level the land to prevent water accumulation, implement a crop rotation system of at least 5 years, and maintain proper planting density to improve ventilation and light penetration. During planting, cultivation, and harvesting, minimize wounds on the plants. Remove infected and damaged roots during harvest, control soil pests to reduce injuries, and thoroughly clear diseased residues from fields to reduce initial infection sources.
2. Soil Treatment in Seedbeds and Fields: Apply 300 g of 20% copper acetate WP or 4 kg of 50% carbendazim WP per mu, mixing with 30 kg of fine soil and spreading evenly on the soil surface, then rake into the soil. Treat planting trenches with medicated soil at planting time.
3. Root Dipping Solution: Before planting, dip the roots in a solution of 3% metalaxyl-thiram at 700-fold dilution, 50% carbendazim sulfonate at 500-fold dilution, or 20% copper acetate WP at 900-fold dilution for 10 min, then air dry before planting.

13.2 Powdery Mildew Disease in Astragalus

Powdery mildew is a major disease affecting Astragalus throughout its entire growth cycle, with occurrences widespread across major Astragalus-producing provinces, including Jilin, Inner Mongolia, Heilongjiang, Gansu, Hebei, Beijing, Shanxi, Shandong, Shaanxi, and Hubei.

Symptoms

Leaves, petioles, young stems, and pods are all susceptible. White mildew appears on both surfaces of the leaves, initially as small white spots that gradually expand to cover the entire leaf. This dense, felt-like fungal layer consists of the conidiophores and conidia of the pathogen. In later stages, small black particles (cleistothecia) appear within the powdery mildew layer (Fig. 13.4). Affected plants exhibit yellowing, wilting, and leaf drop, and in severe cases, the entire plant may die. Tiny black particles, representing a parasitic fungus of the powdery mildew, are also commonly observed within the mildew layer.

Pathogen

The pathogen responsible for powdery mildew in Astragalus is *Erysiphe astragali* DC., a fungus belonging to the phylum Ascomycota, genus *Erysiphe*. Synonyms include *Microsphaera astragali* (DC.) Trevis. and *Trichocladia astragali* (DC.) Neger. The cleistothecia are spherical to nearly spherical, dark brown to black, with diameters of 94.6–161.1 μm (average 116.4 μm). The asci are oval to elongated, typically numbering six to eight, measuring (62.7–85.0) μm × (31.3–53.7) μm, with a stalk. Ascospores are oval to elongated, usually numbering four, pale yellow to bright yellow, and measure (24.6–35.8) μm × (15.7–20.2) μm. The appendages are filamentous, colorless, unsegmented, often curved, with one to two branches at the tip, measuring (165.7–645.0) μm × (17.9–31.4) μm (average 450 × 22.4 μm), with 9–24 appendages (Fig. 13.5). The same pathogen infects *Astragalus mongholicus,*

Fig. 13.4 Symptoms of Powdery Mildew Disease in Astragalus

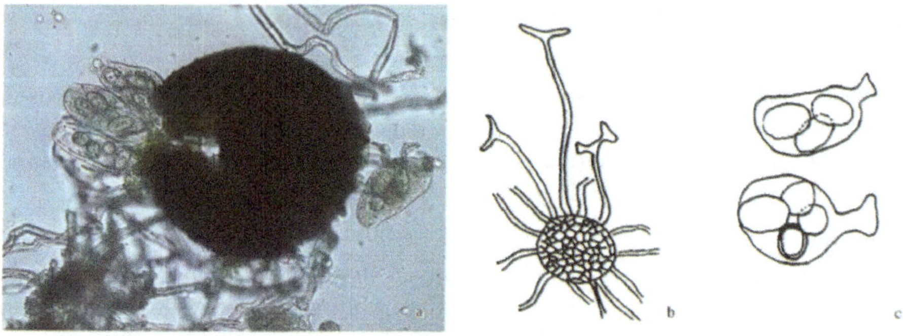

Fig. 13.5 Morphology of *Erysiphe astragali.* (**a**) Cleistothecium micrograph. (**b**) Cleistothecia. (**c**) Ascus and Ascospores

A. membranaceus, and Astragalus chrysopterus. Additionally, *Erysiphe pisi* DC., the powdery mildew pathogen of peas, is also reported to infect Astragalus.

Occurrence Pattern

The pathogen overwinters as cleistothecia on infected plant debris at the soil surface or as mycelia on root buds. In the following year, when temperature and humidity conditions are favorable, ascospores are released for initial infection. Conidia produced on infected areas are dispersed by wind and rain, allowing multiple cycles of reinfection. Observations in Gansu Province indicate that powdery mildew in Astragalus begins sporadically in late June, develops slowly in July, and peaks from mid-August to early September, persisting until harvest. The disease is more severe in areas below 2000 meters in elevation. At temperatures of 20–26 °C and relative humidity of 51–65%, the incubation period is around 6 days. Mycelial growth is rapid at temperatures above 20 °C, and temperature significantly influences disease progression and spread. *Hedysarum polybotrys* Hand.-Mazz.exhibits minimal infection, with an incidence rate below 5%, indicating resistance to powdery mildew.

Control Measures

1. Cultivation Practices: Apply balanced basal fertilizers with a proper nitrogen, phosphorus, and potassium ratio to avoid excessive nitrogen, which can lead to excessive growth. Proper plant spacing improves air circulation and light exposure. In early winter, thoroughly remove infected plant debris from the field to reduce sources of initial infection.
2. Chemical Control: Apply fungicides early in the disease cycle for effective control. Recommended options include 62.25% chlorothalonil WP, 75% azoxystrobin-triazole WP, 25% triadimefon WP, and 10% difenoconazole WP, all of which provide excellent efficacy.

13.3 Downy Mildew Disease in Astragalus

Downy mildew is a primary disease impacting Astragalus production, predominantly found in Gansu Province. In key production areas, the infection rate ranges from 43% to 100%, with severity levels from 1 to 4. Six-year-old seedlings show a 23% to 25% infection rate, all being systemic infections. Infection rates range from 14.5% to 41%, with severity levels from 1 to 3. This disease primarily affects the leaves, significantly impacting the yield and quality of Astragalus.

Symptoms

Astragalus downy mildew exhibits both local and systemic infection characteristics. In 1- to 2-year-old plants, it appears as localized infections, primarily affecting the leaves. Early symptoms include faint, polygonal, or irregular lesions along the leaf edges, light brown to brown. A white to pale gray mildew, consisting of the pathogen's sporangiophores and sporangia, develops on the corresponding undersides of the leaves. In advanced stages, this mildew layer turns dark gray; severe cases lead to yellowing, wilting, curling, and defoliation of lower leaves, leaving only the upper leaves intact. In perennial plants, systemic infection is more common, causing stunted growth to about one-third of normal height and yellowing of smaller leaves. Other symptoms are consistent with localized infections (Fig. 13.6).

Pathogen

The pathogen causing downy mildew in Astragalus is *Peronospora astragalina* Syd., a member of the Oomycota phylum and genus *Peronospora*. Sporangiophores emerge from stomata, mostly as single branches, occasionally with multiple branches, colorless, measuring (224.0–357.4) μm × (6.1–8.2) μm (average 285.6 μm × 7.5 μm). The main axis occupies two-thirds of the total length, branching dichotomously four to six times at the upper portion, ending in straight or slightly curved tips that extend at acute or right angles, measuring (7.7–15.9) μm × (1.5–2.5) μm (average 10.5 μm × 2.2 μm) (Fig. 13.7a). Sporangia are oval, with a protrusion on one end, colorless, and measure (18.0–28.3) μm × (14.1–20.6) μm (average 20.4 μm × 19.1 μm) (Fig. 13.7b). Oogonia are nearly spherical, pale yellow-brown, measuring (43.7–61.7) μm × (43.7–61.7) μm (average 51.2 μm × 47.3 μm). Antheridia are rod-shaped, lateral, solitary, and measure (30.8–39.8) μm × 9.0 μm. Oospores are spherical, pale yellow-brown, with a diameter of 23.1–36.0 μm (average 31.5 μm) (Fig. 13.7c). Sporangia begin to germinate in water droplets after 10–12 h at temperatures between 5 °C and 30 °C, with an optimum of 20 °C. At 100% relative humidity, the germination rate is only 5.5% after 36 h, with no germination below 95% relative humidity. The pH range for germination is 4.98–9.18, with an optimum pH of 7.28. Extracts from Astragalus leaves strongly stimulate sporangia germination. The pathogen can also infect *A. mongholicus*, *A. membranaceus*, and *A. chrysopterus*.

Occurrence Pattern

The pathogen overwinters as oospores in plant debris on the soil surface or as mycelia within perennial plants. In the following year, when environmental conditions

Fig. 13.6 Symptoms of Downy Mildew Disease in Astragalus. (**a**) Local infection symptoms. (**b**) Systemic infection symptoms

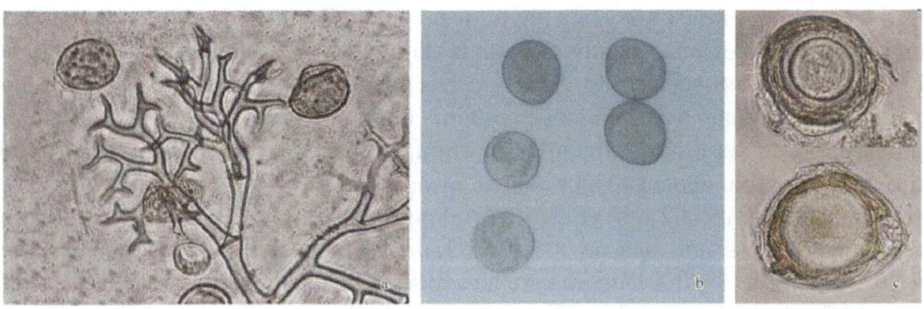

Fig. 13.7 Morphology of *Peronospora astragalina.* (**a**) Sporangiophores. (**b**) Sporangia. (**c**) Oospores

are favorable, these overwintering pathogens germinate and infect the host, causing initial infection. In Longxi County, Gansu Province, symptoms appear shortly after regrowth in early May in systemically infected plants, which serve as the primary centers of infection and sources of initial spread. Sporangia produced on infected plants are dispersed by wind and rain, leading to reinfection. At 20–26 °C and 65–77% relative humidity, the incubation period is approximately 5 days. The disease first appears in early to mid-July, develops slowly in late July, and peaks from early August to mid-September, persisting until harvest. High rainfall, extended dew periods, and high humidity favor disease development. In high-altitude regions such as Min County, Zhang County, and Weiyuan County, dew periods often last 9–11 h at night in July and August, creating a water film on leaf surfaces that facilitates sporangia germination and infection. Consequently, the disease is more severe in areas above 2000 meters. Typically, the middle and upper leaves are more severely affected, while the lower leaves are less impacted. In the later stages, large numbers of oospores form within infected tissue, falling into the soil with diseased leaves to overwinter and serve as the initial infection source for the following year.

Control Measures

1. Cultivation Practices: Adequate plant spacing improves air circulation and light exposure; increased phosphorus and potassium fertilization enhances host resistance. After harvest, thoroughly remove plant debris to reduce sources of initial infection.
2. Chemical Control: At the onset of disease, spray with a 400-fold dilution of 70% propineb WP, which is effective against downy mildew in Astragalus. For combined downy and powdery mildew infections, apply a mixture of 200-fold dilution of 40% fosetyl-aluminum WP and 2000-fold dilution of 15% triadimefon WP.

Disease (Pathogen)	Symptoms	Occurrence pattern
Leaf Blight [*Septoria psammophila*]	Near-circular to polygonal brown to dark brown lesions develop between the lateral veins on the leaf surface. On the underside of the leaf, these lesions are polygonal and bear numerous small black dots, while the upper surface has fewer black dots. These small black dots are the conidiomata of the pathogen	The pathogen overwinters on the soil surface in plant debris as conidiomata. The following year, when conditions are favorable, it releases conidia, which spread via wind and rain to cause initial infection, and secondary infections is frequent. Disease typically begins in late June and spreads rapidly in mid to late August. High rainfall and prolonged dew periods intensify the disease
Gray Spot Disease [*Phyllosticta astragalicola*]	Circular to near-circular lesions appear on the leaf surface, with pale brown edges and a grayish--white center. In later stages, small black dots develop, which are the conidiomata of the pathogen	The pathogen overwinters on the soil surface in plant debris as conidiomata. The following year, under suitable temperature and humidity, conidia cause initial infection, and spores produced on lesions can cause secondary infections
Ring Spot Disease [*Ascochyta* spp.]	Circular brown lesions form on the leaf surface, with faint concentric rings. In the later stages, small black dots appear, representing the conidiomata of the pathogen	The pathogen overwinters on the soil surface in plant debris as conidiomata. The following year, under favorable conditions, conidia cause primary infection. Scattered occurrences are reported in Longxi County, Gansu Province

(continued)

Disease (Pathogen)	Symptoms	Occurrence pattern
Leaf Spot Disease [*Cladosporium* spp.]	Lesions spread inward from the leaf edge, forming elongated to semicircular, yellow-brown to brown lesions that are slightly raised, with faint concentric rings in the center. In later stages, a gray-black mold layer develops on the lesions	The overwintering situation of the pathogen is unclear. The disease is present in all counties of Dingxi City, Gansu Province, with an incidence rate of 4% to 17.5% and severity levels of 1 to 2
Brown Spot Disease [*Alternaria* spp.]	Small, irregular yellow-brown spots emerge on the leaf surface, with indistinct edges and sparse black mold on the spots	The pathogen overwinters as mycelium and conidia on the soil surface or in soil. The following year, under favorable conditions, it infects the host, and conidia produced on lesions can cause secondary infections
Stem Base Rot [*Rhizoctonia solani*]	The plant exhibits weak growth, gray-green leaves, and irregular dark brown lesions on the root and stem base, with light brown filamentous material on the surface	The pathogen overwinters as sclerotia on plant debris and various hosts. It has strong saprophytic capability, surviving in soil for 2–3 years. Under suitable conditions, sclerotia germinate, producing mycelium that infects seedlings, or the mycelium directly infects seedlings. The pathogen spreads through water flow, farm tools, and contaminated fertilizers
Stem Nematode Disease [*Ditylenchus destructor*]	The above-ground parts of the plant exhibit weak growth, with stunted stems, smaller leaves, and yellowing foliage. In severe cases, the plant may die. The root crown and the middle to upper sections of the root show areas of rough, transversely wrinkled, brown epidermis, with inner tissues displaying dry rot, resembling chaff. This decay gradually spreads inward and downward, leading to extensive root rot. It often co-occurs with root rot disease	Reference: Stem Nematode Disease of Angelica
Bacterial Angular Leaf Spot [*Pseudomonas viridiflava*]	Large, angular yellow-green, oil-soaked lesions form on the leaf surface	The overwintering situation of the pathogen is unclear. The disease occurs in the southern counties of Dingxi City, Gansu Province, with an incidence rate of 16.6% to 25.5% and severity levels of 1 to 2

Chapter 14
The Diseases of Alpinia (Yizhi)

Alpinia (*Alpinia oxyphylla* Miq.), a plant in the ginger family (Zingiberaceae), produces dried mature fruits that are used in traditional Chinese medicine. It is one of China's renowned four southern medicines, valued for its effects in warming the kidneys, consolidating essence, reducing urination, warming the spleen, stopping diarrhea, and controlling salivation. It is used to treat symptoms such as enuresis due to kidney deficiency, frequent urination, seminal emission with cloudy discharge, cold-induced diarrhea, abdominal cold pain, and excessive salivation. Modern pharmacological and clinical studies indicate that Alpinia has sedative, hypnotic, analgesic, anticancer, antidiarrheal, anti-ulcer, antiaging, antioxidant, antiallergic, anti-dementia, memory-enhancing, cardiovascular-protective, and neuroprotective properties. In China, Alpinia is primarily cultivated in Qiongzhong, Wuzhishan, Tunchang, and Baisha counties of Hainan; Yangchun, Xinyi, Gaozhou, and Guangning in Guangdong; Napo, Jingxi, Debao, and Wuming in Guangxi; Jinghong and Mengla in Yunnan; and in parts of Fujian. The main diseases affecting Alpinia include ring spot leaf blight, anthracnose, root-knot nematode disease, wilt disease, and viral diseases.

14.1 Ring Spot Leaf Blight of Alpinia

Ring spot leaf blight is currently the most common foliar disease in Alpinia cultivation. It has been reported in Hainan, Guangdong, Guangxi, Yunnan, and Fujian. Surveys indicate that when the disease is severe, the infection rate can reach up to 70%, significantly impacting normal growth.

Symptoms
Ring spot leaf blight in Alpinia typically begins at the leaf tips or edges, forming large, irregularly shaped reddish-brown lesions with a gray-brown center. These lesions display distinct concentric rings of alternating light and dark brown and are

© China Agriculture Press Co., Ltd. 2026

T. Wang et al., *Atlas of Chinese Medicinal Plant Diseases*,
https://doi.org/10.1007/978-981-95-6326-5_14

dotted with numerous small black spots, which are the conidiomata of the pathogen. A distinct yellow halo surrounds each lesion. Under favorable conditions, lesions expand continuously, often covering one-third to half of the leaf area. In severely affected plants, all the diseased leaves turn brown and wither (see Fig. 14.1).

Pathogen

The pathogen responsible for ring spot leaf blight in Alpinia is *Pestalotiopsis palmarum* (Cooke) Steyaert, an asexual fungus from the genus Pestalotiopsis in the Ascomycota phylum. On PDA medium, the colonies appear white and woolly, circular to oval, with regular edges and a pale yellow underside featuring rings (see Fig. 14.2a). The conidiomata are disc-shaped, black, and measure 185–192 μm in diameter. The conidiophores are short, unbranched, and colorless. The conidia are spindle-shaped, measuring (17–24) μm × (4.5–7.5) μm, with four septa dividing the spore into five cells. The middle three cells are light brown, while the terminal cells are colorless. The top of each conidium has two to three colorless appendages, occasionally branched, measuring 10–19 μm in length (see Fig. 14.2b).

Occurrence Pattern

The pathogen overwinters as mycelium or conidiomata on infected plants and plant debris. Under favorable environmental conditions, it produces conidia, which is spread by wind and rain, primarily entering the host through wounds. Ring spot leaf blight can occur at any stage from seedling to fruit-bearing, but it is more commonly observed on older leaves of mature plants. The disease is more likely to develop during the high-temperature, rainy season, with its peak incidence in August and September.

Control Measures

1. Enhanced Cultivation Management: Alpinia is a shade-loving plant and can be intercropped with areca, rubber, or oil tea trees to maintain a well-ventilated environment. Promptly remove diseased plant debris from the field and destroy it to reduce infection sources for the following year. Apply organic fertilizers, which improve soil structure through humus formation and enhance Alpinia's disease resistance by enriching soil nutrients.

Fig. 14.1 Symptoms of Ring Spot Leaf Blight in Alpinia

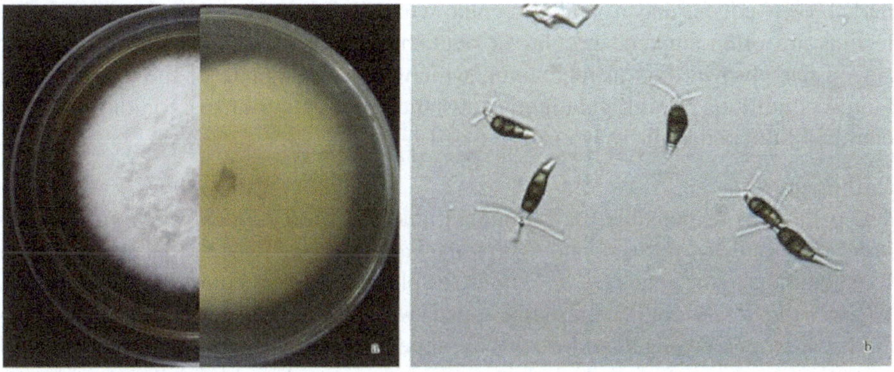

Fig. 14.2 Morphology of the Ring Spot Leaf Blight Pathogen in Alpinia. (**a**) Colony characteristics; (**b**) Conidia

2. Biological Control: Biological agents are recommended for disease control. Per mu (approximately 0.067 hectares), apply 150–200 g of 20 billion spores/g *Bacillus cereus* wettable powder, 100–125 g of 200 million spores/g Trichoderma water-dispersible granules, or 300–400 g of 3% polyoxin wettable powder by spraying. Combine with chemical agents as needed for enhanced control.
3. Chemical Control: Spray both sides of the leaves evenly and in a timely manner. Use 65% zineb WP at a 500-fold dilution, 10% difenoconazole WG at a 1000-fold dilution, 50% carbendazim WP at an 800–1000-fold dilution, or 250 g/L azoxystrobin SC at a 2000-fold dilution. Spray every 7–10 days for two to three consecutive applications. Rotate these fungicides to prevent resistance development, strictly adhering to the recommended safety intervals.

Here is the refined and carefully proofread version, ensuring accuracy, elegance, and adherence to the specified format:

14.2 Anthracnose of Alpinia

Anthracnose primarily affects the leaves and fruits of Alpinia, occurring from the seedling stage through to maturity. This disease has recently emerged as a significant issue in Alpinia cultivation, with cases reported in areas like Ledong Li Autonomous County and Wuzhishan City in Hainan Province. If left untreated when infection is severe, anthracnose can lead to extensive leaf and fruit infections, rendering the fruit unsuitable for medicinal use.

Symptoms

The pathogen typically invades from the leaf tip or edge, initially causing pinhead--sized spots with yellow halos. As lesions enlarge, they may become circular, oval, or irregularly shaped, appearing dark brown to grayish-white with concentric rings, dark brown edges, and a sunken center with slightly raised margins. On fruits, lesions are often sunken, circular or oval, black-gray in color, with scattered or ringed dark brown dots in the center, which are conidiomata of the pathogen. In humid conditions, a pink gelatinous substance, consisting of the conidiomata and conidia of the pathogen, may emerge (see Fig. 14.3).

Pathogen

The pathogens responsible for Alpinia anthracnose are *Colletotrichum gloeosporioides* (Penz.) Penz. & Sacc. and *Colletotrichum siamense*. The sexual stage of both fungi belongs to the genus *Glomerella* in the phylum Ascomycota. On PDA medium, *C. gloeosporioides* colonies initially appear white, gradually turning light gray, circular, with regular edges and a cottony appearance, featuring sparse white aerial mycelium on the surface and a light orange reverse (see Fig. 14.4a). The conidia are cylindrical or clavate, single-celled, colorless, measuring (8–17.5) μm × (4.5–7.5) μm, with a single oil droplet in the center (see Fig. 14.4b). *C. siamense* colonies are also cottony, with smooth, circular edges, initially white and later turning light gray, with a light orange spore mass in the central mycelial clump and light gray at the periphery, with dense gray-white aerial mycelium on the surface (see Fig. 14.4c). The conidiophores are cylindrical, and the conidia are single-celled, colorless, oblong with rounded ends, arranged in a palisade-like pattern within the conidiomata, measuring (7–18) μm × (4–6) μm (see Fig. 14.4d).

Occurrence Pattern

The pathogen overwinters in the soil and plant debris. The following year, it spreads through air currents, wind, rain, and human activity. The pathogen can invade through wounds, with infections occurring at temperatures between 10 and 30 °C. The disease is favored at temperatures of 20–24 °C and relative humidity levels of 85–98%, with low temperatures and continuous rainfall further promoting infection. The peak infection period is from May to July each year. In fields intercropped with rubber trees, anthracnose severity often increases with the incidence of anthracnose in the rubber trees.

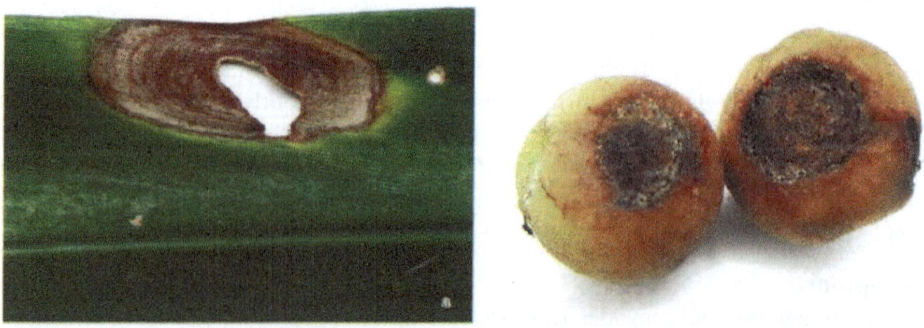

Fig. 14.3 Symptoms of Anthracnose in Alpinia. (**a**) Symptoms on leaves; (**b**) Symptoms on fruits

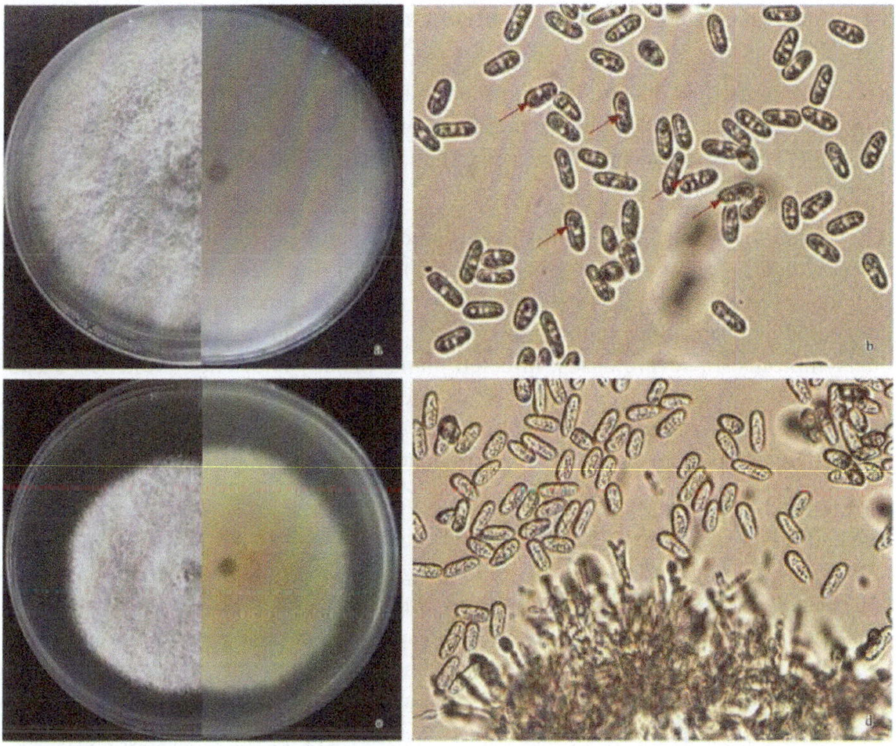

Fig. 14.4 Morphology of the Anthracnose Pathogen in Alpinia. (**a**) Colony characteristics of *Colletotrichum gloeosporioides*; (**b**) Conidia of *C. gloeosporioides* with internal oil droplet (indicated by arrow); (**c**) Colony characteristics of *C. siamense*; (**d**) Conidia and conidiomata of *C. siamense*

Control Measures

1. Enhanced Field Management: Implement rational crop rotation and appropriate planting density to improve ventilation and reduce humidity between plants. Apply adequate base fertilizer to promote vigorous plant growth, as high concentrations of potassium and nitrogen fertilizers may exacerbate anthracnose. Promptly remove and destroy diseased leaves and fruits.
2. Biological Control: Use biological agents such as *Trichoderma* spp., *Bacillus subtilis*, and *Bacillus velezensis* for controlling Alpinia anthracnose. In recent years, nanotechnology has gained attention for preparing plant protection agents, and nano-magnesium oxide has demonstrated effective antimicrobial action against *C. gloeosporioides*, inhibiting mycelial growth and spore germination.
3. Chemical Control: Spray both sides of the leaves evenly and in a timely manner. Use 10% difenoconazole WG at a 1500-fold dilution, 25% pyraclostrobin SC at a 1500-fold dilution, 45% prochloraz EC at a 1000-fold dilution, or 25% azoxystrobin SC at a 1500–2000-fold dilution. Rotate these fungicides, applying every 10 days as needed based on disease progression, and strictly observe the recommended safety intervals between applications.

14.3 Root-Knot Nematode Disease of Alpinia

Root-knot nematode disease is a common issue affecting Alpinia, primarily damaging the roots and causing wilting and death of the above-ground parts. This disease has been reported in the provinces of Hainan, Guangdong, and Yunnan.

Symptoms
After invading the host's roots, the pathogenic nematodes stimulate excessive tissue growth, resulting in numerous nodules (galls) of varying sizes and irregular shapes. Initially white, these galls turn light brown and may appear as single nodules or connected in a bead-like chain. In mild cases, the above-ground parts show no obvious symptoms. In severe cases, infected plants display shortened branches, weakened shoots, smaller leaves with loss of green color and luster, curled leaf edges, and symptoms of dehydration and nutrient deficiency. Eventually, the roots turn brown and rot, leading to the wilting and death of the entire plant (see Fig. 14.5).

Pathogen
The causative nematode of Alpinia root-knot disease is the southern root-knot nematode (*Meloidogyne incognita*), belonging to the phylum Nematoda, class Secernentea, order Tylenchida, and family Meloidogynidae. The species is sexually dimorphic. Female nematodes are enlarged, spherical or pear-shaped, with a prominent neck and slightly cap-like lip region. The perineal pattern varies significantly, generally showing a high dorsal arch. The pattern is distinctly oval or square, with a rounded or flat top on the dorsal arch, sometimes trapezoidal, with tightly packed striae that are wavy to serrated on the dorsal and lateral surfaces, occasionally smooth, and often with unclear or branched lateral lines. The body length is 493.5–676.5 μm, with a maximum body width of 331.5–526.4 μm. Stylet length is 13.1–15.6 μm, with a DGO (distance from stylet to opening of the gland orifice) of 2.4–4.5 μm, and the stylet knobs measure 2.1–3.0 μm high and 3.2–4.5 μm wide. The metacorpus (middle esophageal bulb) measures 42.0–61.2 μm high and

Fig. 14.5 Symptoms of Root-Knot Nematode Disease in Alpinia. (**a**) Above-ground symptoms; (**b**) Below-ground symptoms

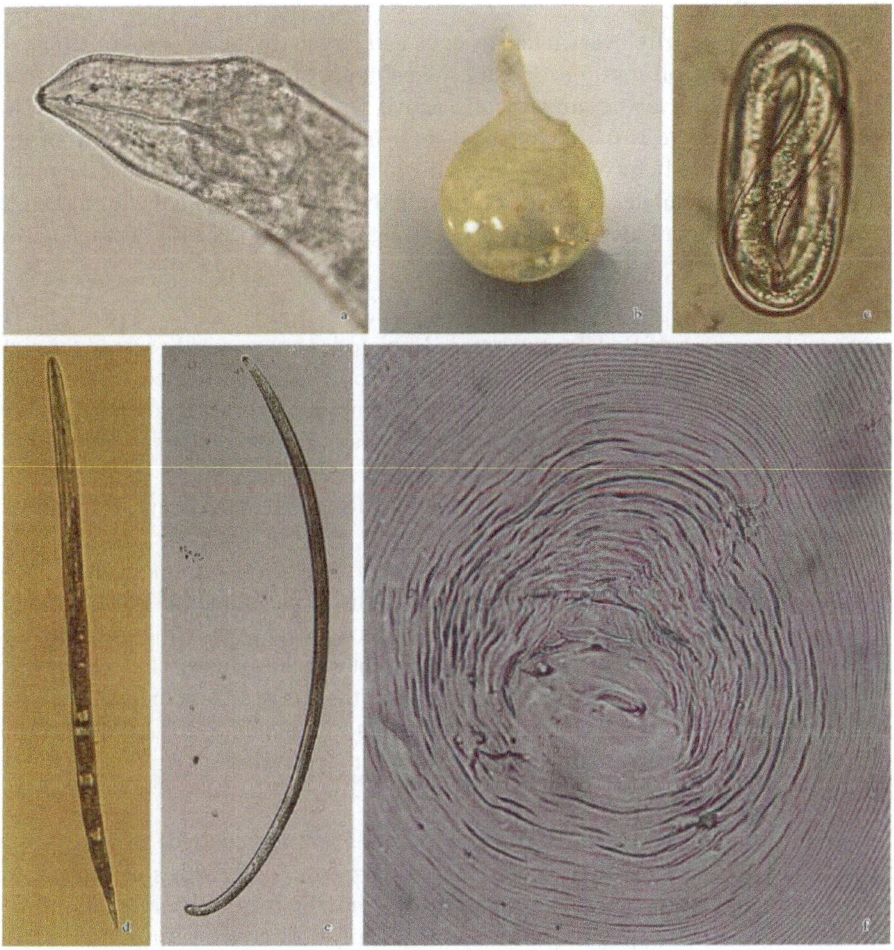

Fig. 14.6 Morphology of Southern Root-Knot Nematode. (**a**) Head of female; (**b**) Female nematode; (**c**) Single egg mass; (**d**) Second-stage juvenile; (**e**) Male nematode; (**f**) Perineal pattern. Note: Figures provided by Fu Meiying from the Plant Protection Institute, Hainan Academy of Agricultural Sciences

29.0–47.5 μm wide. Males are elongated, with a flattened or concave lip region and typically two to three incomplete annuli; the conical part of the stylet is bluntly rounded, often cylindrical near the basal knob, which is round. The body length is 1047.5–1681.5 μm, with a maximum width of 33.4–41.2 μm. Stylet length is 22.5–40.0 μm, with a DGO of 1.7–3.0 μm; stylet knobs measure 2.5–3.5 μm high and 4.5–5.1 μm wide. Second-stage juveniles (J2) measure 355.4–479.6 μm in body length, with a maximum width of 13.5–18.2 μm, stylet length of 10.5–14.0 μm, DGO of 2.0–3.6 μm, tail length of 43.1–58.6 μm, and a transparent tail length of 11.5–22.5 μm (see Fig. 14.6).

Occurrence Pattern

The pathogen primarily overwinters as eggs and larvae. In the absence of a host, it can survive in the soil for up to 3 years. When temperatures reach 10 °C or higher, eggs hatch, and the larvae are typically active within a 5–30 cm soil depth. The second-stage juvenile is the infective stage, usually invading through the root tip, moving intercellularly by compressing cell wall spaces, thereby infecting the plant. The nematodes stimulate accelerated host cell division, resulting in the formation of root galls. At temperatures of 25–30 °C, a single generation can be completed in approximately 25 days.

Control Measures

1. Strict Prevention of Soil Transmission: Nematodes are primarily spread through infested soil, manure, and farm equipment. Tools used in infected fields should be thoroughly cleaned and disinfected before moving to other fields. Sterilize the seedling medium before sowing, and remove any infected plants promptly, disinfecting the surrounding soil.
2. Rational Crop Rotation and Intercropping: Rotate or interplant with rubber, areca, oil tea, or pepper.
3. High-Temperature Solarization: In Hainan, for non-forest Alpinia planting, prepare the field in the high-temperature months (June–August). Plow, irrigate, cover the soil with plastic film, and expose it to the sun to raise the soil temperature to 45–48 °C (or even 50–60 °C) at a depth of 20–25 cm. This, along with high humidity (90–100% relative humidity), effectively kills nematodes.
4. Water-Flooding Control: Flood infected fields to a depth of 20–25 cm for over 20–30 days induces nematode death by oxygen deprivation.
5. Chemical Control:

   (a) Pre-Planting Soil Disinfection: Apply 35% formaldehyde solution at a 200–300-fold dilution in the planting furrows, cover with soil and plastic film, and fumigate for 7–15 days after removing the film, aerate the soil for one week before planting. Alternatively, apply 0.5% avermectin granules(GR) at 2–3.5 kg per mu (approximately 0.067 hectares) for soil treatment.

   (b) Post-Planting Soil Disinfection: Apply 1.8% avermectin EC at a 2000-fold dilution by root drenching twice, or spray soil with 41.7% fluopyram SC at 0.03 mL/plant (diluted in 400–500 mL of water) or apply 40% fosthiazate EC at 7500–9000 mL per hectare.

14.4 Fusarium Wilt Disease in Alpinia

Fusarium wilt is a common systemic disease that severely impacts Alpinia, particularly during its mature growth stage. This disease is widespread across China, with occurrences in Hainan, Guangdong, Guangxi, Yunnan, and Fujian provinces.

Symptoms

Fusarium wilt is a vascular disease with distinct internal symptoms. Infected vascular bundles display reddish-brown spots or streaks, darkening near the stem base and eventually turning black and dry. Mature plants show characteristic yellowing in the lower leaves and outer leaf sheaths, beginning at the leaf edges and gradually spreading inward. The boundary between diseased and healthy tissue is distinct. Further infection causes the entire leaf to yellow and wither. Affected leaves wilt rapidly, and in severe cases, the entire plant may die (Fig. 14.7).

Pathogen

The pathogen responsible for Fusarium wilt in Alpinia is *Fusarium oxysporum* Schltdl. ex Snyder et Hansen, an ascomycete fungus in its asexual form. On PDA medium, it produces dense, velvety mycelium, often displaying purple, yellow, or brown pigments. The pathogen produces three types of spores: microconidia, macroconidia, and chlamydospores. Microconidia are colorless, single-celled or two-celled, kidney-shaped or ovoid, often forming in false-head clusters, measuring (5–12) μm × (2.5–4) μm. Macroconidia are colorless, multicellular, slightly curved, and fusiform with tapering tips, with a distinct foot cell at the base, usually having three to seven septa (commonly three), and measuring (10–60) μm × (2.5–6) μm. Chlamydospores are light yellow, near-spherical, 6–8 μm in diameter, with thick, smooth walls, occurring singly, in pairs, or in short chains on short monophialides (Fig. 14.8).

Occurrence Pattern

The pathogen overwinters as mycelium and chlamydospores within plant debris in the soil. Infected plant residues, compost, and infested soil are primary sources of initial infection. The pathogen enters the host through the stem base, fibrous roots, root hairs, or wounds, proliferating within the host's vascular bundles and spreading through the conducting tissue. It releases toxins that impair cell function, blocking vessels and causing leaf wilt. The pathogen's optimal growth temperature is 24–28 °C, with a range of 17–37 °C. Acidic soil (pH 5–5.6), root injuries from transplanting or cultivation, and poor plant health favor disease development. Conidia spread via rain splash, leading to secondary infections.

Fig. 14.7 Symptoms of Fusarium Wilt Disease in Alpinia

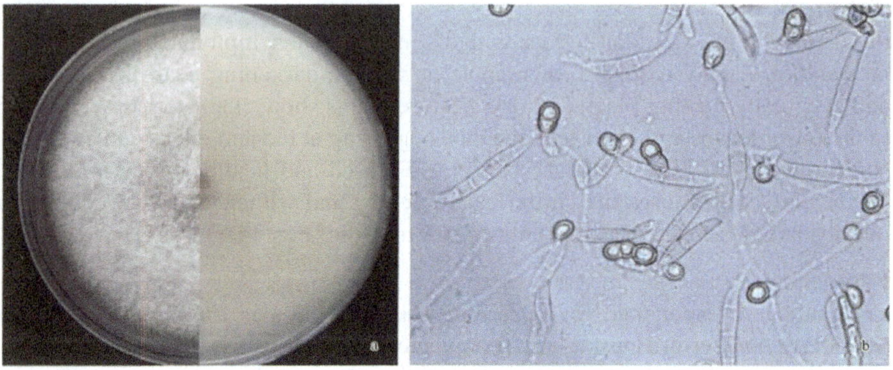

Fig. 14.8 Morphology of Fusarium wilt pathogen. (**a**) Culture characteristics; (**b**) Macroconidia and Chlamydospores

Control Measures

1. Deep Plowing and Soil Exposure: Before planting, deep plowing and soil exposure reduce pathogen levels. Apply well-composted organic fertilizers to enhance disease resistance.
2. Seedbed Treatment: Mix Trichoderma powder with the seedling medium in a 1:10 ratio, or apply 8–10 g/m^2 of 50% carbendazim WP or 1–2 g/m^2 of 70% thiophanate-methyl WP mixed with 1 kg of fine soil or sand as a seedbed covering, or spray the seedbed with a 1000-fold dilution of 70% thiophanate-methyl WP.
3. Soil Sterilization: In areas with severe disease occurrence, treat the soil before transplanting. Mix Trichoderma powder with the seedling medium in a 1:500 ratio, then proceed with seeding. Alternatively, for each hectare, mix 60 kg of 50% carbendazim wettable powder or 15 kg of 70% thiophanate-methyl powder with 225–300 kg of dry fine soil. Spread this mixture on the soil surface and rake it in, or apply it directly to planting holes during transplanting.
4. Field Management: Regularly clean the field, removing infected plants to prevent pathogen spread to healthy plants. If Fusarium wilt symptoms appear, remove affected plants and sterilize the surrounding soil. During rainy seasons, create drainage ditches to prevent water accumulation, increase sunlight exposure, and strengthen cultivation practices to enhance disease resistance.
5. Chemical Control: For biological control, apply 200–300 g per acre of *Bacillus subtilis* WP (10 billion CFU/g) per acre as a root drench or spray 150–200 g on foliage. Alternatively, use 50–60 mL of 5% chitosan oligosaccharide solution for foliar spray or a 1000–1200-fold dilution of 10% Jinggangmycin-*Pseudomonas fluorescens* WP as a root drench. Combine with chemical control using an 800-fold dilution of 70% thiophanate-methyl WP, a 500-fold dilution of 50% carbendazim WP, or an 800-fold dilution of 30% thiophanate-methyl AS for root drenching and foliar spraying. Rotate treatments every 7–10 days as needed based on disease severity.

14.5 Viral Disease in Alpinia

Viral disease in Alpinia is rare, with symptoms first observed in only two plants in Lingjiao Village, Wanling Town, Qiongzhong, Hainan Province, in July 2020.

Symptoms
Infected leaves display alternating yellow and green streaks or general yellowing. In severe cases, plants exhibit stunted growth, significantly inhibiting development (Fig. 14.9).

Pathogen
The causative virus has not yet been identified.

Occurrence Pattern
The virus can spread through seeds, seedlings, insect vectors, surrounding weeds, and field operations. It is essential to screen seedlings before transplanting, promptly clear surrounding weeds, and avoid injuring plants during field operations to prevent infection. Effective control of insect vectors is also critical. Hot, dry weather conditions favor disease spread. Overuse of ammonia fertilizers, tender plant tissue growth, thin or compacted soil, and poor drainage contribute to higher disease severity.

Control Measures

1. Field Management: Control nitrogen fertilizer application, maintain adequate soil moisture, and use organic and microbial fertilizers to promote vigorous plant growth and enhance resistance. Avoid injuring plants during field operations and prevent virus transmission through tools. Immediately remove and destroy infected plants. Avoid intercropping with Solanaceae, Cucurbitaceae, or Brassicaceae crops, which are also prone to viral diseases.
2. Disinfection: Use virus-free materials for propagation and pathogen-free substrates for seedling growth. Disinfect seeds by soaking them in a 10% trisodium phosphate solution for 20–30 min, in warm water at 50–55 °C for 10–15 min, or by treating them in a temperature-controlled environment.

Fig. 14.9 Symptoms of Viral Disease in Alpinia Leaves

3. Chemical Control: For preventive treatment, use a 1000-fold dilution of 6% oligosaccharide-streptomycin WP, a 1500-fold dilution of 0.5% glucan SP, an 800-fold dilution of 2% ningnanmycin AS, or a 700-fold dilution of 5% chitosan oligosaccharide solution. Major insect vectors, including aphids, thrips, and whiteflies, can be managed with a 3000-fold dilution of 10% imidacloprid WP, a 2500-fold dilution of 4% abamectin-thiamethoxam ultra-low-volume liquid, a 2500-fold dilution of 30% spirotetramat-flonicamid SC, or a 2000-fold dilution of 60 g/L spinetoram SC.

14.6 Occasional Diseases of Alpinia Oxyphylla (Table 14.1)

Table 14.1 Characteristics of common diseases in Alpinia oxyphylla

Disease (Pathogen)	Symptoms	Occurrence pattern
Damping-Off Disease (*Rhizoctonia solani*)	Primarily affects young leaves, where small, circular, reddish-brown spots appear initially. These spots gradually expand, forming irregular brownish--black patches. A grayish-green cloud-like pattern appears on the undersides, eventually covering the entire leaf, leading to wilting and death. Sclerotia, appearing as granular black particles, may be observed on the lower parts of the leaves. The infected areas turn brownish-red, and the mycelial strands may connect the lesions	The fungus overwinters as sclerotia or mycelium on infected plants or plant debris. Under favorable conditions, at 23–28 °C, the fungus germinates, producing mycelium that spreads through air or directly invades host plants. Transmission is facilitated by rain, contaminated farm tools, or fertilizers containing fungal fragments
Ring Spot Disease (*Pestalotiopsis alpinae*)	Primarily affects leaves, with symptoms initially appearing on the lower and middle portions of the plant. The infection typically begins at the leaf tips or edges, with early-stage spots manifesting as dark brown to grayish-brown, which expand into irregular shapes with darker margins. In later stages, black specks may appear within the central lesion. Severe infections can cause the affected leaves to wither and die rapidly	The pathogen overwinters as conidia or pycnidia on plant debris. The following year, when temperature and humidity conditions are favorable, the spores germinate, initiating infection at the leaf surface

Chapter 15
The Diseases of *Codonopsis*

Codonopsis is a perennial herbaceous vine of the genus *Codonopsis* in the Campanulaceae family. The medicinal *Codonopsis* mainly refers to the dried roots of *Codonopsis pilosula* (Franch.) Nannf., *Codonopsis pilosula* var. *modesta* (Nannf.) L.T. Shen, and *Codonopsis pilosula* subsp. tangshen (Oliver) D.Y. Hong. With mild nature and sweet taste, it enters the spleen and lung meridians. It has the functions of invigorating the spleen and benefiting the lungs, nourishing blood, and promoting the production of body fluids. It is mainly used to treat spleen and lung qi deficiency, poor appetite, lassitude, cough with deficiency asthma, qi and blood deficiency, sallow complexion, palpitations, shortness of breath, and thirst due to fluid impairment.

Codonopsis is widely cultivated across China, with major production areas in southeastern Tibet, western Sichuan, northwestern Yunnan, eastern Gansu, southern Shaanxi, Ningxia, eastern Qinghai, Henan, Shanxi, Hebei, Inner Mongolia, and the northeast regions. It is further categorized by regional origin as follows: Eastern *Codonopsis*: Primarily produced in Fengcheng and Kuandian of Liaoning, Yanbian and Tonghua of Jilin, and Shangzhi, Wuchang, and Bin County of Heilongjiang.

Lu *Codonopsis*: Primarily grown mainly in southeastern Shanxi, including Pingshun, Lingchuan, Changzhi, Huguan, Jincheng, and Xinxiang of Henan. Tai Codonopsis: Mainly from the Wutai Mountain region in Shanxi. Western Codonopsis: Primarily from Minxian, Wenxian, Lintan, Zhuoni, and Zhouqu of Gansu; Nanping, Pingwu, Songpan, and Zoige of Sichuan; and Hanzhong and Ankang of Shaanxi; and the Wutai Mountain area in Shanxi. Special local varieties include Wen Codonopsis and Jing Codonopsis from Wenxian, Gansu, Min Codonopsis from Minxian, Gansu, and Feng Codonopsis from Fengxian, Shaanxi, and parts of Gansu. Codonopsis is also distributed internationally in the Democratic People's Republic of Korea, Mongolia, and the Russian Far East.

Reported diseases of *Codonopsis* in China mainly include rust, gray mold, powdery mildew, leaf blight, root rot, and violet root rot.

© China Agriculture Press Co., Ltd. 2026

T. Wang et al., *Atlas of Chinese Medicinal Plant Diseases*,
https://doi.org/10.1007/978-981-95-6326-5_15

15.1 Rust of *Codonopsis*

Rust is a significant issue in Codonopsis cultivation, commonly affecting production areas across northern and southern provinces. Currently, rust disease in Codonopsis has been reported in Jilin, Hubei, Sichuan, Guizhou, and Chongqing. The severity of the disease varies across major production areas, with severe cases causing leaf necrosis and hindering plant growth.

Symptoms

Rust disease primarily affects the leaves, though it can also infect stems and flower peduncles. Symptoms appear on both leaf surfaces. Initially, yellow-brown chlorotic spots with indistinct edges form on the upper leaf surface. Corresponding reddish-brown pustules appear on the underside, concealed beneath the epidermis. Over time, these pustules rupture, releasing a brownish powder. Lesions on stems and flower peduncles are generally larger. In advanced stages, slightly raised, blackish-brown teliospore clusters form, leading to premature leaf senescence (Fig. 15.1).

Fig. 15.1 Symptoms of *Codonopsis* Rust. (**a**) Symptoms on upper surface of leaves, (**b**) Symptoms on the lower surface of leaves, (**c**) Spore clusters, (**d–f**) Spore clusters on flower receptacles

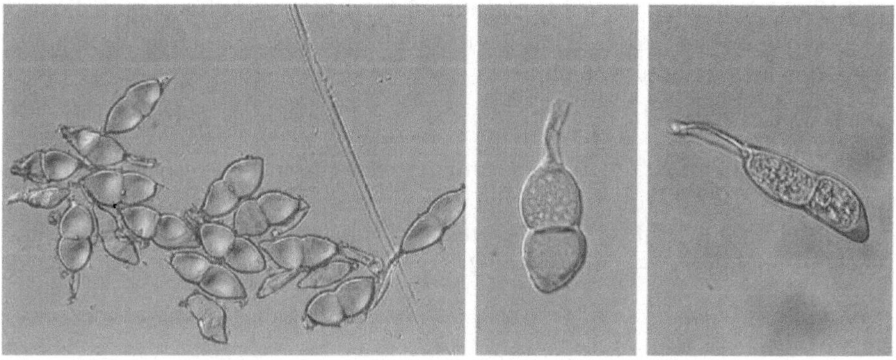

Fig. 15.2 Teliospore Morphology of *Codonopsis* rust pathogen. (Note: Specimens were collected from Dunhua, Jilin. Uredinial piles and urediniospores were not found)

Pathogen

The pathogen responsible for Codonopsis rust disease is *Puccinia campanumoeae* Pat., a basidiomycete fungus of the genus Puccinia. Teliospore clusters form on the underside of leaves, occurring singly or in clusters and occasionally merging. They are firm and yellow-brown in color. Teliospores are oblong, two-celled, brown, with smooth walls and constricted at the septa. They have a rounded papilla at the apex, and the stalk is colorless, reaching up to 70 μm in length, with spores measuring (33–46) μm × (15–17) μm (Fig. 15.2).

Occurrence Pattern

In northeastern and northern regions, *Codonopsis* rust disease is particularly severe in autumn, especially around the Mid-Autumn Festival and before seed maturation. The pathogen overwinters as teliospores on plant branches and leaves, and spreads through air currents. The disease typically appears in May, with severe outbreaks from June to July. Optimal temperatures for disease development are 23–26 °C, and it is prone to occur under conditions of frequent rain, strong winds, dense fog, heavy dew, and high relative humidity.

Control Measures

1. Agricultural Control.

    1. Crop Rotation: Rotate with leguminous or gramineous crops for over 3–4 years. Avoid continuous cropping to reduce pathogen accumulation and lower the probability of disease occurrence.
    2. Proper Planting Density: Maintain adequate plant spacing to control light transmittance in the field to the humidity of the microclimate in the field.
    3. Strengthen Water and Fertilizer Management: Apply sufficient basal fertilizer, top-dress early, and increase the application of phosphorus and potassium fertilizers to promote robust growth with thick and dark green leaves, and enhance disease resistance.

4. Field Sanitation: Regularly clean the field, remove and disposing of diseased and dead plant material outside the field. Spray the soil surface with a 2–3° Baumé lime sulfur solution.

2. Chemical Control.

Apply 30–45 mL of 29% pyraclostrobin · kresoxim-methyl suspension concentrate, 50–83 g of 10% difenoconazole · cyproconazole water-dispersible granules, or 15–30 g of 30% trifloxystrobin wettable powder per mu (Chinese acre). Spray regularly, once every 7–10 days, with 2–3 consecutive applications recommended.

15.2 Gray Mold of *Codonopsis*

Gray mold of *Codonopsis* is a disease that has emerged in recent years. It was first discovered and reported in Gansu Province in 2009. In July 2020, it was found to be widespread in Codonopsis cultivation areas in Dunhua City, Jilin Province. It is currently a relatively serious disease in production.

Symptoms
Gray mold primarily affects the leaves, stems, withered flowers, and receptacles of *Codonopsis*. The leaves and stems near the ground are severely affected, showing extensive soft rot covered with a gray-brown mold layer. Infected leaves may initially show symptoms starting at the tips or edges, spreading inward to form V-shaped lesions. Circular lesions may also develop within the leaf tissue and expand outward. On withered flowers and receptacles, light brown to gray-brown irregular lesions occur at the tips, covered with a layer of grayish-brown or brown mold (see Fig. 15.3).

Pathogen
The pathogen causing gray mold of *Codonopsis* is *Botrytis cinerea* Pers.: Fr., which is an asexual morph of the ascomycete genus Botrytis. On PDA medium, the fungus initially forms white colonies that gradually change to light brown to brown, and it may form small black sclerotia can be prodeced. The mycelium is brown, thick, and septate, with a diameter of 14.1–18.2 μm. Conidiophores are gray-brown, erect or slightly curved, septate, and may be clustered, single-branched, or have a tree-like branching structure, with swollen, head-shaped tips covered with numerous small protrusions. These protrusions have small stalks densely bearing a large number of conidia, forming a grape-like cluster. Conidia are unicellular, colorless to light brown, nearly round or oval, and measure (7.9–17.1) μm × (6.2–9.5) μm (see Fig. 15.4).

Occurrence Pattern
The pathogen overwinters as mycelium, sclerotia, and conidia on diseased plant debris or in the soil. In the following spring, under suitable conditions, sclerotia germinate to produce conidiophores with abundant conidia. After maturing, conidia

Fig. 15.3 Symptoms of *Codonopsis* Gray Mold

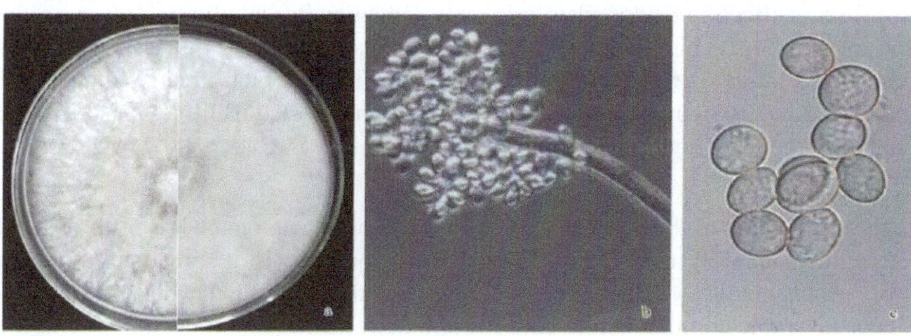

Fig. 15.4 Morphology of the *Codonopsis* Gray Mold Pathogen. (**a**) Colony characteristics, (**b**) Conidiophores and conidia, (**c**) Conidia

are dispersed through wind, rain, air currents, and agricultural activities. Conidia can enter the host through stomata, wounds, senescent tissues, or directly into young tissues. Secondary infections occur frequently. The disease is more severe under conditions of low temperature and rainy weather, especially when plants are densely planted.

Control Measures

1. Agricultural Control.

    1. It is recommended to use trellising for vertical cultivation, which enhances air circulation and light penetration, reduces field microclimate humidity, promotes healthy plant growth, and alleviates the occurrence of gray mold.
    2. Maintain field sanitation by promptly removing diseased tissues after harvest or plant removal to reduce sources of primary infection.

2. Biological Control

    At the onset of disease, biological agents such as polyoxin, carvacrol, *Bacillus subtilis*, *Trichoderma* spp, eugenol, *Bacillus amyloliquefaciens* QST713, *Pseudomonas fluorescens*, and berberine hydrochloride can be used for treatment.

3. Chemical Control

    At the early stage of infection, chemical agents such as boscalid, pyrimethanil procymidone, cyprodinil, pyrimethanil, or their combination formulations can be applied for effective control.

15.3 Root Rot Disease in *Codonopsis*

Root rot of *Codonopsis*, also known as root decay disease, is a common and recurring disease affecting *Codonopsis*, primarily damaging its underground fibrous roots and lateral roots. This disease significantly reduces yield and quality, sometimes rendering *Codonopsis* entirely unfit for medicinal use, which greatly impacts the industry. Known occurrences of root rot have been reported in Jilin, Gansu, Hubei, Guizhou, and Chongqing Municipality.

Symptoms

In the early stages of the disease, dark brown lesions appear on the lower fibrous and lateral roots, causing mild decay. As the disease progresses, it spreads to the main root, which gradually develops a water-soaked decay from the bottom upward. Under favorable conditions, the entire root may rot. If infection occurs later in the season or under unfavorable conditions, the disease may stop spreading, leaving the lower part of the root decayed and disintegrated, while the upper half near the ground will not rot in the cerrent year. Such "half-rotted roots" show a clear boundary between healthy and infected part, with the decayed area appearing dark brown and the vascular bundles turning deep brown. In following spring, the upper portion

Fig. 15.5 Symptoms of *Codonopsis* Root Rot

may sprout, but rot resumes under suitable temperature and humidity. Above-ground symptoms begin with yellowing leaves, moving upward from the lower leaves. When most of the root is mostly decayed, the entire plant withers. In wet soil, white, cottony fungal growth appears on decaying roots, representing the pathogen's mycelium and spores (Fig. 15.5).

Pathogen

The pathogen causing *Codonopsis* root rot belongs to the genus Fusarium (*Fusarium* spp.), which are asexual morphs of ascomycete fungi. The reported species include *Fusarium oxysporum* Schltdl. ex Snyder et Hansen, *Fusarium solani* (Mart.) Appel & Wollenw. ex Snyder et Hansen, *Fusarium tricinctum* (Corda) Sacc., and *Fusarium acuminatum* Nirenberg et O'Donnell, among which *Fusarium oxysporum* is the dominant species. On PDA medium, *F. oxysporum* forms circular, raised colonies with smooth edges. The dense aerial mycelium is initially white to pale pink, and later pigments ranging from light purple to dark purple are produced, making the reverse side of the colony appear pale purple. Macroconidia are slender, sickle-shaped, slightly curved, and evenly tapered at both ends. They have 1–5 septa, with constrictions at the septa, and measusre (23–56.6) μm × (3–5) μm. Microconidia are numerous, single-celled, and ovoid, measuring (5–12.6) μm × (2.6–2.7) μm. Chlamydospores are spherical, 6.1–8.2 μm in diameter, and occur singly, in pairs, or in chains. Based on morphological characteristics, culture traits, and molecular sequences of the pathogen, and with reference to Flora Fungorum Sinicorum and Fusarium Flora of Zhejiang, it was identified as Fusarium oxysporum, and the colony and spore morphology are shown in Fig. 15.6.

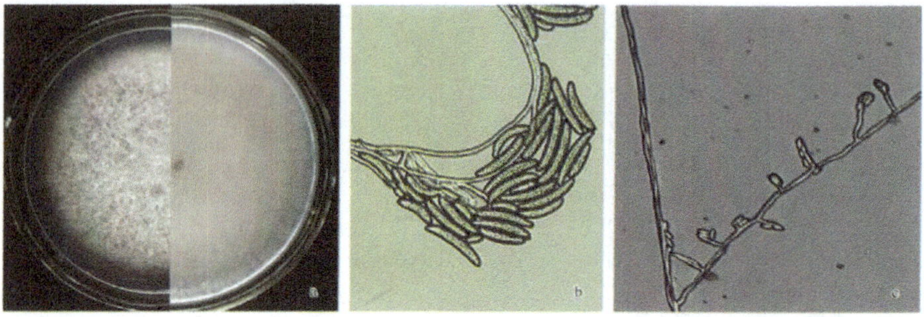

Fig. 15.6 Morphological characteristics of *Fusarium oxysporum*. (**a**) Culture characteristics, (**b**) Macroconidia, (**c**) Microconidia

Occurrence Pattern

The pathogen overwinters in the soil and on infected Codonopsis roots. The infected Codonopsis roots from the previous year start to show symptoms in mid-May of the next year, and the disease reaches its peak from June to July. The disease is severe in continuous cropping fields with high temperature, heavy rainfall, waterlogging due to low-lying terrain, dense vine growth, high humidity, and abundant underground pests. It mostly occurs on the roots of two - to three - year - old plants. In addition, root injuries increase susceptibility to root rot, which spreads to neighboring plants when soil moisture increases.

Control Measures

1. Agricultural Control

    1. Use Disease-Free Seeds and Seedlings: Before sowing, discard infected seeds and disinfect remaining seeds. Select healthy *Codonopsis* plants for transplanting.
    2. Field Selection: Choose well-drained, elevated plots with loose soil and avoid continuous cropping.
    3. Land Preparation and Bed Formation: Deep till and prepare high beds before planting.
    4. Field Management: Install trellises to improve air circulation and light penetration, preventing dense vine coverage on the soil. After rain, promptly drain the field to reduce moisture levels.
    5. Field Sanitation: Remove infected plants promptly and disinfect affected spots with quicklime. In late autumn, thoroughly clear the field of all diseased plant debris to reduce initial sources of primary infection.

2. Chemical Control

    1. Seed Treatment: Apply seed coatings with fludioxonil, metalaxyl-fludioxonil, and thiram-fludioxonil, etc.
    2. Soil Treatment: During land preparation, apply 3 kg of 50% carbendazim wettable powder mixed with 20–30 kg of fine soil per mu, and apply it along

furrows; alternatively, mix 200 g of 20% copper acetate with 20 kg of fine soil, spread it evenly on the ground and harrow it into the soil.

3. Root Drenching: Upon identifying infected plants, drench the root area and neighboring plants with 25% carbendazim or 50% thiophanate-methyl to control disease spread. Metalaxyl-mancozeb and polyoxin can also be used for root drenching.

Figure 15.5, symptoms of root rot of *Codonopsis*
Figure 15.6, morphology of *Fusarium oxysporum* (root rot pathogen)

Disease (pathogen)	Symptoms	Occurrence pattern
Violet root rot (*Helicobasidium mompa*)	The root surface shows purplish–red or dark brown filamentous mycelia, accompanied by cobweb-like mycelial membrane, and purple hemispherical sclerotia. Mildly infected roots become hard, short and thin, and grayish-brown, while severely infected roots rot completely, eventually forming black-brown hollow structures	It is prone to occur in plots with acidic soils (pH 4.7–6.5), shallow soil layer and poor water and fertility retention capacity. Continuous cropping fields have severe disease, and years with heavy precipitation and high humidity have severe disease. Sticky and compacted fields also have severe disease
Powdery mildew (*Sphaerotheca codonopsis*, *Podosphaera erigerontis–canadensis*)	Leaves, petioles, and fruits are all affected. At the initial stage, small white powdery spots appear on both sides of the leaves, and then spread to the entire leaf. The leaf surface is covered with a sparse white powdery layer. In the later stage, small black particles, which are the cleistothecia of the pathogen, are produced in the white powder. The diseased plants grow weakly, and the leaves may turn yellow and curl	The disease begins in late April, and the peak period is from June to early September. It commonly occurs under both dry and humid conditions. The disease is severe in places where plants are dense, leaves intertwined and ventilation is poor
Leaf spot (*Septoria codonopsidis*)	Polygonal, circular to near-circular brown lesions form on the leaf surface with purple edges, reaching diameters of 1–5 mm. In the later stages, black specks form in the center of the lesion	The disease occurs severely under the conditions of low temperature and continuous rainy and dewy weather
Downy mildew (Peronospora sp.)	Irregular, grayish-brown spots appear on the leaf surface, and leaf edges may curl upward. When the humidity is high, gray mold-like substances are produced on the back of the leaves	The disease is prone to occur under rainy and high-humidity condition

Chapter 16
The Diseases of Platycodon

Platycodon grandiflorus (Jacq.) A.DC., commonly known as balloon flower or bell-flower, or monk's hat flower is a perennial herbaceous plant in the Campanulaceae family. It is widely distributed in northeastern, northern, eastern, and central provinces of China, as well as in Guangdong, northern Guangxi, Guizhou, southeastern Yunnan (Mengzi, Yanshan, Wenshan), Sichuan (Pingwu, eastern Liangshan), and Shaanxi. Internationally, it is found in the Democratic People's Republic of Korea, Japan, the Russia's Far East, and southern East Siberia. The root of Platycodon is used in traditional Chinese medicine, with effects such as relieving cough and reducing phlegm, benefiting the lung and draining pus. It is harvested in spring and autumn, cleaned, fibrous roots removed, and either peeled fresh or left unpeeled before drying.

The major diseases affecting *Platycodon* include root rot, southern blight, wilt, sclerotinia rot, black spot, anthracnose, root-knot nematode disease, and dodder disease, with root-knot nematode and dodder diseases being particularly severe.

16.1 Root Rot Disease in *Platycodon*

Root rot is a major disease impacting *Platycodon* production, occurring in all major cultivation areas. Reported cases are mainly in Liaoning, Jilin, and Anhui provinces. This disease primarily affects the root, causing decay, which can lead to wilting or death of the above-ground parts.

Symptoms
Root rot primarily damages the roots of *Platycodon*. Initially, brown necrotic lesions appear near the root crown, gradually spreading downward. The root crown and stem base begin to dry and turn brown, and the decay progressively extends upward into the stem, leaves, and leaf veins, causing yellowing of leaves and branches. A characteristic symptom is a reddish discoloration on the root surface, and under

© China Agriculture Press Co., Ltd. 2026

T. Wang et al., *Atlas of Chinese Medicinal Plant Diseases*,
https://doi.org/10.1007/978-981-95-6326-5_16

high humidity, a pinkish fungal layer forms on the roots and stems, representing the conidia of the pathogen's spores. In severe cases, the entire plant wilts, and within 15–20 days, the entire root may decay. Secondary invasion by saprophytic microorganisms often leads to soft rot and disintegration, leaving only the outer skin of the root. As root decay progresses, stems and leaves gradually yellow and eventually wither and die, while the vascular bundles in the stems remain unchanged in color. When the entire fleshy root turns brown and necrotic, the plant fully wilts and dies (Fig. 16.1).

Pathogen

The pathogen causing *Platycodon* root rot is *Fusarium oxysporum* Schltdl. ex Snyder et Hansen, an asexual morph of the ascomycetes fungus in the genus *Fusarium*. On PDA medium, the colony appears raised and flocculent, with dense, white mycelia. Initially, colonies are white, but later produce pigments that range from pinkish-white to light pink, with a slight purple-red tinge. Microconidia are single-celled, oval to ovoid, measuring (5.04–15.93) μm × (1.64–6.06) μm. Macroconidia are sickle-shaped, slightly curved, with one to three septa, measuring (24.39–44.54) μm × (3.26–7.31) μm.Most of them have 3 septa, and a few have 1-2 septa (Fig. 16.2).

Occurrence Pattern

The pathogen overwinters as mycelia and chlamydospores in soil and on diseased plant debris, serving as the primary infection source. The disease spreads through irrigation, cultivation practices, and infected seeds. The disease typically starts to occur in late May, with peak incidence from June to July, coinciding with rising temperatures and increased rainfall. Conditions favoring infection include continuous cropping, poor drainage, low-lying fields, and high populations of soil pests.

Fig. 16.1 Symptoms of root rot disease in Platycodon. (**a**) Root decay and peeling, (**b**) Plant wilting and death

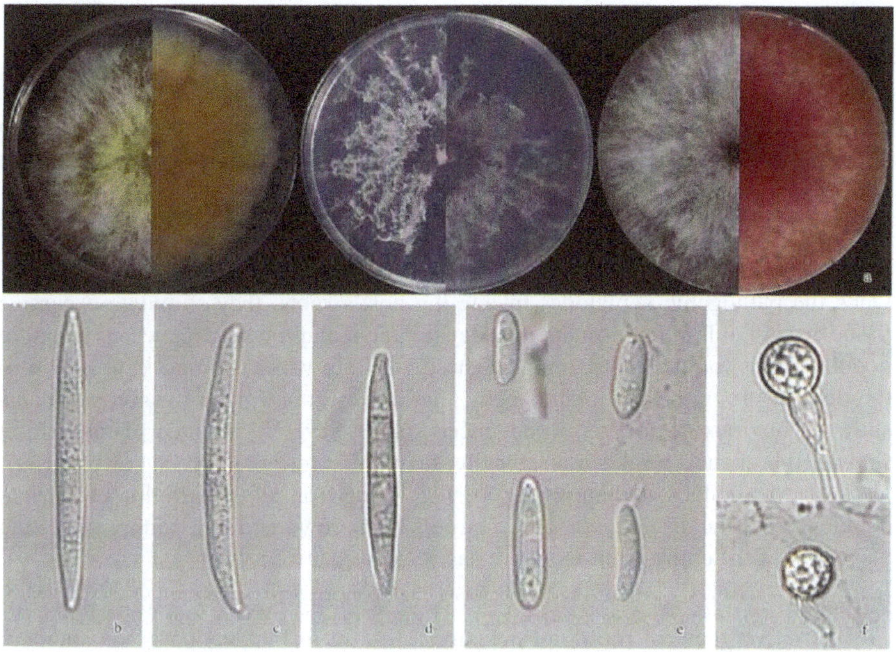

Fig. 16.2 Morphological characteristics of *Platycodon* root rot pathogen. (**a**) Culture characteristics, (**b–d**) Macroconidia, (**e**) Microconidia, (**f**) Chlamydospores

Control Measures

1. Crop Rotation: Practice crop rotation with grasses, especially alternating between dryland and paddy rice cultivation after Platycodon harvest, for effective disease control.
2. Use Disease-Free Seedlings: Select seeds from healthy plants and disinfect with 0.2%–0.5% thiram before planting.
3. Remove Plant Debris: Collect and remove diseased plant residues from the field, destroying or deeply burying them to reduce pathogen levels.
4. Field Management: Ensure proper plant spacing, apply organic fertilizers, and balance nitrogen, phosphorus, and potassium. Avoid waterlogging by promptly draining excess water after rain to reduce soil moisture.
5. Chemical and Biological Control: Use 300 billion CFU/g Bacillus cereus wettable powder, 50% carbendazim wettable powder, or 30% triflumizole wettable powder, all of which effectively inhibit the pathogen's mycelial growth. Additionally, 10% flumorph · 50% mancozeb wettable powder, *Pythium oligandrum* wettable powder (1×10^8 CFU/g), and 10% difenoconazole water-dispersible granules are also effective against the pathogen.

16.2 Root-Knot Nematode Disease of *Platycodon*

Root-knot nematode disease is a significant threat to *Platycodon* production, primarily occurring in the cultivation areas in South China, Central China, and North China. This disease has also been observed in Anhui. It primarily impacts the roots of *Platycodon*, with severe infections causing wilting and death of the above-ground parts.

Symptoms

Once the pathogenic nematodes invade the host's roots, they stimulate excessive tissue growth, inhibiting primary root development and forming numerous irregular nodules (galls) on the roots. Initially, these galls are white, turning light brown as the disease progresses. In early stages, the above-ground parts of infected plants show no obvious symptoms. In advanced stages, severely affected plants exhibit shortened branches, weak shoots, smaller leaves that lose their green color and luster, curled leaf edges, and appearing as if suffering from of dehydration or nutrient deficiency. As the disease worsens, the roots turn brown and rot, leading to wilting and death of the entire plant in severe cases (see Fig. 16.3).

Pathogen

The causative agent of root-knot nematode disease in Platycodon is the southern root-knot nematode (Meloidogyne incognita), belonging to the phylum Nematoda, class Chromadorea, subclass Tylenchida, superfamily Heteroderidae, and family Meloidogynidae. This nematode is sexually dimorphic. The female is enlarged, spherical to pear-shaped, with a prominent neck and a slightly protruding and somewhat cap-like lip region. The perineal pattern varies significantly, often featuring a high dorsal arch. The pattern is distinct, often oval or square. The top of the dorsal arch is rounded or flat, sometimes trapezoidal. The dorsal striae are dense; the

Fig. 16.3 Symptoms of root-knot nematode disease in *Platycodon*. (**a**) Whole plant symptomatic appearance, (**b**) Root galls formed on the roots due to infection

dorsal and lateral surfaces are wavy to serrated, sometimes smooth. The lateral region is often indistinct, and the lateral striae often branched. Body length is 578.2 μm (498.5–654.3 μm), stylet length 14.6 μm (13.3–15.3 μm), DGO (distance from the stylet to the opening of the gland orifice) 3.2 μm (2.5–4.1 μm), stylet knob height 2.6 μm (2.1–3.0 μm), and stylet knob width 3.8 μm (3.2–4.5 μm). The male is elongated, with a flattened to concave lip region, not constricted, and typically 2–3 incomplete annuli. The stylet's conical part has a blunt, rounded tip. The rod-like part is usually cylindrical and narrower near the basal knobs, which are rounded. Body length is 1326.0 μm (1135.5–1541.5 μm), stylet length 24.2 μm (22.5–40.0 μm), and DGO 2.1 μm (1.8–2.6 μm). Second-stage juveniles (J2) measure 386.2 μm (357.3–470.2 μm) in body length, with a stylet length of 13.1 μm (11.5–15.0 μm), DGO 2.8 μm (2.1–3.2 μm), tail length 48.2 μm (43.1–53.6 μm), and a transparent tail length of 14.9 μm (12.6–20.2 μm) (see Fig. 16.4).

Occurrence Pattern

The nematodes overwinters primarily as eggs and larvae. In the absence of host plants, they can survive in the soil for over 3 years. When temperatures rise above 10 °C, eggs hatch, and larvae and eggs are typically active within the 5–30 cm plow layer. The second-stage juveniles are the only infective stage, typically entering through young roots of *Platycodon* to establish infection. Once they finds a suitable site, they induces rapid cell division near their heads, forming multiple giant cells that support nematode development. As the nematodes matures, it creates root galls or knots at the infected sites are formed. At temperatures of 25–30 °C, a full life cycle can be completed within 25 days.

Control Measures

1. Control of Transmission Pathways: Prevent the spread of nematodes via soil and agricultural tools. After working in infected areas, thoroughly clean and disinfect all tools. Use disinfected seedling substrates, and promptly remove infected plants.
2. High-Temperature Solarization: Before planting, take advantage of high temperatures from June to August by plowing, irrigating, and covering the soil with plastic film. This can increase the soil temperature under the film to 45–48 °C or even 50–60 °C at a depth of 20–25 cm, combined with high humidity (90–100% relative humidity), achieving a certain degree of control.
3. Crop Rotation and Residue Management: In fields affected by root-knot nematodes, rotate with non-host crops such as wheat or corn for 3–4 years before replanting *Platycodon*. Avoid rotating with Chinese medicinal plants like *Scutellaria baicalensis* or *Salvia miltiorrhiza*, as these are also suitable hosts for root-knot nematodes, which could worsen the problem.
4. Modification of Cultivation Practices: Switch from broadcast sowing to row sowing. In some regions, *Platycodon* is sown by broadcasting up to 5 kg of seeds per mu (approximately 0.067 hectares), resulting in a final density of over 400,000 plants, which limits ventilation and light penetration. This high density increases the risk of root-knot nematode infestation.
5. Chemical Control: In prepared soil, dig a trench every 20 cm, 20–25 cm deep. Per mu, apply 2.5 kg of 10% thiazophos granules mixed with 10 kg of fine, sieved soil evenly into the trench. Cover the trench slightly, marking it for sowing. After apply-

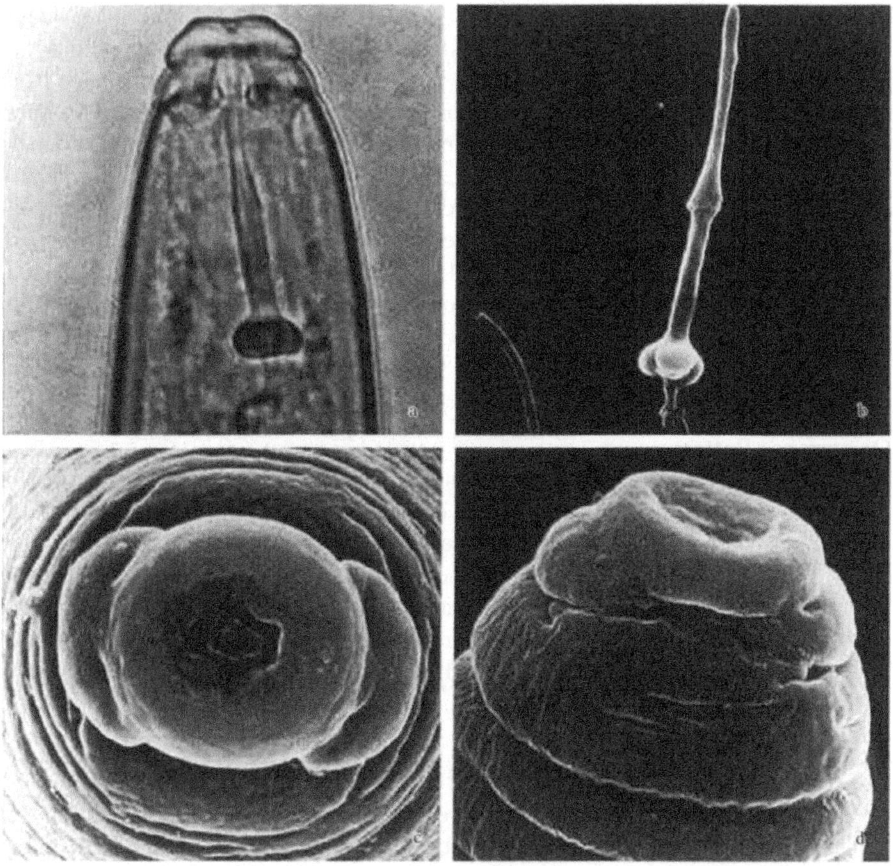

Fig. 16.4 Morphology of Male Southern Root-Knot Nematode. (**a**) Lateral view of male head under a light microscope, (**b**) Stylet, (**c**) Frontal view of male head region under a scanning electron microscope, (**d**) Lateral view of male head under a scanning electron microscope (adapted from Eisenback et al. 1981)

ing the pesticide, sow along the trench line 10 days later, compact with foot pressure, and cover with wheat straw. In late March of the following year, when *Platycodon* sprouts, dig another trench between two rows of *Platycodon* and apply 10% thiazophos granules again. This method achieves an overall control efficacy of over 60%.

16.3 Dodder Infestation in *Platycodon*

Dodder is an important parasitic threat to *Platycodon* production, occurring in all cultivation areas across China. Provinces currently known to experience dodder infestation in Platycodon include Anhui and Liaoning. Dodder primarily harms the above-ground parts of *Platycodon*, and severe infestations can cause wilting and death of these parts. Dodder is widely distributed and mainly spreads through seeds, which can survive in soil for many years.

Symptoms

Dodder is an annual climbing herbaceous parasitic seed plant. *Platycodon* can be parasitized and damaged by dodder during both the seedling and flowering stages. Due to its rapid and lush growth, dodder disrupts *Platycodon*'s photosynthesis, and nutrients are usurped by dodder, resulting in the plant exhibiting yellowing leaves that are prone to dropping, withered branch tips, and weakened growth vigor. Mild infestations affect plant growth, while severe cases lead to the death of the entire *Platycodon* plant (see Fig. 16.5).

Pathogen

The pathogen is Chinese dodder (*Cuscuta chinensis* Lam.), a parasitic seed plant. The seeds are oval, measuring (1–1.5) mm × (0.9–1.2) mm, and light yellow-brown in color. The stem is slender and weak, yellowed, and lacks chlorophyll. Upon contact with the host's stem, dodder produces haustoria that attach to the surface of the *Platycodon* stems and leaves to absorb nutrients. The flowers are white or pale red, clustered, and the fruit is a capsule that splits open upon maturity, releasing 2–4 seeds. Each flower has two capitate styles, and the sepals have longitudinal ridges that run lengthwise, giving the sepals an angular appearance.

Occurrence Pattern

Dodder propagates and spreads through its seeds. After ripening, dodder seeds fall into the soil, and enter dormancy to overwinter. The following spring (April to June), when temperature and humidity are suitable, they germinate. The seedling radicle extends into the soil, while the shoot emerges above the ground, forming thread-like dodder stems that wraps around the Platycodon plant. At the points of contact, haustoria form and penetrate the host. Once haustoria enter the host tissue, some structures differentiate into vessels and sieve tubes, connecting with the host's xylem and phloem to extract nutrients and water.

Control Measures

1. Enhanced Cultivation Management: Select seeds carefully to prevent the mixing of dodder seeds. Combine nursery and field management practices by tilling deeply to bury dodder seeds before they germinate, preventing them from sprouting (generally, dodder seeds buried deeper than 3 cm have difficulty emerging).
2. Manual Removal: Inspect Platycodon nurseries and fields from late spring to early summer. If dodder is detected, remove it immediately, including any affected parts of the host plant. As broken dodder stems can regenerate into new plants, it is crucial to remove them thoroughly. Do not discard pruned stem segments randomly; instead, dry and burn them to prevent further spread. In areas with widespread dodder infestation, remove dodder completely before seed maturity to prevent seeds from falling and increasing the primary infection source for the following year.
3. Chemical Control: During dodder's growth period (May to October), apply quizalofop-p-ethyl herbicide. Spray once every 10 days, for a total of two applications. Apply the herbicide before dodder flowers and produces seeds.

Fig. 16.5 Symptoms of Dodder infestation in *Platycodon*. (**a**) Extensive dodder infestation, (**b**) Dodder parasitizing *Platycodon*, (**c**) Platycodon stems and leaves affected by dodder, (**d**) *Platycodon* fruit stalks affected by dodder

Chapter 17
The Diseases of *Artemisia argyi*

The dried leaves of *Artemisia argyi* H. Lév. & Vaniot, a plant in the Asteraceae family, are known in Chinese as "Ai Ye." The leaves are soft, with a fresh fragrance. *A. argyi* leaves are believed to have the properties of bitterness, warmth, and pungency, which are associated with the liver, spleen, and kidney meridians. Therefore, *A. argyi* leaves are commonly used to treat conditions such as hematemesis, uterine bleeding, excessive menstruation, cold pain in the lower abdomen, menstrual irregularities, infertility due to uterine cold, and skin itching. *A. argyi* is widely distributed across China, and major production areas include Hubei, Henan, Shandong, and Hebei. Common diseases affecting *A. argyi* production include powdery mildew, sooty mold, and leaf spot.

17.1 Powdery Mildew of *A. argyi*

Powdery mildew is a major disease of *A. argyi* and is commonly observed in Hubei and Henan, with the incidence of 10–30%. It affects the above-ground parts of the plant, spreads rapidly, and can quickly expand from the initial infection center, making it difficult to control and seriously impacting the growth of *A. argyi*.

Symptoms
Powdery mildew primarily infects leaves, petioles, and stems of *A. argyi*, with symptoms most evident on the leaves, especially on the upper surfaces. In the early stages of infection, white, amorphous spots appear on the leaf surface, along with small powdery patches. These spots gradually enlarge and merge into a continuous white powdery layer, eventually covering the entire leaf. In the later stages of infection, the leaves turn yellow, wilt, and, in severe cases, the entire plant dies (see Fig. 17.1).

© China Agriculture Press Co., Ltd. 2026

T. Wang et al., *Atlas of Chinese Medicinal Plant Diseases*,
https://doi.org/10.1007/978-981-95-6326-5_17

Fig. 17.1 Symptoms of powdery mildew in *A. argyi*

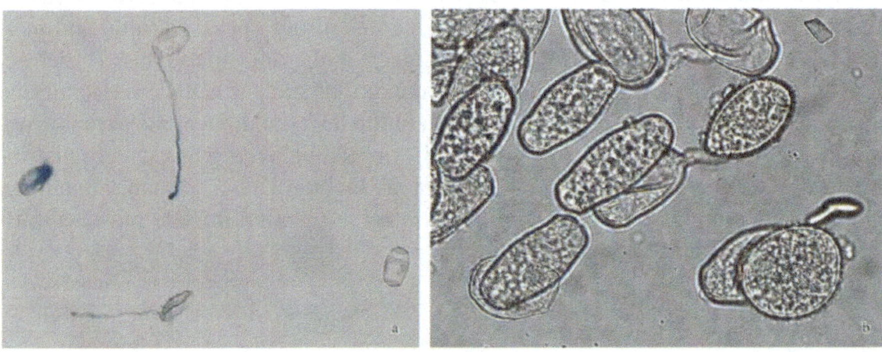

Fig. 17.2 Morphology of powdery mildew pathogen. (**a**) Spore morphology (400x). (**b**) Spore morphology (1000x)

Pathogen

The pathogen responsible for powdery mildew in *A. army* is known in its asexual form as *Oidium artemisiae* var. *artemisiae*, which produces conidia on the host plant. Its sexual form is classified as *Erysiphe artemisiae*, which forms fruiting bodies and produces sexual spores on the mycelium. This fungus is an ectoparasitic organism with well-developed conidia, commonly found on the epidermis of the plant. The conidia are oval to ellipsoidal, with a smooth or slightly warty surface, measuring approximately 34.57 μm in length and 18.36 μm in width. Germ tubes emerge from the apex (see Fig. 17.2).

Occurrence Pattern

The timing and severity of powdery mildew disease outbreaks vary by region due to differences in geography and climate. In Hubei, powdery mildew on *A. argyi* primarily occurs in winter, with infection beginning around mid-November and peaking from December to February of the following year. Temperature is a major factor influencing disease occurrence; lower winter temperatures is conducive to the occurrence of the disease, while higher temperatures inhibit the occurrence of the disease. The pathogen's spores are dispersed by wind and rain, quickly spreading to surrounding plants, often resulting in severe outbreaks. The pathogen overwinters in dormant fruiting bodies within plant debris and infects host plants with ascospores.

Control Measures

1. Seedling Selection: Use healthy seedlings from noninfected areas, and implement strong management practices to enhance plant resistance.
2. Cultivation Management: Maintain appropriate plant spacing to avoid overcrowding, promptly remove infected plants, and control powdery mildew at an early stage to reduce disease spread and prevent large-scale outbreaks. Carefully manage water and nutrients, reducing nitrogen and increasing phosphorus to improve disease resistance. Regularly disinfect soil, cultivate and loosen it, and reduce residual of infected plants.
3. Chemical Control: In the early development stage of powdery mildew disease, the effect of using fungicide is better, strengthen the investigation of the condition, and determine the time of fungicide spraying. In the early stages, 0.5% rhein methyl ether solution diluted 416-fold can be effective. For severe infections, 20% mepronil EC diluted 2,500-fold or 42% benzothiazole SC diluted 3,000-fold is recommended. Additionally, 15% triadimefon WP diluted 1,000-fold also provides good control.

17.2 Sooty Mold Disease of *A. argyi*

Sooty mold disease significantly influences the leaves of *A. argyi*, occurring in all areas of *A. argyi* production fields. This disease causes extensive formation of gray to black mold spots or even a full mold layer on the surfaces of *A. argyi* leaves. This inhibits normal photosynthesis, adversely affecting both the yield and quality of *A. argyi* wool.

Symptoms

Sooty mold initially forms as small black mold spots on leaves and shoots, which later expand and merge to cover the entire leaf surface and tender shoots with a black mold layer. The appearance of a black moldy or soot-like layer on the leaf surface is a distinctive feature of this disease (Figs. 17.3 and 17.4).

Pathogen

A. argyi sooty mold disease is caused by fungi from the genus Meliola (*Meliola* sp.), belonging to the Ascomycota phylum, Microthyriales order. These fungi feed on the secretions from aphids and their own exudates, forming a gray-brown or ink-brown fungal mat on the leaf surface. In severe cases, this mycelium can envelop entire leaves, severely impairing photosynthesis. Conditions that exacerbate the disease include high humidity, poor ventilation, and the presence of honeydew-secreting pests such as aphids.

Fig. 17.3 Symptoms of Sooty Mold Disease on *A. argyi* leaves

Fig. 17.4 Comparison of Healthy and Diseased *A. argyi* Leaves. (**a**) Front of *A. argyi* leaves, (**b**) Back of *A. argyi* leaves

Occurrence Pattern
The pathogens spread via wind, rain, and insects, with infections typically manifesting in spring and autumn. The development of sooty mold is closely linked to poor management, high planting density, weak plant growth, and the activity of aphids.

Control Measures
1. Strengthen Cultivation Management: Avoid overly dense planting of *A. argyi* to ensure adequate ventilation and light penetration, reducing humidity levels and preventing an overly moist environment.
2. Chemical Control: The disease's prevalence is closely linked to honeydew--secreting insects; therefore, controlling aphids, scale insects, and similar pests through spraying is crucial for disease mitigation. Applications may include 25g of bromoamine chrysanthemum ester EC at 300 mL/ha or 10% imidacloprid WP at 450g/ha. For sooty mold caused by parasitic fungi, it is advisable to apply a 45% Daisen ammonium AS diluted to 200-400-fold or 80% mancozeb WP diluted to 1,000-fold.

17.3 Leaf Spot Disease of *A. argyi*

Leaf spot disease of *A. argyi* is a prevalent fungal disease that occurs across all cultivation regions. The disease is particularly severe in areas with dense planting, poor ventilation, and high humidity, with infection rates exceeding 30%.

Symptoms
Leaf spot disease in *A. argyi* typically initiates at the lower parts of the plant. In the early disease stages, small yellow spots appear on the leaf surfaces. As the disease progresses, these spots enlarge and their centers darken to brown. Eventually, the spots coalesce, causing significant leaf wilt and shedding (Fig. 17.5).

Pathogen
The pathogen responsible for *A. argyi* leaf spot disease is *Alternaria alternata* (Fr.) Keissl., a fungus from the phylum Ascomycota, Microthyriales order. On PDA media, the fungus forms brownish-black colonies and produces a black pigment. The conidia, mostly solitary and brown, are typically rod-shaped, ovate, or elongated oval, often with a conidial stalk. They contain 1–5 transverse septa and 0–4 longitudinal or oblique septa, most commonly 2–3 transverse and 1–2 longitudinal or oblique septa (Fig. 17.6). Conidia measure 9.68–35.49 μm in length and 6.4–18.26 μm in width, with a length-to-width ratio of 1.04–4.49.

Occurrence Pattern
The onset of mugwort leaf spot disease occurs in early March as new leaves emerge. The spores are dispersed by wind and rain, and the disease rapidly spreads following rainfall, affecting entire fields. Conditions such as high planting density and poor ventilation exacerbate the severity of the disease. Leaves closest to the ground are typically the first and most severely affected. Fungal mycelia and conidia

Fig. 17.5 Symptoms of *A. argyi* Leaf Spot Disease

overwintering in plant debris and soil serve as primary sources of infection for the following year. They are spread by wind and rain, infecting through wounds or natural openings. The optimal conditions for spore germination are temperatures between 15 °C and 25 °C and relative humidity above 95%. Furthermore, fields with prolonged monoculture are susceptible to disease outbreaks.

Control Measures

1. Enhance cultivation management: Regulate planting density, ensure adequate field ventilation, and maintain timely drainage during the rainy season to lower humidity levels.

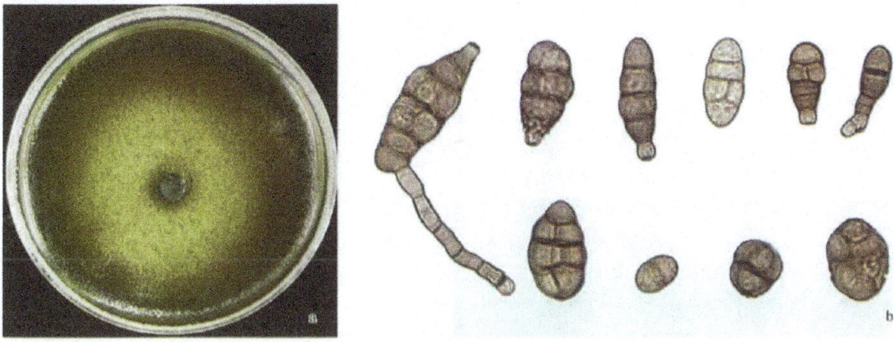

Fig. 17.6 Morphology of the *A. argyi* Leaf Spot Disease pathogen (**a**) Cultural characteristics (**b**) Conidia morphology

2. Chemical control: In laboratory tests, fungicides such as quinoxyfen, propiconazole, difenoconazole, thiophanate-methyl, and tebuconazole have proven effective in inhibiting mycelial growth and are recommended for use.

Chapter 18
The Diseases of Atractylodes (Bai Zhu)

Atractylodes macrocephala Koidz., commonly known as Bai Zhu, belongs to the Asteraceae family and is a medicinal plant unique to China. It is a prominent traditional Chinese herb, one of the famous "Zhejiang Eight Flavors," and is celebrated with the epithet "Southern Bai Zhu, Northern Ginseng." Bai Zhu is a perennial herb, and its rhizome is used medicinally. It is primarily distributed in Jiangsu, Zhejiang, Fujian, Jiangxi, Anhui, Sichuan, Hubei, and Hunan. The rhizome is thick, with scattered brown-yellow oil spots on the cross-section, and has a fresh fragrance. Bai Zhu is known for its benefits in tonifying the spleen and qi, drying dampness, promoting urination, stopping sweating, and calming the fetus. Major diseases affecting Bai Zhu include southern blight, mosaic disease, rusty leaf disease, sclerotial blight, root rot, and leaf spot. Among these, root and rhizome diseases like sclerotial blight, root rot, and damping-off are particularly severe.

18.1 Sclerotial Blight of Bai Zhu

Sclerotial blight, also known as "sugar rot" or "white mold disease," is prevalent across most Bai Zhu production areas, with severe outbreaks in the hot, humid regions of the south. Typical field infection rates range from 15% to 20%, reaching up to 100% in severely affected areas.

Symptoms
Sclerotial blight primarily affects the rhizomes of mature Bai Zhu plants. In the early stages, dark brown, irregular, water-soaked lesions appear on infected roots and stems, accompanied by reduced leaf size and yellowing, which weakens the overall plant growth. As the disease progresses, the cortical tissues in the rhizomes decay, severely damaging the vascular tissues, which disrupts water transport from the roots to the upper parts. This leads to wilting and browning of the upper leaves. In severe cases, rhizome rot manifests in two forms: under low temperatures, the

© China Agriculture Press Co., Ltd. 2026

T. Wang et al., *Atlas of Chinese Medicinal Plant Diseases*,
https://doi.org/10.1007/978-981-95-6326-5_18

parenchyma tissue within the rhizome gradually decays, leaving only the lignified vascular bundles in a matted form, making the plant easily detachable from the soil; under high temperatures and humidity, the rhizome is covered with white mycelium and exhibits wet rot, resembling a rotting potato. Affected rhizomes and the surrounding soil and leaves display white mycelium, and numerous sclerotia—initially milky white, later turning pale yellow to light brown—develop as the pathogen's overwintering structures, similar in size to canola seeds (see Fig. 18.1).

Pathogen

The pathogen responsible for sclerotial blight is *Sclerotium rolfsii* Sacc., an asexual form of a basidiomycete, with its sexual stage identified as *Athelia rolfsii* (Curzi) C.C. Tu et Kimbr. Sclerotia are mostly round, with some elongated or irregular shapes, appearing full or flattened, and ranging in color from milky white to tea brown (on culture media, sclerotia appear round and vary from tea-brown to black). The pathogen produces abundant white aerial mycelium with a diameter of 6.40–7.36 μm. Mycelial growth is rapid and radiates outward, covering a 9 cm Petri dish in about 3 days on PDA medium. After 5 days, numerous milky white, small, grain-like sclerotia form on the culture surface, nearly spherical or ellipsoidal, measuring (1.45–1.76) mm × (1.29–1.61) mm. After 9 days, the sclerotia turn tea brown. The sclerotia have a smooth surface, with dense and firm texture; the outer cells are darker and smaller, while the inner cells are lighter and larger, or sometimes colorless (see Fig. 18.2).

Occurrence Pattern

High temperatures and humidity favor this disease, with the optimal infection temperature being 30–35 °C. Initial infected plants typically appear in late April, with peak infections from June to August, gradually declining after August. The pathogen primarily spreads through sclerotia that are disseminated via seeds, seedlings, rainwater, contaminated soil, and infected plant debris. Mycelium from infected plants also spreads the disease to neighboring plants over short distances. The sclerotia produced at the end of the disease cycle serve as the pathogen's overwintering structure, capable of surviving in soil or plant debris with strong resilience and long--term viability.

Control Measures

1. Seed Selection: Use high-quality, disease-free Bai Zhu seeds. Before sowing, disinfect the seeds by soaking them in a 0.1% solution of 50% carbendazim wettable powder diluted at 1:1000 for 5–10 min, then allow them to air dry before planting.
2. Preplanting Soil Treatment: Deeply till the soil to improve its structure and aeration. Apply quicklime across the entire field for sterilization at a rate of 1200 kg/ha, ensuring even distribution. Leave the field to air out for 2 weeks after treatment.
3. Crop Rotation for Disease Prevention: Rotate with nonhost gramineous crops for 3–5 years. Avoid rotating with susceptible plants such as Scrophularia, Aconitum, Paeonia, Glycine max (soybean), or Rehmannia. Where possible, alternating wet and dry crops yields better disease control results.

Fig. 18.1 Symptoms of Sclerotial Blight in Bai Zhu. (**a**) Early stage of infection, (**b**) Advanced stage of infection, (**c–e**) Sclerotia on the roots and nearby soil surface

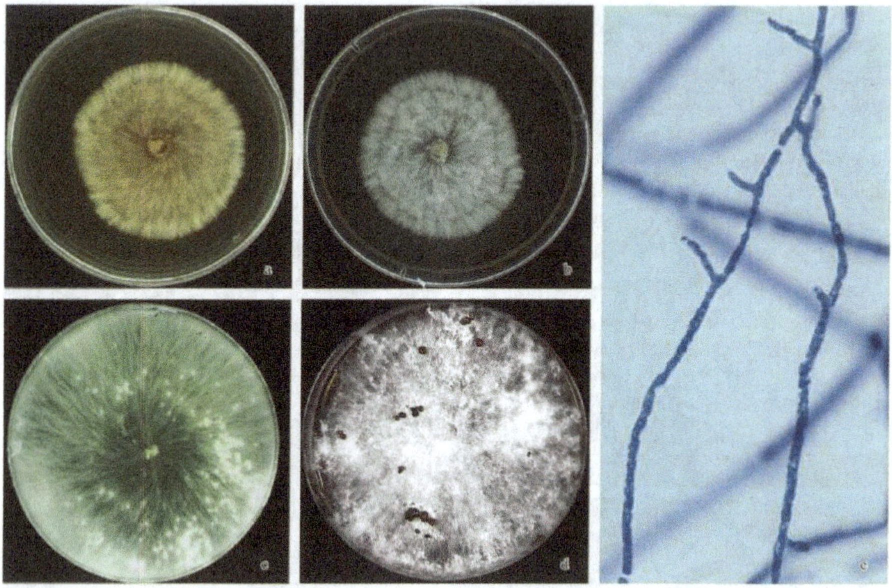

Fig. 18.2 Morphology of Sclerotial Blight Pathogen in Bai Zhu. (**a**, **b**) Early stage of culture, (**c**, **d**) Late stage of culture, (**e**) Hyphal morphology

4. Field Sanitation: Promptly remove and destroy infected plants, and apply lime to disinfect the soil in affected areas. Clear any remaining mycelium to prevent further disease spread.
5. Chemical Control: In the early stages of infection, apply 50% procymidone WP or 50% carbendazim WP, each diluted 1,000-fold, as a foliar spray to control the disease.

18.2 Root Rot of Bai Zhu

Root rot is a significant soil-borne disease affecting Bai Zhu production, primarily damaging the vascular bundles in the roots, making it challenging to control. Since the rhizome of Bai Zhu is used medicinally, infection by root rot leads to a rapid decline in quality and yield, sometimes resulting in a complete loss of economic value. In Enshi, Hubei, root rot causes an average yield loss of up to 40%, with severe cases exceeding 90%, posing a major threat to Bai Zhu cultivation.

Symptoms

Root rot in Bai Zhu is a systemic vascular disease. Initially, the infection typically begins in the root hairs, then spreads to the main roots, stem bases, and even lower leaves close to the ground. It may also directly infect the main root, causing disease throughout the vascular system. The infected areas show yellow-brown, sunken lesions, which gradually turn dark brown, dry out, and rot. The aboveground symptoms include wilting and yellowing of the leaves. Cross-sections of the stem reveal significant browning of the vascular bundles, forming a ring of brown lesions that disrupt nutrient conduction. In advanced stages, leaves and fine roots wither and detach, and the entire rhizome becomes spongy, with dark brown dry rot. The cortex of the rhizome shrinks, wrinkles, and detaches from the woody tissue, ultimately leading to plant death, with the whole plant easily pulled from the soil (see Fig. 18.3).

Pathogen

Root rot in Bai Zhu is primarily caused by fungi from the Fusarium genus, an asexual form of ascomycetes. Pathogen types vary across Bai Zhu production regions, with *Fusarium oxysporum* (Schltdl. ex Snyder et Hansen) and *Fusarium solani* (Mart.) Appel et Wollenw. ex Snyder et Hansen as the dominant strains. *Fusarium oxysporum* forms dense, cottony, aerial mycelium on PDA medium, with colors ranging from white, light red, and pale purple to waxy yellow; in late growth stages, mycelial balls may form. *Fusarium solani* produces yellowish-white concentric rings on PDA medium with less aerial mycelium. Fusarium species produce both macroconidia and microconidia, with spore size and shape varying by strain. Macroconidia typically measure (5.64–36.56) μm × (1.69–6.81) μm with two to six septa, while microconidia measure (5.65–11.19) μm × (1.30–4.46) μm with zero or one septum. Spore shapes range from ovoid and elliptical to fusiform (see Fig. 18.4).

Occurrence Pattern

The pathogen overwinters as mycelium and chlamydospores in infected seedlings, soil, and plant debris, serving as the primary infection source the following year. The pathogen can enter through wounds caused by pests or directly infect the plant, with a high transmission rate. Infected plants typically appear in mid-April, with peak infection from May to June, gradually subsiding after August and lasting until October. High temperatures and humidity promote disease development, especially following extended rain, as sunny conditions can cause large-scale outbreaks in Bai Zhu's late growth stages. The disease spreads through wind, rain, agricultural practices, and soil pests, with long-distance transmission mainly through contaminated soil and seeds.

Control Measures

1. Seed Selection: Choose healthy, pathogen-free seeds with high disease resistance. For large-scale cultivation, establish sterile nurseries. Soak selected seedlings in 0.1% 25% carbendazim wettable powder solution, diluted to 1:400–500, for 12–24 h, then air-dry before planting.
2. Seedling Disinfection: Disinfect seedlings to prevent the spread of the disease.
3. Scientific Site Selection: Select high, well-drained sandy loam soil for planting, preferably in fields that have not previously grown Bai Zhu or in newly cultivated land to reduce the initial pathogen load in the soil and lower the risk of disease.
4. Enhanced Field Management: Rotate with shallow-rooted crops such as maize or cabbage for 3 or more years, avoiding consecutive Bai Zhu planting. Practice

Fig. 18.3 Symptoms of Bai Zhu Root Rot. (**a**, **b**) Plant symptoms, (**c**, **d**) Rhizome symptoms

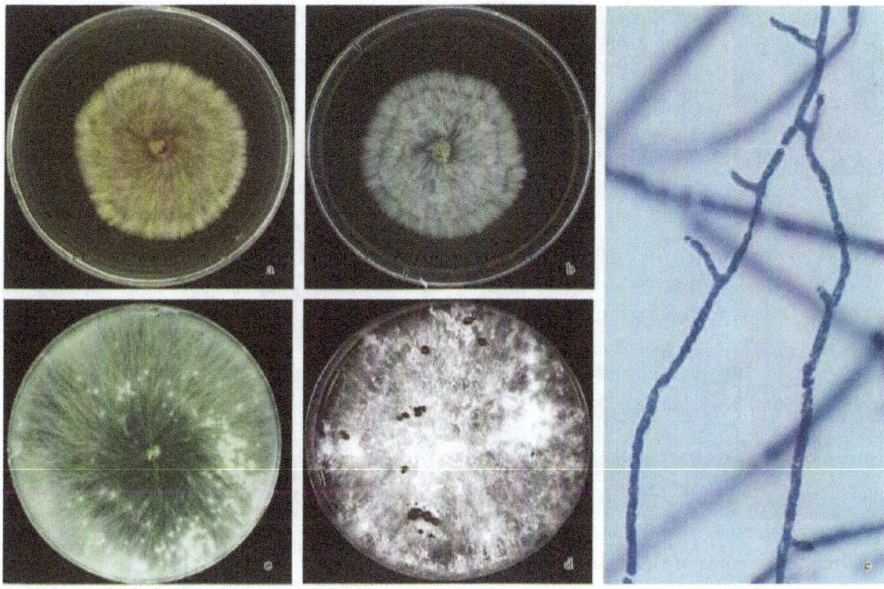

Fig. 18.4 Morphology of Bai Zhu Root Rot pathogen. (**a**, **b**) Culture characteristics of Fusarium oxysporum, (**c**) Conidial morphology of *Fusarium oxysporum*, (**d**) Culture characteristics of *Fusarium solani*, (**e**) Conidial morphology of *Fusarium solani*

reasonable planting density, promptly drain fields after rain to reduce humidity, and immediately remove and destroy infected plants to prevent disease spread within the planting area.

5. Chemical Control: In the early stages of infection, remove centrally infected plants and apply a drench treatment with 50% carbendazim WP or 70% thiophanate-methyl WP diluted to 500–1,000-fold around the diseased plants and surrounding area. Alternatively, use 3% kasugamycin WP or 98% pyrazole WP diluted to 1,000-fold as a foliar spray for disease prevention.

18.3 Damping-Off Disease in Bai Zhu

Damping-off disease, commonly referred to as "stem rot," is a major disease affecting Bai Zhu seedlings. It primarily targets seeds before germination, young shoots, seedlings, and larger transplanted seedlings. Severe outbreaks can result in total crop loss. This disease is found across most Bai Zhu production regions in China, with typical field infection rates ranging from 5% to 20%.

Symptoms

Once infected, young seedlings develop yellow-brown lesions at the base of the stem, often with faint concentric rings and darker edges. The lesions then lose

Fig. 18.5 Symptoms of Bai Zhu Damping-Off Disease

moisture and become sunken, turning from yellow-brown to dark brown. In the early stages, infected seedlings may appear wilted under sunlight but recover in the shade. As the lesions expand and encircle the stem, the stem dries and constricts, forming what is known as a "wire stem," leading to wilting, collapse, and ultimately death of the seedling. Under high-humidity conditions, brown, web-like mycelium may develop at the stem base, along with many yellow-brown, granular sclerotia resembling soil particles. Sometimes, moist leaves near the soil surface are also infected, developing water-soaked brown lesions at the edges that eventually spread to cover the entire leaf (see Fig. 18.5).

Pathogen

The pathogen responsible for Bai Zhu damping-off is *Rhizoctonia solani* Kühn, an asexual form of a basidiomycete fungus from the genus Rhizoctonia. On PDA medium at room temperature, it grows rapidly, forming uniform, thin, felt-like brown colonies, with the reverse side appearing yellow to yellow-brown. The colony edges may be irregular or smooth. The hyphae are septate, initially colorless and later turning pale brown to yellow-brown, with right-angle branching and slight constriction at the branch base. In later growth stages, the colony develops brown sclerotial masses. The sclerotia appear as irregular brown granules with a rough surface and a soft texture when moist (see Fig. 18.6).

Occurrence Pattern

Damping-off disease in Bai Zhu tends to occur during extended periods of cool, rainy weather, thriving in low-temperature, high-humidity environments. In early spring, when newly emerged Bai Zhu seedlings have not yet developed protective cork tissue, they are especially vulnerable to infection, leading to seedling and seed rot. The first signs of infection typically appear from late March to early April, with

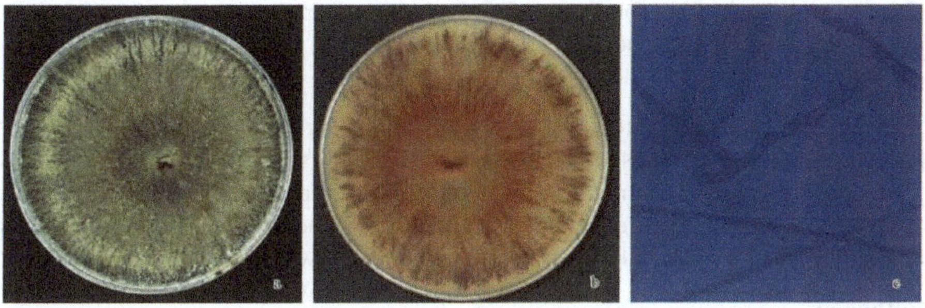

Fig. 18.6 Morphology of Bai Zhu Damping-Off pathogen, (**a**, **b**) Culture characteristics, (**c**) Hyphal structure

peak infection from late April to early May, then gradually decrease. The pathogen overwinters in the form of mycelium or sclerotia in soil and plant debris, where it can survive for 2–3 years. Under favorable conditions, the pathogen infects seedlings through wounds on the epidermis, causing root and stem rot, and spreads via agricultural tools, irrigation water, and rainwater. *Rhizoctonia solani* prefers low temperatures and high humidity, with an optimal growth temperature range of 13–22 °C and an ideal pH range of 3–9.5. Continuous cropping or planting in previously infected fields significantly increases disease incidence.

Control Measures

1. Seedling Selection: Use pathogen-free, healthy, and high-quality Bai Zhu seedlings.
2. Scientific Site Selection: Choose sandy loam fields with good drainage and ventilation, ideally on slopes.
3. Soil Treatment: Disinfect the soil to reduce initial sources of pathogen infection.
4. Enhanced Field Management: Avoid continuous cropping or rotation with other hosts of *Rhizoctonia solani*. After rainfall, create drainage ditches, loosen soil, and reduce field humidity to minimize pathogen spread. Promptly remove and destroy infected plants to prevent pathogen accumulation. Ensure adequate field ventilation and increase organic fertilizer application.
5. Chemical Treatment: In the early stages of infection, apply a drench treatment using 50% carbendazim WP at a 1,000-fold dilution, 50% metalaxyl-thiram wettable powder at an 800–1,000 dilution, 5% lime water solution, or 20% pencycuron EC at a 1,000-fold dilution. To avoid phytotoxicity, mix the chemical with soil and apply clean water to Bai Zhu leaves when treating the soil.

18.4 Atractylodes Leaf Spot Disease

Atractylodes leaf spot disease occurs in most production regions nationwide, with severe outbreaks in the hot and humid southern areas. Typical field infection rates range from 10% to 30%, with rates in severely affected plots exceeding 80%. The disease leads to leaf perforation and decay, impairing photosynthesis and consequently reducing Atractylodes yield.

Symptoms

The disease usually initiates at the leaf tips and margins, presenting as irregular brown to dark brown spots with clear boundaries that spread quickly. These spots gradually expand from the margins and tips across the entire leaf, causing the leaves to turn black, dry up, and eventually wilt and die. In addition to the leaves, stems and branches can also be affected. Spots tend to crack when dry (Fig. 18.7).

Pathogen

The pathogens responsible for Atractylodes leaf spot disease vary by region. In Pan'an County, Zhejiang Province, the disease is mainly caused by *Alternaria longipes* [*Alternaria longipes* (Ellis et Everh.) E.W. Mason]. This fungus forms round colonies on PDA media with dense aerial mycelium that appears light gray on the surface and dark green on the reverse. On SNA media, the conidia range from nearly oval to inversely club-shaped, possessing three to eight transverse septa and zero to two longitudinal septa, measuring (14.2–58) μm × (98.0–200) μm. Occasionally, multiple conidia connect to form chains. Conidiophores are light brown with few branches and one or several septa. The optimal growth temperature is 25–30 °C. In

Fig. 18.7 Symptoms of Atractylodes Leaf Spot Disease

Guizhou Province, the pathogen is *Alternaria tenuissima* [*A. tenuissima* (Nees: Fr.) Wiltshire]. Colonies on PDA media are sparse and gray-white, later becoming velvety, dense, and ranging from dark green to black-brown. Conidia are brown, inversely club-shaped, with four to seven transverse and two to four longitudinal septa, measuring (20.0–44.5) μm × (75–130) μm (Fig. 18.8).

Occurrence Pattern

The pathogens overwinter in plant debris or soil as mycelia or conidia, producing new conidia the following spring. These are disseminated by rainwater and air currents, infecting plants through stomata or by directly penetrating the epidermis. The pathogens prefer high temperature and humidity, with peak disease incidence occurring from mid to late May. The disease spreads mainly through insects, rainwater, and improper farming practices.

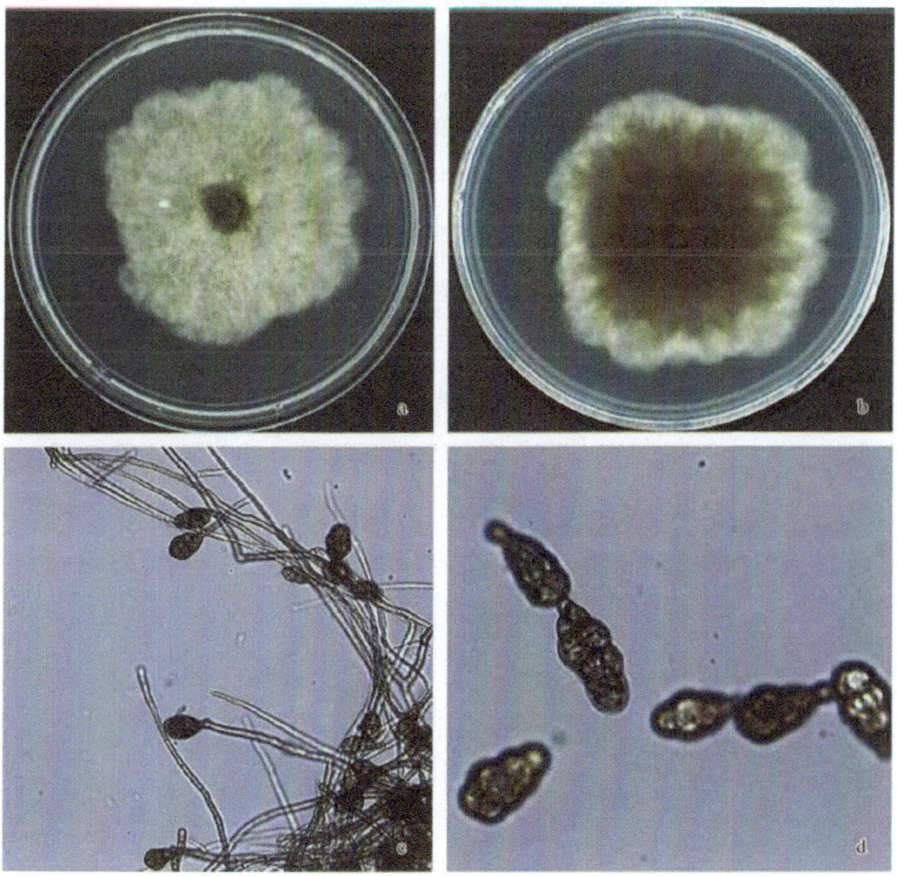

Fig. 18.8 Morphology of the pathogen causing Atractylodes Leaf Spot Disease (**a**) Cultural characteristics (**b**) Conidial morphology

Control Measures

Seedling Selection: Opt for robust, sterile seedlings.

Field Management: Maintain appropriate planting densities, ensuring good field ventilation and prompt drainage after rain; remove and destroy diseased plants immediately to maintain field cleanliness; avoid agricultural activities when dew is heavy.

Chemical Control: At the onset of disease, treat Atractylodes leaves with a spray of *Bacillus subtilis* WP at 800-fold dilution, containing 100 billion viable spores per gram.

18.5 Occasional Disease of Bai Zhu (Table 18.1)

Table 18.1 Characteristics of Bai Zhu Alternaria Diseases

Disease (pathogen)	Symptoms	Occurrence pattern
Rust disease [*Puccinia atractylodis*]	Small, round, yellowish-green spots appear initially, with a yellow halo around the edges. As the spots gradually enlarge, the undersides develop yellow depressions (teliosori), which eventually rupture to release yellow powder (teliospores). Leaves may develop perforations	Generally begins in May to June, with peak infection in June to July, and reduction starting from late July to August
Damping-Off disease [*Rhizoctonia solani*]	After the stem is infected, shallow brown lesions with cloudy edges appear, resembling water-soaked areas, similar to those seen in damping-off diseases caused by water mold. The lesions gradually spread to the entire stem, leading to the death of affected tissue. Subsequently, smaller sclerotia may develop on the decayed tissue	Often occurs in September in fields with excessive moisture. Over-fertilization with nitrogen during this period can contribute to increased disease severity

Chapter 19
The Diseases of Atractylodes

Plants of the genus Atractylodes in the Asteraceae family are unique to East Asia. They are generally believed to include species such as Hubei Atractylodes [*Atractylodes carlinoides* (Hand.-Mazz.) Kitam.], Southern Atractylodes (also commonly referred to as Maocangshu) [*A. lancea* (Thunb.) DC.], and Northern Atractylodes [*A. chinensis* (Bunge) Koidz.]. Except for *Atractylodes carlinoides*, the rhizomes of all other species within the genus can be used medicinally. The Flora of China categorizes Southern and Northern Atractylodes as a single species under the collective name Atractylodes [*A. lancea*].

Atractylodes is a perennial herb, and its rhizomes are used in traditional medicine. The plant is primarily distributed across Hubei, Jiangsu, Hebei, Inner Mongolia, Shanxi, Henan, Liaoning, Jilin, Heilongjiang, Gansu, and Qinghai. Its rhizomes are thick, with red-brown oil spots visible in cross-section, and they possess a distinct aroma. Atractylodes has recognized medicinal properties, including drying dampness, strengthening the spleen, dispelling wind-cold, and improving eyesight.

The primary diseases affecting Atractylodes include root rot, leaf spot, anthracnose, sclerotinia disease, southern blight, black spot, rust, and viral infections. Among these, root rot and other root diseases are particularly severe.

19.1 Root Rot of *Atractylodes*

Root rot is the most common soil-borne fungal disease in *Atractylodes* cultivation, with complex pathogenic factors and significant impact, making it challenging to control. The incidence rate of root rot in Atractylodes is between 15% and 40%, with certain regions experiencing severe outbreaks where the incidence can reach as high as 70%, leading to total crop loss.

© China Agriculture Press Co., Ltd. 2026

T. Wang et al., *Atlas of Chinese Medicinal Plant Diseases*,
https://doi.org/10.1007/978-981-95-6326-5_19

Symptoms

In the early stages of the disease, leaves turn yellow and wilt, while the primary and fibrous roots display yellowish-brown discoloration that gradually deepens to dark brown. The disease spreads from the roots to the stems. In advanced stages, leaves wilt and fall, stems rot, and the cortex separates from the woody tissues, leaving only woody fibers and fragments, or the roots show water-soaked decay (Fig. 19.1).

Pathogen

The pathogen responsible for *Atractylodes* root rot is a fungus of the Fusarium genus (*Fusarium* spp.) in the Ascomycota phylum, including species such as

Fig. 19.1 Symptoms of *Atractylodes* Root Rot. (**a**, **b**) Field symptoms; (**c**) Plant symptoms; (**d**, **e**) Root symptoms

Fusarium solani (Mart.) Appel et Wollenw. ex Snyder et Hansen, *Fusarium oxysporum* Schltdl. ex Snyder et Hansen, *Fusarium striatum* Sherb., and *Fusarium proliferatum* (Matsushima) Nirenberg. Based on pathogen identification in samples from major Atractylodes production regions, *F. oxysporum* and *F. solani* are the predominant strains. *F. oxysporum* produces dense, cotton-like aerial mycelium, with colors varying from white, dark red, purple, pink, to waxy yellow. In later growth stages, some strains may form mycelial balls. *F. solani* has sparse, yellow-white to light gray aerial mycelium with concentric rings and a central yellow--brown spore mass. Spore morphology and size vary significantly among strains, with shapes including fusiform, oval, and ovoid. Microconidia are mostly zero to one septate, measuring (5.65–11.19) μm × (1.30–4.46) μm, while macroconidia are typically two to six septate, measuring (5.64–36.56) μm × (1.69–6.81) μm (Fig. 19.2).

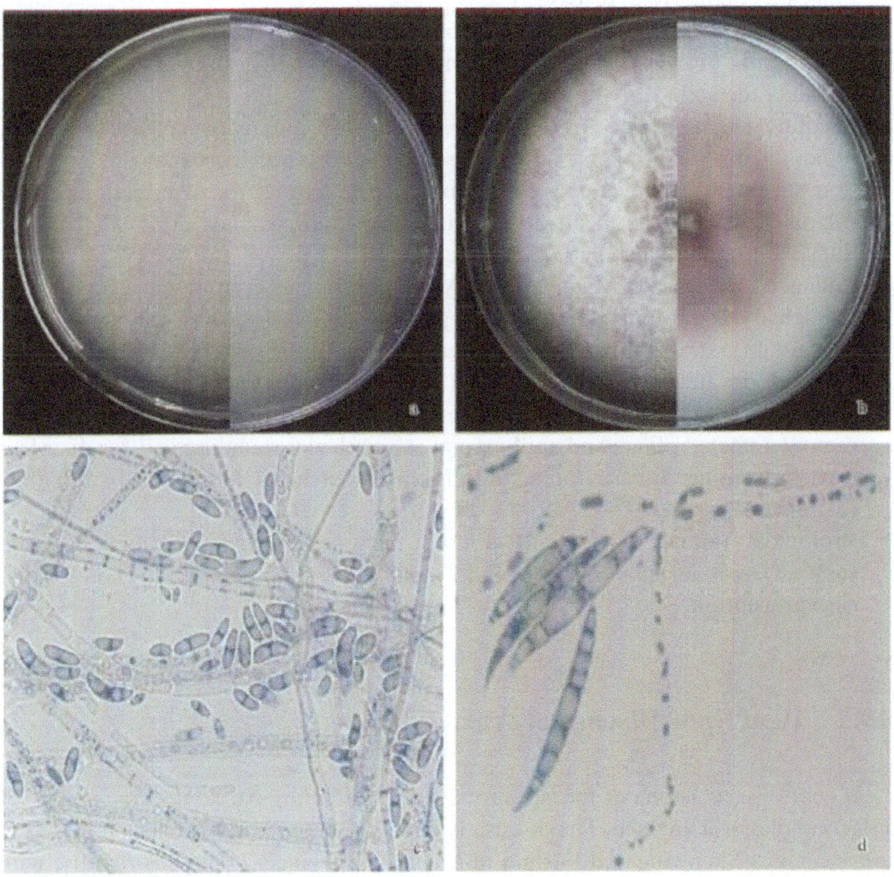

Fig. 19.2 Morphology of *Fusarium oxysporum*. (**a**, **b**) Colony characteristics; (**c**, **d**) Conidia morphology

Occurrence Pattern

The occurrence of Atractylodes root rot varies across regions and cultivation sites. The pathogen overwinters in soil as mycelium, conidia, or chlamydospores within diseased residues, or in storage rhizomes indoors. The disease typically begins in mid-May, with some areas experiencing cases as early as April. The peak period is from June to September, especially following the rainy season, as high temperature and humidity, coupled with poor soil drainage, can lead to large-scale outbreaks. Wounds facilitate pathogen invasion, and long-term continuous cropping increases the risk of the disease. The effectiveness of rhizome and seed disinfection is closely related to disease prevention.

Control Measures

1. Scientific Site Selection: Atractylodes cultivation is best suited to hilly areas or mountain slopes with partial shade. Avoid high temperatures, strong sunlight, waterlogged lowlands, clayey soils, and poorly drained fields, as these conditions are conducive to disease.
2. Soil Disinfection: Apply 50% carbendazim WP at 7–8 g/m^2 or a disinfection powder mix (50% wood ash, 50% lime, 2% sulfur powder) at 7.5×10^3 g/m^2 to disinfect the soil.
3. Seedling Disinfection: For sprouting, select rhizomes with at least two buds, removing diseased rhizomes. Soak the rhizomes in a 50% carbendazim WP solution at 800–1000 times concentration or garlic extract for disinfection. Allow rhizome cut surfaces to air-dry for 1–2 days until wounds are healed before transplanting to reduce pathogen infection risks.
4. Enhanced Cultivation Management: Avoid continuous cropping and high planting density, as prolonged consecutive planting increases disease incidence. New and rotation fields have lower disease rates. It is recommended to rotate Atractylodes fields with cereal crops (such as wheat or corn) for at least 3 years and avoid previous crops susceptible to root rot. During the growing season, remove infected plants and apply wood ash to the surrounding soil for disinfection, preventing disease spread. Apply sufficient basal fertilizer to strengthen resistance.
5. Biological Control: During cultivation or nursery stages, apply beneficial fungi such as *Trichoderma harzianum* and *Trichoderma asperellum* to the soil for disease prevention.

19.2 Leaf Spot Disease of Atractylodes

Leaf spot disease is one of the primary diseases affecting Atractylodes, capable of infecting plants at any age. The disease is widespread, with occurrences reported in areas such as Shenyang and Fushun in Liaoning Province. It spreads rapidly, with infection rates in affected plants often exceeding 50%, and in later stages, reaching

Fig. 19.3 Symptoms of Leaf Spot Disease of Atractylodes. (**a**) Field symptoms; (**b**) Leaf symptoms

over 80%. This leads to extensive leaf death, adversely impacting the normal growth of the plant.

Symptoms

Leaf spot disease primarily affects older leaves on the lower part of the Atractylodes plant, eventually spreading to upper leaves and, in some cases, infecting stems. In the early stages, small white spots with black edges appear on the leaf surface, surrounded by a yellow halo at the margin of the lesion. As the disease progresses, the spots enlarge, forming visible black pycnidia on the lesions, either embedded or partially embedded in the lower epidermis of the leaf. In the field, infections are usually scattered, forming multiple centers of infection. Disease severity increases noticeably after rainfall (Fig. 19.3).

Pathogen

The causative agent of Atractylodes leaf spot disease is the fungus *Paraphoma chrysanthemicola*, an anamorphic ascomycete. Colonies on PDA medium are dark brown with grayish-white margins and appear roughly circular with irregular edges. Hyphae are light black-brown, septate, with some thickened outer walls. Conidia are hyaline, single-celled, short, ellipsoid or cylindrical, measuring 6.5–9 μm in length and 1.75–2.95 μm in width. Conidiogenous cells are ampulliform or flask--shaped, with a diameter of 2.5–3 μm. Pycnidia are near-spherical, composed of four to six layers of pseudoparenchymatous tissue, measuring 80–253 μm in diameter and 65–140 μm in height, with black outer walls and ostioles shaped like bottles or ampullas (Fig. 19.4).

Occurrence Pattern

The pathogen overwinters in the field within infected plant debris or soil, forming pycnidia that serve as the primary inoculum for the following year. Under suitable conditions, conidia are spread by rainwater or insects. In Liaoning Province, initial infections can occur as early as late June, persisting through mid-October, with peak incidence from July to August, influenced by rainfall and temperature. Disease

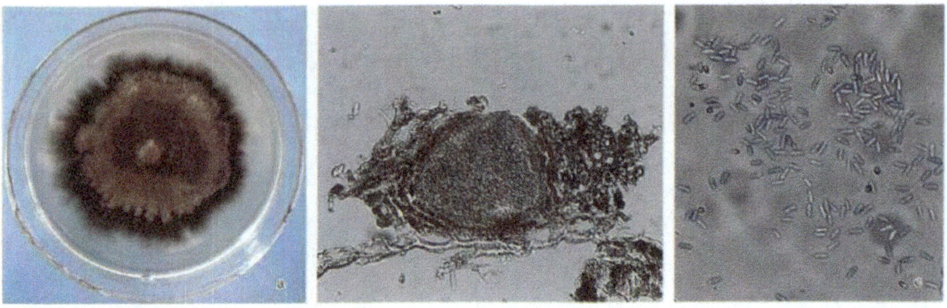

Fig. 19.4 Morphology of Leaf Spot Pathogen of Atractylodes. (**a**) Colony characteristics; (**b**) Pycnidia; (**c**) Conidia

severity decreases in cold and dry conditions. Dense planting increases humidity between leaves, facilitating the spread of the pathogen.

Control Measures

1. Enhanced Cultivation Management: Promptly remove infected plants in the field. Ensure proper plant spacing to facilitate air circulation and reduce humidity between plants. After rainfall, drain the fields promptly and apply organic fertilizers along with phosphorus and potassium to enhance plant resistance. In winter, clear the field of fallen leaves and debris, then bury or burn them deeply to reduce overwintering inoculum.
2. Chemical Control: Apply fungicides at the onset of disease. Recommended chemicals include 75% chlorothalonil WP at a concentration of 600 times, 65% zineb WP at 500 times, 80% mancozeb WP at 600–800 times, or 70% thiophanate-methyl WP at 600 times. Spray at intervals of 7–10 days, repeating 2–3 times.

19.3 *Atractylodes Lancea* Anthracnose

Atractylodes lancea anthracnose is a significant foliar disease that affects *A. lancea*. This disease is widespread and commonly found in most cultivation areas. Typically, the infection rate among plants ranges from 10% to 50%, but in severe cases, it can exceed 80%. The disease leads to extensive leaf necrosis, impacting the plant's normal growth.

Symptoms

Atractylodes lancea anthracnose mainly affects the leaves, beginning as small, nearly circular or irregularly shaped brown spots at the leaf tips and edges. These spots are initially brown, with a clear demarcation between healthy and infected tissue. As the disease progresses, the spots enlarge and merge into deep brown, slightly concave lesions. The surface of these lesions may exhibit scattered or concentrically arranged small black granules, which are the pathogen's conidiomata. Under humid conditions, large red mucilaginous masses of conidia become evident

Fig. 19.5 *Atractylodes lancea* Anthracnose symptoms. (**a**) Field symptoms (**b**, **c**) Symptoms on leaves (**d**). Sporulation on leaves

on the lesions. The mature leaves at the base yellow, wilt, and detach. In severe cases, the disease also affects the young leaves and flowers at the top of the plant (Fig. 19.5).

Pathogen

The pathogen responsible for *Atractylodes lancea* anthracnose, *Colletotrichum anactylodicola*, varies by region. In Pan'an County, Zhejiang Province, the disease is predominantly caused by *Alternaria longipes* [*Alternaria longipes* (Ellis et Everh.) E.W. Mason]. This fungus forms circular colonies on PDA media, showing dense aerial mycelium that appears light gray on the surface and dark green on the reverse. On SNA media, conidia range from nearly oval to inversely club-shaped and contain three to eight transverse septa and zero to two longitudinal septa, measuring (14.2–58) μm × (98.0–200) μm. Occasionally, multiple conidia form chains. Conidiophores are light brown, with few branches and one or several septa. The optimal growth temperature is 25–30 °C. In Guizhou Province, the identified pathogen is *Alternaria tenuissima* [A. tenuissima (Nees: Fr.) Wiltshire]. On PDA media,

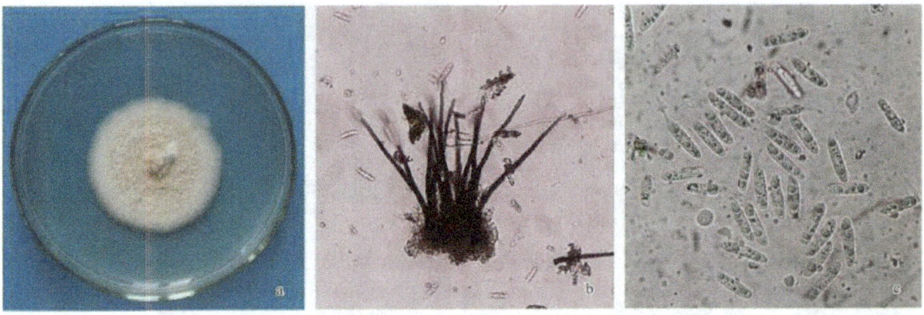

Fig. 19.6 Pathogen morphology of *Atractylodes lancea* Anthracnose. (**a**) Cultural characteristics (**b**) Conidiomata and setae (**c**) Conidia

colonies are sparse, gray-white, later becoming velvety, dense, and ranging from dark green to black-brown. Conidia are brown, inversely club-shaped, with four to seven transverse and two to four longitudinal septa, measuring (20.0–44.5) μm × (75–130) μm (Fig. 19.6).

Occurrence Pattern

The pathogens overwinter in plant debris or soil as mycelial or conidial forms, and mature conidia in the following spring become the primary source of new infections. These are disseminated by rain and wind, as well as through irrigation water, insects, and farming activities. The development of the disease is significantly influenced by temperature and humidity, with peak incidences occurring from late May through July to August. The disease tends to have a focal point of infection, spreading rapidly and often causing leaf spots that lead to batch leaf necrosis.

Control Measures

1. Field Management: Enhance sanitary conditions to reduce the source of pathogens. Remove and destroy fallen leaves and branches at the end of autumn or early winter to reduce sources of infection for the following year. Apply fertilizers timely and maintain proper irrigation and drainage, especially during rainy autumn periods, to reduce humidity. Early weeding and the application of organic fertilizers can improve plant resistance.
2. Chemical Control: Early application of fungicides is crucial. Options include a 600–800× dilution of 70% thiophanate-methyl WP or 250 g/L myclobutanil EC, or 80% Antracol WP at 600× dilution.

19.4 Sclerotinia Disease of Atractylodes

Sclerotinia disease is one of the major diseases affecting Atractylodes. This disease occurs widely across various production regions, with an average incidence rate of 10–30%, reaching over 50% in severe cases.

Symptoms

Sclerotinia disease primarily affects the roots and rhizomes of Atractylodes, though it can also impact the base of the stem. Infected plants show yellowing and wilting of the older leaves at the base, which gradually spreads upward, ultimately leading to the death of the entire plant. The stem and rhizome develop brown or dark brown rot, with indistinct boundaries between healthy and diseased tissue. The cortex decays easily, exposing the fibrous tissues underneath. In high humidity, a white, cottony layer of mycelium can be observed on the affected areas. Over time, black sclerotia, measuring 0.8–6.9 mm in diameter and varying in shape from oval to irregular, form on the diseased parts (Fig. 19.7).

Pathogen

The pathogen responsible for sclerotinia disease in Atractylodes is *Sclerotinia nivalis* I. Saito, a fungus from the phylum Ascomycota. On PDA medium, the pathogen forms circular colonies with abundant, fluffy, aerial mycelium that initially appears white and later turns pinkish. After 7 days, concentrated hyphae near the edge of the petri dish form small, white protrusions. These protrusions gradually enlarge and darken, forming black sclerotia interspersed with white mycelium. The central area of the colony does not show sclerotia formation. Sclerotia are spherical or

Fig. 19.7 Symptoms of Sclerotinia Disease in Atractylodes. (**a**) Aboveground symptoms. (**b, c**) Root symptoms. (**d**) Basal stem symptoms

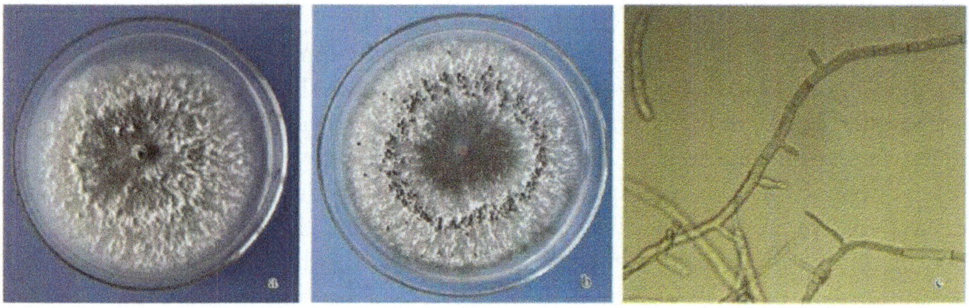

Fig. 19.8 Morphology of the sclerotinia disease pathogen in Atractylodes. (**a**) Colony morphology after 7 days. (**b**) Colony morphology after 14 days. (**c**) Mycelium morphology

near-spherical, varying in diameter from 0.5 to 4.5 mm. Some sclerotia elongate or merge into irregular shapes, adhering closely to the medium's surface. The sclerotia have a smooth texture, compact structure, and are firm (Fig. 19.8).

Occurrence Pattern

The pathogen overwinters in the soil or in plant debris as sclerotia and mycelium, which serve as primary infection sources for the following year. During the growing season, sclerotia germinate, spreading via rainwater or irrigation, or they may produce ascospores that disperse by wind and rain, leading to secondary infections and further spread. In the late stages of disease development, sclerotia form near the infected plant's rhizome and on the soil surface. Field infections typically begin in mid-April, with peak occurrence from late April to May. High nitrogen levels, poor drainage, minimal management, and waterlogging after rain can all promote disease spread.

Control Measures

1. Scientific Site Selection: Choose fields that have not previously been used for Atractylodes cultivation, and avoid planting adjacent to previous year's Atractylodes fields.
2. Enhanced Cultivation Management: During spring rains, loosen the soil promptly after rainfall and ensure adequate drainage to reduce field humidity. From early April, regularly inspect Atractylodes fields; remove and destroy infected plants immediately, and apply lime powder around infected areas for disinfection.
3. Chemical Control: At the onset of disease, spray with 65% mancozeb WP at a dilution of 400–600 times, using 75–100 kg per acre. Repeat spraying every 7 days for two to three applications.

19.5 *Atractylodes Macrocephala* White Scourge Disease

White scourge disease ranks as a significant root disease in *Atractylodes macrocephala* cultivation, only surpassed by root rot. It affects all authentic production regions, with severe infection rates reaching 40–50%.

Fig. 19.9 *Atractylodes Macrocephala* white scourge disease symptoms. (**a**) Field symptoms (**b**, **c**) Sclerotia (**d**, **e**) Root symptoms

Symptoms

Initially, the aerial parts of the plant show no obvious symptoms. As the disease progresses, leaves wilt and eventually die, remaining attached to the plant, a symptomatology similar to soft rot. Infection of the *Atractylodes rhizomes* or stem bases leads to water-soaked rot that turns brown. In advanced stages, only a net-like vascular tissue remains visible, with white mycelium adhering to the diseased plant or its surroundings, forming spherical or oval sclerotia about 0.4–1.2 mm in diameter, resembling the size of rapeseeds. The ease with which plants can be uprooted is notable (Fig. 19.9).

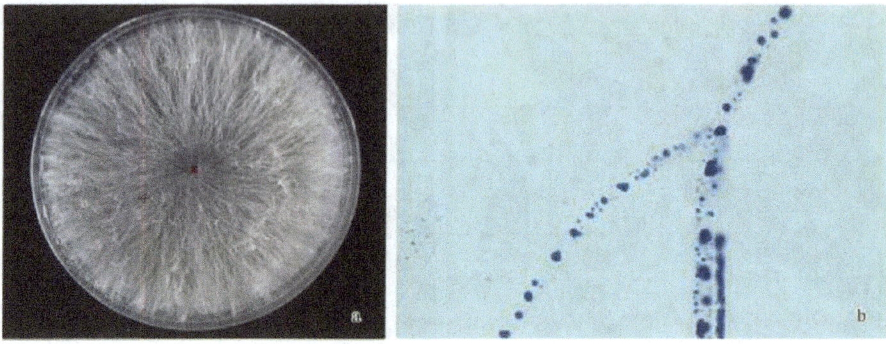

Fig. 19.10 Pathogen morphology of *Atractylodes macrocephal.* (**a**) White Scourge Disease a. Cultural characteristics (**b**) Mycelial form

Pathogen
The causative agent of *Atractylodes macrocepha*la white scourge disease is *Sclerotium rolfsii* Sacc., the asexual stage, and its sexual form is *Athelia rolfsii* [*Athelia rolfsi* (Curzi) C.C. Tu & Kimbr.]. The fungus's mycelium is white and forms spherical or oval sclerotia. Sclerotia are initially milky white, transitioning to beige, yellowish-brown, and eventually to reddish-brown. On PDA medium, the fungus displays white, silky mycelium that spreads radially, producing numerous sclerotia in later stages. These sclerotia initially appear milky white, gradually changing to beige and eventually to tea-brown, spherical or irregularly shaped (Fig. 19.10).

Occurrence Pattern
The pathogen primarily overwinters as sclerotia in the soil or as mycelium on seeds or plant debris. Under favorable conditions, sclerotia germinate to produce mycelium, which infects the rhizomes and stem bases, initiating infection. Mycelium on diseased plants continually produces new sclerotia, which are disseminated by water, soil movement, or infected plant material, causing further plant infections. The pathogen thrives in high temperatures and high humidity, with optimal growth at 30–35 °C in moist environments. Peak disease occurrence spans from early June to mid-August; under suitable conditions, the disease rapidly expands to surrounding areas.

Control Measures
1. Seed Selection: Choose disease-free, robust seedlings or rhizomes and prefer resistant varieties.
2. Cultivation Management: Avoid consecutive planting with susceptible crops such as Solanaceae, legumes, and cucurbits. Implement proper field management, maintain appropriate planting density, ensure effective drainage especially during rainy seasons, promptly remove and destroy diseased plants and contaminated soil, and apply lime to infected pits.
3. Chemical Control: Apply appropriate fungicides early in disease development or after detection. Suitable treatments include a 1000–1,500-fold dilution of 25%

Fig. 19.11 Symptoms of black spot disease in Atractylodes

fungicide SC, a 1200–1,500-fold dilution of 10% benzimidazole WG, or a 400–500-fold dilution of 70% mancozeb and zinc WP. Additionally, treatments with thiophanate-methyl and carbendazim are recommended.

19.6 Black Spot Disease of Atractylodes

Black spot disease is a significant disease affecting Atractylodes, with a widespread distribution and high incidence rate. Primarily impacting the leaves, this disease can result in wilting and leaf drop in later stages, which impairs photosynthesis in the plant. It can also lead to seedling mortality, with general yield reductions of 10–15%, and in severe cases, over 90%.

Symptoms

In the early stages of infection, black spot disease begins on the leaves at the base of the Atractylodes stem, gradually spreading to the upper leaves. Lesions appear as circular or irregular spots, typically starting at the leaf edges and tips, then expanding quickly. Black mold is often visible on both sides of the affected leaves. In advanced stages, the lesions turn grayish-brown and coalesce, leading to leaf wilting and shedding, leaving only the stems intact (Fig. 19.11).

Pathogen

The pathogen responsible for black spot disease in Atractylodes is *Alternaria alternata*, an ascomycete fungus from the genus Alternaria. The hyphae are dark brown, and the majority of conidia have three transverse septa and one longitudinal septum, with an inverted club shape. Under the microscope, conidia measure (35–102) μm × (11–26) μm. On PDA medium, the colony's underside appears bluish-green to black, with flat, firm hyphae. Optimal growth temperature for hyphae is 20–25 °C, and they thrive in slightly acidic conditions, with a pH range of 5.0–6.0 (Fig. 19.12).

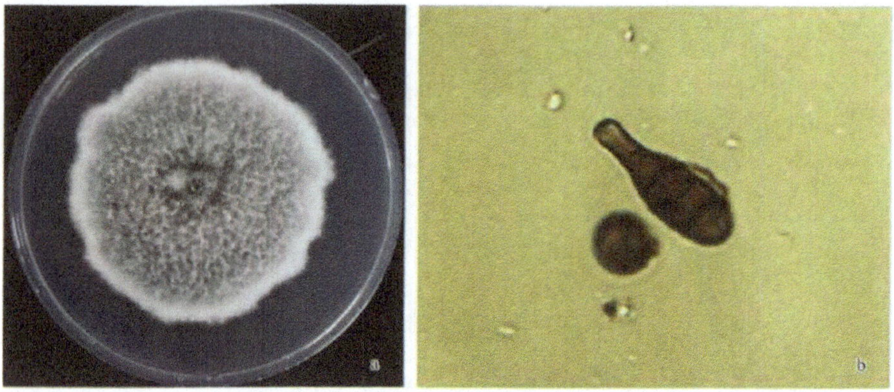

Fig. 19.12 Morphology of black spot disease pathogen in Atractylodes. (**a**) Colony morphology. (**b**) Conidial morphology

Occurrence Pattern

The pathogen overwinters as mycelium or conidia on infected plant debris, serving as the initial infection source for the following year. Under favorable conditions, conidia are produced and spread by wind, rain, or insects, infecting Atractylodes leaves. In the region of Liaoning, the disease typically begins in mid-May, with an average temperature of around 20 °C and relative humidity of 85%. Peak infection occurs in July and August when relative humidity is approximately 90%. As temperatures drop in late September, disease progression slows considerably.

Control Measures

1. Seedling Selection: Use robust, disease-free seeds for nursery and transplanting.
2. Enhanced Cultivation Management: After harvesting, deep-plow the soil and remove plant debris. Rotation with rice can help reduce pathogen levels. For infected plants, apply 30% benzo-epoxiconazole EC or 10% prothioconazole WG to inhibit pathogen spread and minimize damage.
3. Chemical Control: Initiate treatment at the early stages of infection for best results.

19.7 Atractylodes Rust Disease

Atractylodes rust disease primarily affects the foliage and is prevalent in all the recognized growing regions across the country. In advanced stages, the disease can lead to leaf wilting and defoliation, significantly affecting the plant's photosynthesis.

Symptoms

The disease predominantly manifests on leaves, particularly affecting the younger, more tender leaves more than the older ones. Initially, pale yellow spots appear on the upper surface of the leaves. On the underside, these spots change to discolored patches. As the disease progresses, the upper surface of the leaves develops slightly sunken, chlorotic marks, while the underside shows pale yellow, blister-like protrusions, representing the aecia (Fig. 19.13).

Pathogen

The causal organism of Atractylodes rust disease is *Puccinia atractylodis* P. Syd. et H. Sydow, a member of the Basidiomycota phylum. Aeciospores are spherical or oval, measuring (20–26) μm × (19–29) μm, with a thick, yellowish-brown wall. Teliospores are solitary, dark brown, and club-shaped with a slanted blunt or flat tip and are slightly constricted at the septa, measuring (56–62) μm × (15–19) μm (Fig. 19.14).

Occurrence Pattern

The pathogen overwinters as teliospores in the soil or on plant debris and reinfects by producing aeciospores in the following season. The disease starts in mid-May, peaking in June and July. It spreads through wind and rain, with high humidity facilitating the disease's proliferation.

Fig. 19.13 Symptoms of Atractylodes rust disease

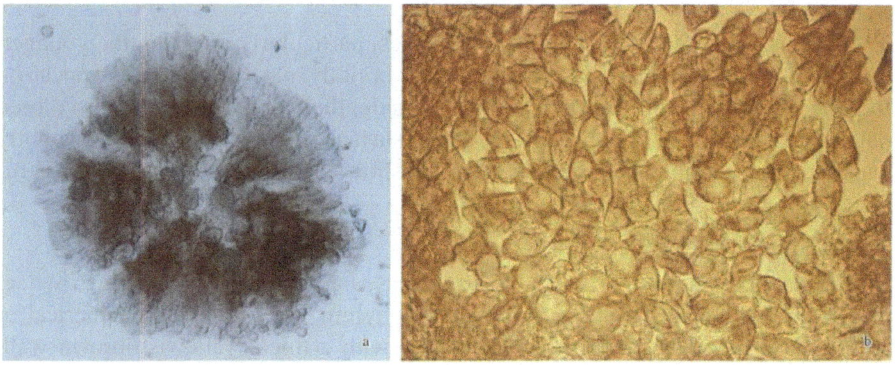

Fig. 19.14 Morphology of the pathogen causing Atractylodes Rust Disease. (**a**) Aecia (**b**) Telia

Control Measures

1. Cultivation Management: Maintain appropriate planting densities and ensure adequate ventilation to prevent excessive field humidity.
2. Chemical Control: Early application of fungicides is recommended for effective disease management. Suitable options include a 1,000–1,500-fold dilution of 15% triadimefon WP or an 800-fold dilution of 50% propineb WP.

19.8 Atractylodes Virus Disease

Atractylodes virus disease is prevalent across all recognized production areas in the country, typically affecting 5–30% of plants.

Symptoms

Infected plants exhibit symptoms such as mosaicism, dwarfing, and wrinkling. Early symptoms include leaves displaying shades of deep green, light green, or pale yellow. As the disease progresses, severe deformities such as wrinkling, curling, and even death of leaves occur. The plant experiences stunted growth, and the subterranean tubers become deformed, undersized, and of poor quality (Fig. 19.15).

Pathogen

The causative agent of Atractylodes virus disease is the Cucumber mosaic virus (CMV), classified within the Cucumovirus genus of the Bromoviridae family. CMV is a classic tripartite single-stranded positive-sense RNA virus (Fig. 19.16).

Occurrence Pattern

CMV is widely transmitted by aphids in a nonpersistent manner, with the cotton aphid and peach aphid as primary vectors. It can also spread through sap contact or seeds, with a seed transmission rate generally between 4% and 8%. The disease typically begins in plants from germination to the four-to-five leaf stage, peaking

Fig. 19.15 Symptoms of atractylodes virus disease

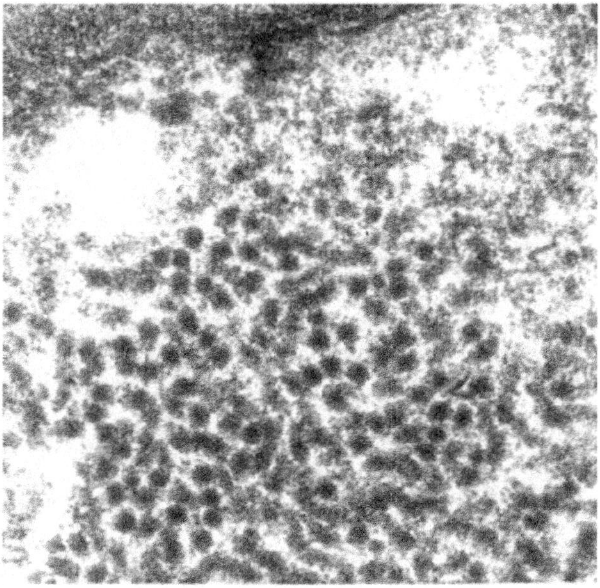

Fig. 19.16 Morphology of cucumber mosaic virus

between May and June. Aphids play a crucial role in secondary infections, and disease incidence significantly increases with large aphid populations.

Control Measures

1. Choosing Resistant Varieties: Employing resistant varieties is an effective strategy to manage viral diseases.
2. Cultivation Management: Promote vigorous seedlings and timely planting after the last local frost, adjusting dates appropriately for protected cultivation environments.
3. Proper Fertilization: Apply bio-organic fertilizers such as seaweed extracts and adhere to tailored fertilization practices while enhancing overall plant management.
4. Vector Management: Utilize insect nets to control aphid vectors responsible for virus transmission (Table 19.1).

Table 19.1 Common diseases of Atractylodes

Disease (Pathogen)	Symptoms	Occurrence pattern
Wilt disease [(*Fusarium solani* and *F. Equiseti*)]	Infection initially appears in the lower leaves of Atractylodes, which turn yellow-green and gradually wilt upward along the stem, eventually spreading to the entire plant. Leaves may turn yellow and die without falling off. Infected plants may show "half-wilted" yellowing on individual branches, which can progressively extend to the whole plant	Disease onset begins around mid-June, with a peak occurrence from July to august, and diminishes by mid-October
Soft rot disease [(Erwinia carotovora pv. Carotovora)]	Infected roots and rhizomes exhibit decay, becoming mushy or soft with a sour odor. Early symptoms include browning and softening of the roots. Upper parts of the plant may show yellowing, which progressively leads to decay of roots and stems, ultimately causing wilting and death of the entire plant. The affected vascular bundles turn brown, impairing water transport functions and leading to plant death	Generally starts in early June, peaking in mid-October. Lower humidity during this period reduces disease severity
Gray spot disease [(*Cercospora atractylodis*)]	Primarily affects leaves, forming circular spots 2–4 mm in diameter. Spots have a light gray center, dark brown edges, and a black mildew layer on the upper surface	The pathogen overwinters in plant residues and infected roots. In the following year, it produces conidia, which initiate primary infection, followed by secondary infections

Chapter 20
The Diseases of *Erigeron breviscapus*

Erigeron breviscapus (Vant.) Hand.-Mazz., also known as short-calyx fleabane and Dengzhanxixin, is a perennial herbaceous plant of the genus Erigeron in the Asteraceae family. The whole plant is used as medicine. It grows on sunny hillsides, grasslands, forest edges, and in sparse forests at an altitude of 1,100–3,500 meters. It is distributed in Yunnan, Sichuan, Guizhou, Guangxi, and Hunan provinces. Yunnan produces more than 95% of *E. breviscapus* medicinal materials in the country, and the annual cultivation area is maintained at around 10,000 mu, primarily in regions including Honghe, Qujing, Dali, Kunming, Yuxi, and Chuxiong, with Honghe and Qujing as the main production areas.

Historically utilized by the Miao, Yi, and De'ang ethnic groups in Yunnan, *E. breviscapus* has a warm property, tastes pungent and slightly bitter. It has the effects of promoting blood circulation and dredging collaterals, relieving pain, dispelling wind and eliminating dampness. Traditionally, it has been mostly used for treating stroke-induced hemiplegia, chest-stuffiness and heartache, rheumatic arthralgia, headache, toothache, etc. Since being recognized by medical professionals in the 1970s, this herb has developed into a natural special-effect medicine for treating occlusive cerebrovascular diseases and sequelae of cerebral hemorrhage. It has been successively included in 1974 edition, 1996 edition of the "Yunnan Provincial Standards for Drugs", and the 1977 edition, 2005 edition, 2010 edition, 2015 edition, and 2020 edition of the "Pharmacopoeia of the People's Republic of China".

The main diseases of *E. breviscapus* include root rot, rust, downy mildew, and root-knot nematode infections, with root rot causing the most serious losses.

© China Agriculture Press Co., Ltd. 2026

T. Wang et al., *Atlas of Chinese Medicinal Plant Diseases*,
https://doi.org/10.1007/978-981-95-6326-5_20

20.1 Root Rot Disease in *Erigeron breviscapus*

Root rot is the most important disease of *E. breviscapus*, which can occur at all growth stages and is a common root disease. It appears in almost all planting areas in Yunnan Province and has a severe impact on the plant's yield and economic value. *Erigeron breviscapus* is cultivated in a system where it is planted once and harvested multiple times. Typically, three harvests are possible, with up to five in ideal conditions. However, root rot directly reduces the number of viable harvests, with severely affected areas achieving only one or two harvests, leading to considerable yield loss. The incidence of this disease is related to factors such as soil, climate, management level, and harvest frequency. Fields that have not been harvested show a disease incidence rate of less than 1%, while areas harvested three times or more can exceed an 80% infection rate.

Symptoms

The pathogen typically invades the root-stem junction close to ground level. Early infection stages often show no obvious symptoms. However, as the pathogen spreads, it infects the vascular tissues, disrupting water transport and causing wilting of the plant's above-ground parts. As the disease progresses, leaves and flowers wilt and die, roots turn black and rot, and a white mold layer may sometimes appear on the roots (Fig. 20.1).

Fig. 20.1 Root Rot symptoms of *Erigeron breviscapus*. (**a**) Field symptoms. (**b**, **c**) Leaf wilting and withering (**d**) Root rot

Pathogen

Root rot of *Erigeron breviscapus* is caused by multiple *Fusarium* species (asexual morph of Ascomycota), including *Fusarium oxysporum* Schltdl. ex Snyder et Hansen, *F. solani* (Mart.) Appel et Wollenw. ex Snyder et Hansen, *F. sporotrichioides* Sherb., *F. semitectum*, *F. subglutinans*, and *F. graminearum* Schwabe, with *F. oxysporum* and *F. solani* being the most common.

When *F. oxysporum* is cultured on PDA medium at 25°C for six days, colonies reach a diameter of 8.5 cm. The aerial mycelium is relatively well-developed, velvety, and the colony is circular, loose, with colors from white to pinkish-white. Small conidia are borne on single phialides, often aggregating into clusters at the phialide tips. These conidia are unicellular, kidney-shaped, and pseudocapitate on the conidiogenous cells, and measure (6.4–14.0) μm × (2.5–3.5) μm. Large conidia are sickle-shaped, slightly curved, mostly with three septa, measuring (15.0–32.6) μm × (3.5–4.0) μm. Thick-walled chlamydospores are easily produced; they are spherical, single, paired, or in chains, with a diameter of 6.0–8.6 μm (Fig. 20.2a, b).

When *F. solani* is cultured on PDA medium at 25°C for six days, the colony diameter reaches 7.4 cm. The aerial mycelium is velvety, white or light gray, with interspersed light yellow sporodochia, often forming a large number of

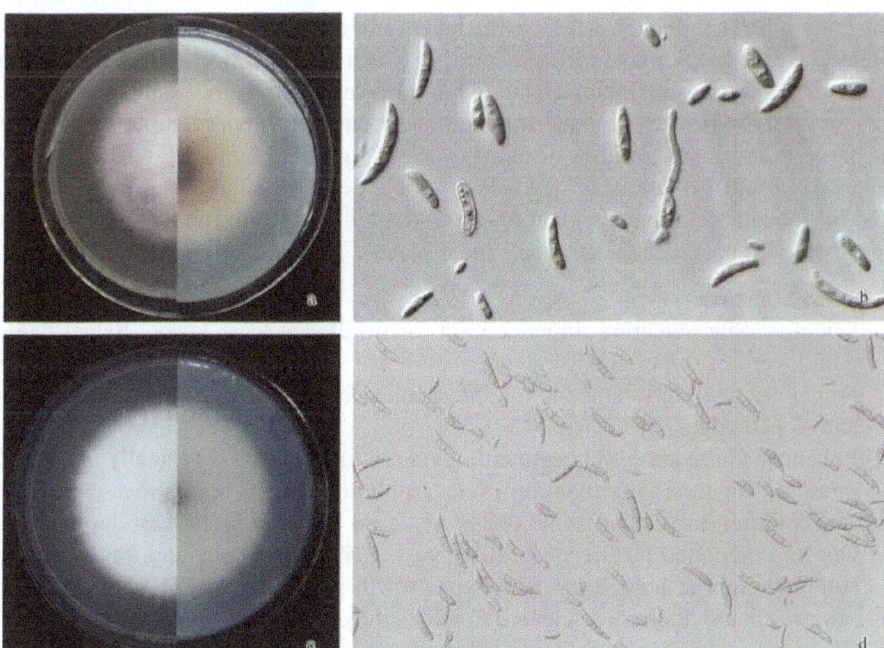

Fig. 20.2 Pathogen morphology of Root Rot in *Erigeron breviscapus.* (**a**) Colony morphology of *Fusarium oxysporum.* (**b**) Conidia of *F. oxysporum.* (**c**) Colony morphology of *F. solani.* (**d**) Conidia of *F. solani*

yellow-brown slimy spore masses, while the culture medium does not change color. Small conidia are oval or kidney-shaped, with one or no septum, measuring (11.5–20.5) μm × (3.5–6.0) μm. Phialides on aerial mycelium are long tubular, forming clusters on sporodochia, branched, with varying lengths, and have septa, measuring (31.2–101.4) μm × (2.6–4.5) μm. Most phialides are longer than 50 μm, and sometimes can reach 200 μm. Large conidia are produced in aerial mycelium, sporodochia, and slimy spore masses, elliptic-curved, with short blunt-rounded terminal cells and blunt-rounded basal cells, having 3–5 septa, with a size of (25.5–40.0) μm × (5.5–6.0) μm. Numerous chlamydospores, oval, borne singly or intercalarily at the spex of middle of hyphae or spores, with a diameter of 6.2–9.8 μm (Fig. 20.2c, d).

Occurrence Pattern
Root rot occurs year-round, particularly from April to October, with its peak during the rainy season from June to October. It affects almost all cultivated areas, with the incidence increasing with each harvest. Severely affected fields can experience a disease incidence rate of over 70%, leading to significant plant loss.

High temperature and high humidity, as well as wounds from harvesting, are the main factors inducing root rot of *E. breviscapus*. *Erigeron breviscapus* is grown using either direct seeding or seedling transplantation. Directly sown plants are often covered with low tunnels during the seedling stage, which are removed during the rainy season, reducing soil moisture and minimizing root rot incidence. In transplanted crops, black plastic film is used for covering to control weeds and retain water in the dry season, increasing field humidity and making the plants more prone to root rot. Disease severity is also linked to the number of harvests; as the number of harvests increases, so does disease severity. High field humidity during harvest (due to rainfall, irrigation, or dew), low stubble height (damaging growth points), and inadequate field management before and after harvesting (non-standard use of fertilizers and pesticides) exacerbate the disease.

Control Measures
1. Scientific Site Selection: Practice crop rotation for at least 1 year. Select fields with a certain slope, good drainage, and loose soil texture. In flat areas, ensure proper drainage and shape the soil into a concave "tile shape," with raised centers and sloped edges.
2. Enhanced Cultivation Management: *Erigeron breviscapus* is typically harvested three to four times per planting cycle. Applying urea increases the severity of root rot after harvest, so avoid using urea until the final preharvest application. Opt for compound fertilizers at other times. Harvest on sunny, dry days, leaving stubble height at least 2 cm to reduce growth point damage. During the rainy season, ensure ditches are cleared to reduce field moisture.
3. Chemical and Biological Control: If root rot has not yet appeared, preventive sprays using *Bacillus subtilis* or fungicides like hymexazol, fenaminosulf, captan, or thiram can be applied alone or in combination. Preventive spraying is effective from April to October, with monthly applications to control root rot and

other fungal diseases. When root rot appears sporadically, remove affected plants and drench the area with a mix of fungicides like hymexazol, fenaminosulf, captan, or thiram to disinfect the roots of healthy plants nearby. Postharvest, before wounds heal, spray plants with a systemic fungicide like difenoconazole, pyraclostrobin, hymexazol, or fenaminosulf, combined with a protective fungicide like captan, mancozeb, or thiram for prevention. If the disease is widespread, 1–2 of the above-mentioned systemic fungicides are alternately used in combination with one protective fungicide for 1–3 applications at 7-day intervals. Follow specific product instructions for concentrations and dosages. New fungicides should be tested on a small scale before large-scale application to ensure no adverse effects.

20.2 Rust Disease of *Erigeron breviscapus*

Rust disease can occur at any growth stage of *E. breviscapus*, with particularly severe outbreaks in the summer. This common foliar disease is widespread and has been documented in key production areas of *E. breviscapus* in Yunnan, such as Honghe and Qujing. Generally, it appears in localized areas, with infection rates in greenhouse cultivation reaching over 75% in some cases, leading to complete crop failure. In contrast, infection rates for open-field cultivation are usually below 1%.

Symptoms

Rust disease primarily affects the basal leaves of *E. breviscapus*, but it can also infect scapes and other parts. In the early stages, affected areas first display discoloration and chlorosis. Raised structures form on the undersides of the leaves, and the leaf epidermis ruptures to produce dark brown powder, which are the telium of the pathogen. As the disease progresses, teliospore masses appear on the upper surface of the leaves as well, eventually covering large portions of the leaves. Infected plants become stunted, with chlorotic and deformed leaves; in severe cases, the entire plant wilts and dies (Fig. 20.3).

Pathogen

The causal agent of rust disease in *E. breviscapus* is *Puccinia dovrensis* Blytt, belonging to the genus Puccinia in the order Pucciniales of Basidiomycota. Only teliospores have been observed. The teliospores are yellow-brown, featuring a single septum that constricts to form a gourd-like shape with nearly equal-sized ends. They measure 29.7–44.9 μm in length (mean 38.8 ± 3.2 μm) and 13.8–23.3 μm in width (mean 20.0 ± 1.4 μm). Each teliospore has a stalk at the base that detaches easily. The spore wall is thick, with lateral walls measuring 1.1–3.5 μm thick (mean 2.3 ± 0.03 μm), and the top wall thickening to form a round or blunt dome, measuring 1.7–7.0 μm in thickness (mean 4.1 ± 0.1 μm) (Fig. 20.4).

Fig. 20.3 Symptoms of Rust Disease in *Erigeron breviscapus*. (**a**) Infected plant. (**b**) Chlorotic lesion on the upper leaf surface. (**c**–**e**) Telium on the underside of the leaf

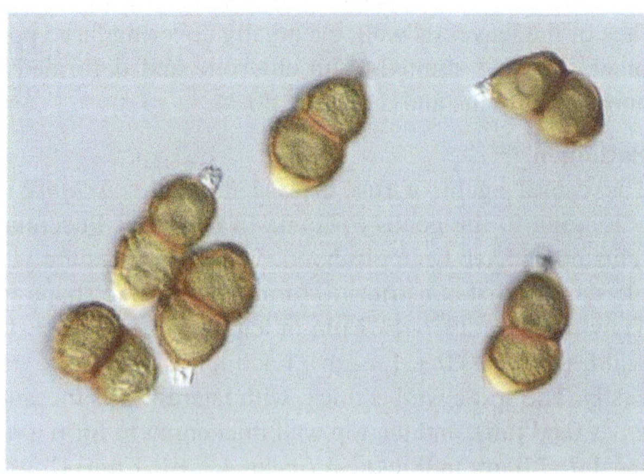

Fig. 20.4 Teliospores of the Rust Pathogen in *Erigeron breviscapus*

Occurrence Pattern

Rust disease can occur year-round, with higher incidence under greenhouse conditions, where the disease incidence in localized areas can reach 100%. Teliospores can be spread via air currents, rain splash, irrigation water droplets, mechanical contact during field operations, or insect vectors. The pathogen produces abundant spores on *E. breviscapus* plants; without timely control, the disease can damage over a large area in a short time.

Control Measures

1. Removal of Infected Plants

   Regular monitoring of disease occurrence in the field is essential. Infected plants should be promptly removed and destroyed to prevent further spread (take care to avoid spore dispersion during removal).
2. Chemical Control

   Chemical treatment is the most effective means of controlling rust disease in *E. breviscapus*. Upon the onset of disease, it is generally recommended to use a combination of a systemic fungicide with a protective fungicide. Systemic fungicides that may be used include triadimefon, carbendazim, difenoconazole, and pyraclostrobin, etc. Protective fungicides include chlorothalonil, thiram, and mancozeb. For mild or sporadic infections, a single spray application is often sufficient. Observe the treated plants for any recurrence. In cases of severe infection or recurrence after a single application, perform three consecutive sprays at 7-day intervals. Alternate systemic fungicides to prevent resistance development in the pathogen. Refer to the specific fungicide instructions for the correct concentration and dosage.

20.3 Section Three: Downy Mildew of *Erigeron breviscapus*

Downy mildew is one of the primary diseases affecting *E. breviscapus* seedlings, and is a common foliar disease. It occurs in key *E. breviscapus* growing regions in Yunnan, including Honghe, Qujing, Yuxi, and Dali. It usually occurs locally and is relatively common at the seedling stage. Due to growers' familiarity with this disease, chemical control measures are generally effective, with the incidence rate typically kept below 5%.

Symptoms

Typical symptoms of downy mildew in *E. breviscapus* include a white, frost-like mildew layer covering the undersides of leaves and petioles. In severe cases, mildew may also appear on the upper leaf surface. Affected areas turn yellow. In dry conditions, infected leaves gradually wilt, and in high humidity, the leaves rot. Severe infections can result in 30% plant mortality (Fig. 20.5).

Fig. 20.5 Symptoms of Downy Mildew in *Erigeron breviscapus*. (**a**) Frost-like mildew covering the underside of leaves. (**b**) Yellowed upper leaf surface covered with mildew

Fig. 20.6 Sporangiophore and Sporangia of Downy Mildew pathogen in *Erigeron breviscapus*

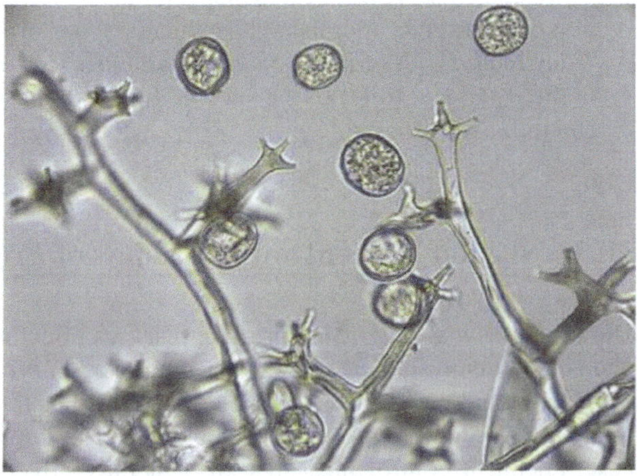

Pathogen

The pathogen causing downy mildew in *E. breviscapus* is Bremia lactucae Regel, which belongs to the genus Bremia in Oomycota. The sporangiophore measures (275–812) μm × (8–15) μm, with 3–6 dichotomous branches at the top. The upper branch tips swell into disc-like structures, each bearing 2–5 smaller branchlets around its upper edge. The branchlets are (4–9) μm × (2–4) μm. The sporangia are nearly spherical, with dimensions of (15–21) μm × (12–18) μm (Fig. 20.6).

Occurrence Pattern
The pathogen overwinters in plant debris. It primarily infects *E. breviscapus* seedlings from April to June, typically starting from the lower leaves and gradually spreading to the entire plant. High seedling density, high field humidity, and large day-night temperature differences in spring easily induce the disease. Following infection, the pathogen forms abundant sporangia on *E. breviscapus* leaves, which spread rapidly through air currents, irrigation water, insect activity, and farming operations.

Control Measures
1. Enhanced Cultivation Management

   Ensure proper plant spacing to facilitate airflow, and irrigate scientifically to avoid waterlogging in the fields. Infected plants should be promptly removed and buried deeply or destroyed to reduce infection sources.
2. Chemical Control

   The systemic fungicides metalaxyl and oxadixyl are highly effective against downy mildew in *E. breviscapus*. To prevent resistance, alternate between applications of metalaxyl-mancozeb, oxadixyl-mancozeb, and metalaxyl-chlorothalonil sprays every 7 days, applying two to three treatments depending on disease severity.

20.4 Root-Knot Nematode Disease in *Erigeron breviscapus*

Root-knot nematode disease is one of the primary diseases during the growth period of *E. breviscapus*, and is a commoon root disease. This disease is widespread, with known occurrences in areas such as Honghe, Qujing, Yuxi, Dali, and Chuxiong in Yunnan Province, primarily occurring in dryland mountainous regions. The disease incidence is directly related to soil conditions in the planting areas, with incidence rate in some local plots being over 50%.

Symptoms
When the damage from root-knot nematodes is mild, there are usually no obvious symptoms on the above-ground parts of the plant. In severe cases, however, leaves turn yellow, plants become stunted, and numerous galls form on fibrous roots. Under dissection microscopy, these galls often reveal scab-like egg masses containing many white, transparent eggs. Removing the egg mass exposes the bodies of female nematodes, with one or more females typically observed on a single gall (Fig. 20.7).

Pathogen
Root-knot nematode disease in *E. breviscapus* is caused by *Meloidogyne incognita* (Fig. 20.8). The nematodes are sexually dimorphic. Female nematodes are enlarged, spherical, or pear-shaped, with a distinct, protruding neck. The lip region is slightly raised, somewhat cap-like, and the excretory pore is at the level of the stylet basal

Fig. 20.7 Symptoms of Root-Knot Nematode Disease in *Erigeron breviscapus*. (**a**, **b**) Root galls formed on the fibrous roots of *E. breviscapus*. (**c**, **d**) Female nematodes (indicated by red arrows) and egg masses (indicated by black arrows) on the root galls

bulb. The perineal pattern varies, generally elliptical or nearly circular. Typically, the dorsal arch is high, with a round or flat top, occasionally trapezoidal. The dorsal striae are dense, and lateral and ventral striae are either wavy or serrated, sometimes smooth, with indistinct lateral lines and occasionally forked lateral lines. The ventral striae are few, smooth, usually arc-shapred, curving from both sides to the middle. Females measure 620.2 (408.0–1019.8) μm in body length, with a maximum body width of 350.9 (283.6–550.1) μm. The stylet is 15.5 (14.0–15.8) μm long, DGO (dorsal gland orifice distance from stylet base) 2.91 (2.5–3.6) μm, stylet basal bulb height 2.6 (1.8–3.6) μm, and stylet basal bulb width 3.9 (3.1–4.3) μm. Males are linear, with a prominent head cap and large, rounded lip disc with a central depression. The stylet cone tip is blunt-rounded, the shaft is often cylindrical, narrowing near the basal bulb. The boundary between the basal bulb and the shaft is distinct. The basal bulb is from oblate to round, with a notch at the front end. Males are 1450.3 (1180–1850) μm long, with a stylet length of 24.5 (22.5–29.3) μm, DGO 2.0 (1.8–2.6) μm, and a spicule length of 33.5 (25.7–40.3) μm, though males are generally difficult to observe. Second-stage juveniles are linear, measuring 385 (350–450) μm in length, with a stylet length of 11.0 (9.0–11.3) μm, DGO 2.5 (2.0–3.0) μm, tail length 40.5 (38.3–45.0) μm, and tail hyaline region length 11.6 (11.3–13.5) μm. Eggs are oval or long-oval, measuring 75.5 (61.3–87.7) μm × 31.0 (23.7–35.0) μm (Fig. 20.8).

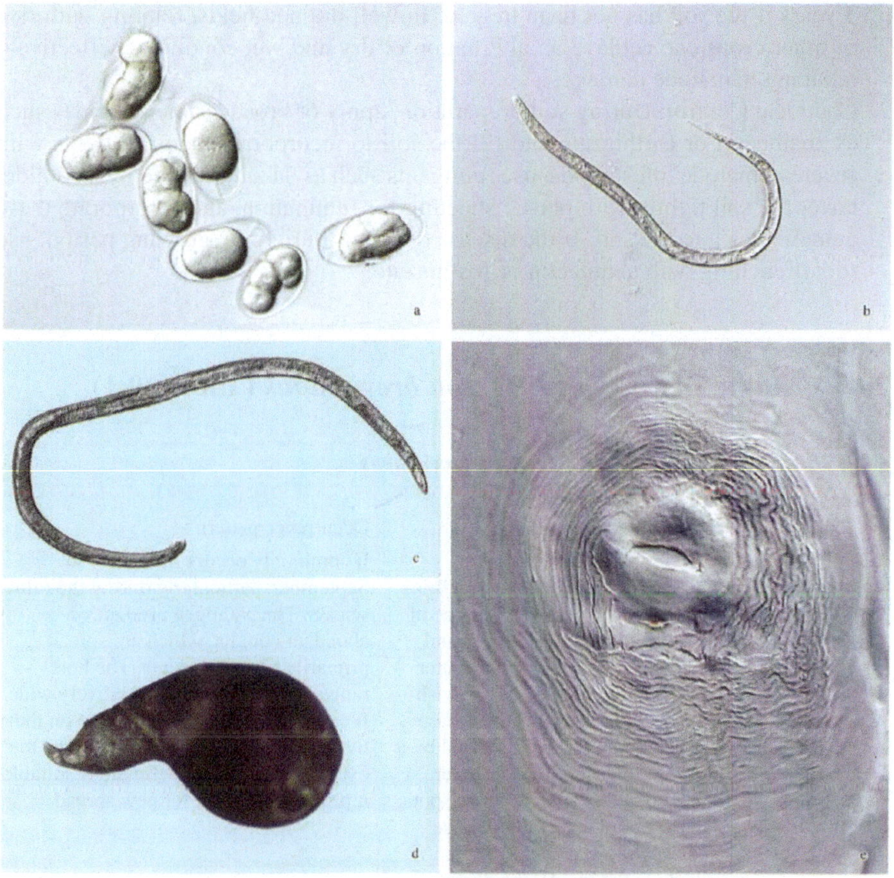

Fig. 20.8 Morphology of *Meloidogyne incognita*. (**a**) Eggs. (**b**) Second-stage juvenile (J2). (**c**) Male nematode. (**d**) Female nematode. (**e**) Female perineal pattern

Occurrence Pattern

Root-knot nematode disease in *E. breviscapus* can occur throughout the year, with a prolonged period of damage and gradual disease progression. Consequently, above-ground symptoms are often subtle, resembling root rot disease symptoms in severe cases. The disease typically manifests as small root galls on fibrous roots, frequently misdiagnosed as root rot. Contaminated soil is the primary cause of this disease. In the absence of host plants, nematode eggs and larvae can survive in the soil for several years. Some fields dedicated to growing *E. breviscapus*, especially those with greenhouses and other facilities, may experience root-knot nematode disease even with crop rotation.

Control Measures

1. Crop Rotation: For fields infected with root-knot nematodes affecting *Erigeron breviscapus* or other crops, avoid planting *Erigeron breviscapus* for at least

3 years if the soil has not been treated. In well-drained fields, rotating with rice or other crops can achieve an alternation of dry and wet conditions, effectively reducing nematode damage.
2. Chemical Control: During soil preparation, apply or broadcast nematicides such as abamectin or fosthiazate and till the soil to incorporate them. In fields with severe nematode infestations, use fumigants such as dazomet or methyl bromide, cover the soil tightly with plastic sheeting for fumigation, and incorporate these nematicides into the soil. If the disease develops during the growing period, use root drenching with abamectin or fosthiazate.

20.5 Major Diseases of *Erigeron breviscapus* (Table 20.1)

Table 20.1 Occasional Diseases of *Erigeron breviscapus*

Disease (Pathogen)	Symptoms	Occurrence pattern
Leaf spot (*Alternaria alternata*)	It infects the leaves of *E. breviscapus*, causing severe wilting and necrosis. The upper surface of the infected leaves is purple, and some have a grayish-white center, while the lower surface is grayish-white or light brown. The disease and healthy parts are separated by a raised purple ring line. There are black granules on the disease spots, and the periphery of the disease spots is yellow-green. From the outside to the inside, the disease spots show a zonate pattern, and the inner part dries up and dies	It commonly occurs from June to September, particularly during the rainy season. The pathogen produces abundant conidia, which are spread primarily via air currents. The host range of the pathogen is relatively wide. It can parasitize or saprophytize on the living bodies or residues of the host to overwinter. When the climate is suitable, it produces conidia for new spread
Verticillium wilt (*Verticillium albo-atrum*)	The disease mainly occurs in the adult plant stage. At the initial stage of the disease, the leaves at the base of the plant start to fade from the leaf tip and mesophyll part. Then the leaf tissue of the whole leaf fades. The area between the leaf margin and the lateral veins turn yellow, showing a mottled appearance, and then turns brown. The leaves become thick and brittle. In the later stage, the disease gradually spreads to the upper leaves. All the lower leaves of the plant wither and die. Severely infected plants die before flowering. When the diseased plant and diseased stem are cut transversely, brown discoloration of the vascular bundles can be seen	It occurs from June to September, especially during the rainy season. The pathogen produces numerous conidia on the diseased parts, which are spread over short or long distances via air currents or rainwater. The hyphae on the diseased plants can also infect adjacent plants at close range. The pathogen overwinters in the form of hyphae on *E. breviscapus* or other host living bodies or diseased residues. When conditions are suitable, it carries out new infections with hyphae or produces conidia

Chapter 21
The Diseases of *Chrysanthemum morifolium*

Chrysanthemum flowers, known as traditional Chinese medicine, is the dried inflorescence of the *Chrysanthemum morifolium* Ramat. in the Asteraceae family. China has a long-standing history of *C. morifolium* cultivation, and through extensive artificial breeding, combined with various environmental and processing techniques, medicinal chrysanthemum flowers in China are classified mainly as Boju, Chuju, Gongju, Hangju, Huaiju, and Qiju.

Chrysanthemum flower is slightly cool in nature, with a sweet, bitter taste, entering the lung and liver meridians. It has the functions of dispersing wind and heat, calming the liver, improving vision, and detoxifying. It is commonly used for treating wind-heat colds, headaches, dizziness, boils, swelling, and eye inflammation. Modern pharmacological studies have confirmed that the chrysanthemum flower improves myocardial nutrition, scavenges reactive oxygen species, enhances capillary resistance, lowers lipid and cholesterol levels in the blood, inhibits tumors, delays aging, and boosts immunity.

C. morifolium is predominantly cultivated in Hubei, Zhejiang, Anhui, Henan, Hebei, and Jiangsu. Common diseases affecting *C. morifolium* production include wilt disease, root rot, southern blight, black spot, leaf blight, brown spot, yellow spot, anthracnose, downy mildew, and viral diseases.

21.1 Wilt Disease of *C. morifolium*

Wilt disease is one of the most prevalent and destructive soil-borne fungal diseases in *C. morifolium* cultivation. It spreads rapidly, causes significant damage, and is challenging to control. It occurs across all major *C. morifolium*-growing areas, with an incidence rate of 10–50%; in some regions, severe outbreaks can result in an incidence rate of up to 80%, even leading to total crop loss.

© China Agriculture Press Co., Ltd. 2026
T. Wang et al., *Atlas of Chinese Medicinal Plant Diseases*,
https://doi.org/10.1007/978-981-95-6326-5_21

Symptoms

Wilt disease significantly impacts *C. morifolium* yield, with distinct symptoms including leaf drooping, yellowing, wilting, and withering. Within a single plant, one side of the stem may exhibit wilted, yellowing leaves while the other side remains healthy. As the pathogenic fungus invades the vascular bundles, the lower stem exhibits hollow and browning, with the vascular bundles turning brown, and the pith may appear hollow. As the infection progresses, the vascular tissue becomes obstructed, preventing the transport of water and essential nutrients, ultimately leading to plant death (Fig. 21.1).

Pathogen

C. morifolium wilt disease is caused by pathogen *Fusarium oxysporum f.* sp. *chrysanthemi* (Snyder et Hansen), which can infect alone or in combination with other *Fusarium* species. On PDA medium, the fungus produces fluffy, abundant

Fig. 21.1 Symptoms of *C. morifolium* Wilt Disease. (**a**, **b**) Field symptoms, (**c**) Plant symptoms, (**d**–**f**) Longitudinal sections of infected *C. morifolium* stems

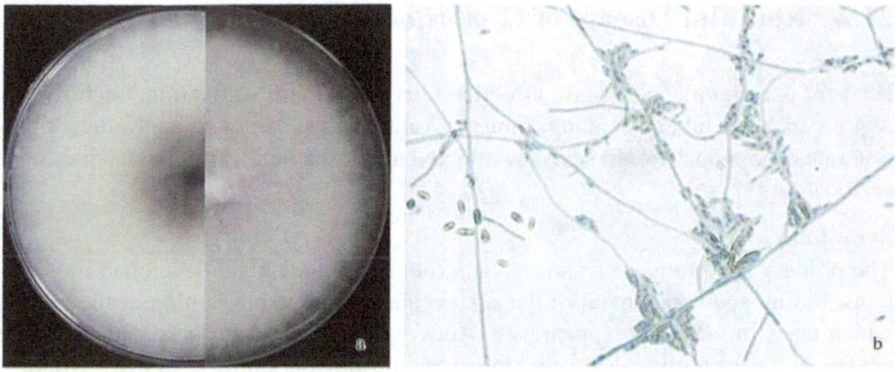

Fig. 21.2 Morphology of *C. morifolium* wilt disease pathogen. (**a**) Cultural characteristics, (**b**) Conidia and mycelium

mycelium, with colony colors ranging from pink to peach, light purple, or white. Macroconidia are spindle-shaped or sickle-shaped with thin walls and pointed ends, usually containing three septa, with some exhibiting four or five, measuring (9.8–32.18) μm × (2.59–5.12) μm. Microconidia form singly on short conidiophores, are kidney-shaped to oval, mostly aseptate though some may have one or two septa, measuring (3.89–9.95) μm × (1.50–4.29) μm. Chlamydospores are spherical to oval, either single- or double-celled, terminal or intercalary, occurring singly, in pairs, or occasionally in chains (Fig. 21.2).

Occurrence Pattern

The pathogen overwinters as mycelium, conidia, and chlamydospores in infected plants and soil. Chlamydospores can survive for several years. The disease spreads through water, soil, and plant debris. Summer temperatures between 24 °C and 32 °C; poor drainage after rain, acidic soil, weak plants, wounds, and dense planting all encourage disease onset and spread.

Control Measures

1. Crop Rotation: Rotate with crops such as rice, wheat, or oilseed rape to reduce disease incidence.
2. Improved Cultivation Practices: Remove and destroy infected plants, select optimal planting densities to improve air circulation, manage soil moisture, choose well-drained fields, and apply fungicides for soil disinfection. Use organic fertilizers as a base and microbial fertilizers during growth to boost plant health.
3. Fertilization: Avoid excessive nitrogen. Use balanced organic and microbial fertilizers to strengthen plant resistance.
4. Chemical Control: Apply 60 g per mu (approximately 667 square meters) of a *Bacillus subtilis* WP (100 billion spores/g) or 400 g of difenoconazole microemulsion (ME) diluted 10,000-fold. Use 250 g/L propiconazole EC diluted 1,000-fold or 10% difenoconazole WG diluted 1,000-fold, applying 0.4–0.5 L of solution per plant. Repeat applications as needed based on disease severity.

21.2 Root Rot Disease of *C. morifolium*

Root rot is a significant disease affecting chrysanthemum cultivation, occurring in regions such as Hubei, Zhejiang, Jiangsu, Anhui, and Hebei, with an average infection rate of around 30%. In severely affected areas, the incidence can reach as high as 100%.

Symptoms
The primary symptoms of *C. morifolium* root rot include underdeveloped root systems, rotting and detachment of the cortex, and complete blackening of the xylem, which takes on a fibrous appearance. Above ground, the stem base rots, with the epidermis easily peeling away, and the xylem turning dark brown. Leaves wilt, turn yellow, and eventually wither; in severe cases, the entire plant dies and can be easily pulled out of the ground. This disease can occur throughout the growth period of *C. morifolium*, with seedlings being particularly susceptible. Root rot spreads and progresses quickly. A distinguishing feature from wilt disease is that the vascular bundles in the infected area turn brown without progressing upward, and the stems show no significant constriction. In practice, root rot often co-occurs with wilt disease (see Fig. 21.3).

Pathogen
The pathogen causing *C. morifolium* root rot is the ascomycete fungus *Fusarium solani* (Mart.) Appel & Wollenw. ex Snyder et Hansen, an asexual species of the genus *Fusarium*. On PDA medium, the pathogen appears as sparse, fluffy aerial mycelium, with colonies that are white on the front of medium and pale yellow on the reverse side. Large conidia are fusiform or sickle-shaped, with thin walls, pointed ends, and typically three to four septa (sometimes five), measuring (19.87–43.51) μm × (3.39–5.42) μm.

Small conidia grow on single or short conidiophores, are abundant, kidney-shaped to oval, with one to two septa, and measure (7.47–20.20) μm × (3.01–6.33) μm. Chlamydospores are spherical to oval, with one to two cells, terminal or intercalary, solitary or in pairs, and occasionally in chains, measuring (6.51–10.57) μm × (4.05–9.39) μm (see Fig. 21.4).

Occurrence Pattern
The pathogen overwinters in infected seedlings, soil, and plant debris, serving as the initial infection source for the following year. It can survive in soil for over 6 years. Planting infected seedlings can lead to direct disease outbreaks. Infections occur from May to November, with peak incidences from June to August and a gradual decrease after September. High soil moisture, especially in waterlogged fields, promotes pathogen reproduction and dissemination. Low-lying, damp fields with poor fertility are more prone to infection. Wounds caused by underground pests and root-knot nematodes also facilitate pathogen invasion, exacerbating the incidence of root rot.

Fig. 21.3 Symptoms of *C. morifolium* Root Rot Disease. (**a–c**) Field symptoms; (**d**, **e**) Root symptoms

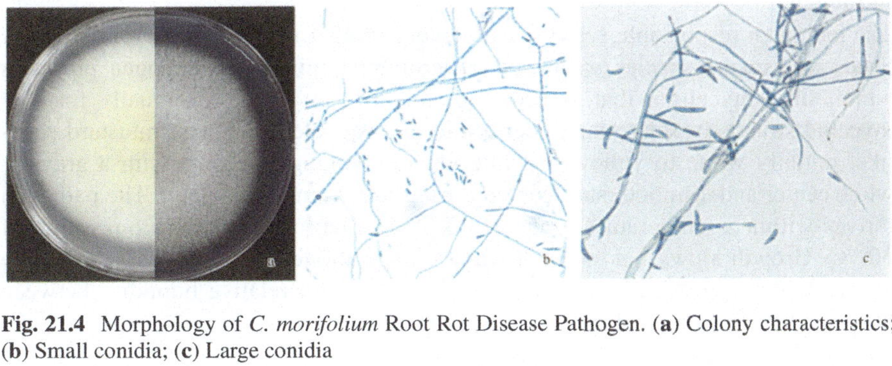

Fig. 21.4 Morphology of *C. morifolium* Root Rot Disease Pathogen. (**a**) Colony characteristics; (**b**) Small conidia; (**c**) Large conidia

Control Measures

1. Crop Rotation: Implement crop rotation or intercropping with other crops, such as rice, wheat, and oilseed rotations.
2. Enhanced Cultivation Management: Disinfect the soil before planting and ensure good drainage. Remove plant debris promptly.
3. Balanced Fertilization: Apply nitrogen, phosphorus, and potassium in appropriate proportions, and increase the application of organic or microbial fertilizers.
4. Chemical Control: Apply 60 g per mu of *B. subtilis* (WP; with 100 billion spores/g), or irrigate each plant with 0.1 L of 400 g/L fluazinam ME diluted 8,000-fold, 250 g/L propiconazole EC diluted 1,000-fold, or 10% difenoconazole WG diluted 600-fold. Depending on disease severity, repeat applications 2–3 times as needed.

21.3 Southern Blight Disease of *C. morifolium*

Southern blight poses a significant threat to *C. morifolium* cultivation, as it can infect plants at any growth stage. This disease is prevalent in regions such as Hubei and Zhejiang, with typical incidence rates between 30% and 50%.

Symptoms

Southern blight primarily affects the rootstock base and stems of mature plants, causing symptoms such as root rot and basal stem rot. When the disease infects the stem base or stem, it results in withering and leaf drop above the affected area. In the early stages, water-soaked, irregular brown lesions appear on the stem base, and the lower leaves and branches discolor and wilt, with symptoms quickly spreading upward. Leaves transition from a healthy dark green to a pale green, eventually covered by a white, cottony mycelial layer that forms mustard seed-sized sclerotia. Initially white, the sclerotia turn yellow and eventually brown as the stem base decays. A cross-section of the stem reveals a hollow pith, with white, thread-like structures lining the inner walls. Southern blight is a destructive disease; mild infections cause root rot, stunted growth, and stem rot with limited branching, while severe cases lead to stem collapse and complete plant death (see Fig. 21.5).

Pathogen

The pathogen responsible for chrysanthemum southern blight is *Sclerotium rolfsii* Sacc., an asexual species within the Sclerotium family. The pathogen produces white, silky mycelium that spreads radially in a fan shape, eventually forming mycelial strands or sclerotia. Under a microscope, sclerotia appear mustard seed-like, initially white to yellowish-white, then darkening to brown with a grayish-white center and a smooth surface, measuring 1–4.2 mm in diameter. The pathogen thrives within a temperature range of 15–35 °C, with an optimal temperature of 30 °C. Growth slows below 15 °C, and ceases altogether below 5 °C or above 40 °C. Optimal growth conditions are 25–35 °C with relative humidity between 80% and 95% (see Fig. 21.6).

Fig. 21.5 Symptoms of *C. morifolium* Southern Blight Disease. (**a**, **b**) Field symptoms; (**c**, **d**) Root symptoms

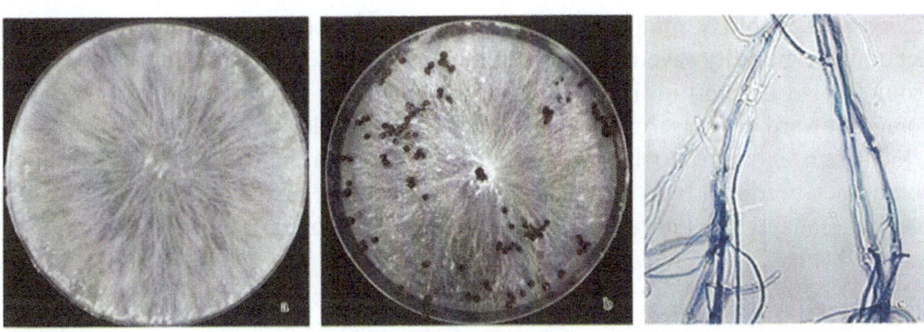

Fig. 21.6 Morphology of *C. morifolium* Southern Blight pathogen. (**a**) Colony characteristics; (**b**) Mycelium characteristics; (**c**) Mycelium

Occurrence Pattern

The pathogen overwinters as sclerotia within plant debris or soil. When favorable temperature and humidity conditions arise in the following season, sclerotia produce mycelium that initiates primary infections. Mycelium from infected plants can spread to adjacent plants upon contact, while sclerotia are dispersed by wind, rain, and insect vectors such as whiteflies and aphids. Disease incidence is higher in fields with continuous cropping, heavy soils, or low-lying areas. The infection period typically extends from mid-June to late September, with peak incidence in July and August due to high temperatures and humidity. Disease progression slows in early September.

Control Measures

1. Crop Rotation: Implement crop rotation with nonhost plants, such as a rice-chrysanthemum rotation.
2. Enhanced Cultivation Management: After harvest, promptly remove plant debris, perform deep plowing, and apply adequate basal fertilizer.
3. Chemical Control: At the onset of infection, spray a 500-fold dilution of 29% lime-sulfur solution, an 8000-fold dilution of 400 g/L fluazinam ME, or a 600-fold dilution of 20% eugenol emulsion. For severe infections, apply every seven days for three to four consecutive treatments, adjusting frequency based on disease severity.

21.4 Black Spot Disease of *C. morifolium*

Black spot disease is one of the most prevalent and severe diseases affecting *C. morifolium*, with widespread incidence across major production regions such as Hubei, Zhejiang, Anhui, Jiangsu, and Henan. In these areas, infection rates can reach up to 100% in some fields, with typical rates ranging from 50% to 80%.

Symptoms

C. morifolium black spot disease can occur throughout the plant's growth cycle, becoming especially severe in warm, humid conditions. The peak outbreak period is from July to August. Initial symptoms appear on the lower leaves and gradually spread upward. Small, irregular, black-brown mold-like spots form on the leaves, which then expand into larger, irregular or nearly circular black-brown lesions, sometimes bordered by a yellow halo. In later stages, lesions coalesce into larger patches, causing the entire leaf to yellow and die, though the dead leaves remain attached (Fig. 21.7).

Pathogen

The causative agents of chrysanthemum black spot disease are fungi from the genus *Alternaria*: *Alternaria alternata* (Fr.) Keissl., *A. tenuissima* (Nees: Fr.) Wiltshire, and *A. chrysanthemi* (Crosier & Heit). On PDA medium, *Alternaria alternata* colonies appear gray-white after 7 days. On PCA medium, they are grayish-brown to

Fig. 21.7 Symptoms of *C. morifolium* Black Spot Disease. (**a**, **b**) Field symptoms, (**c**–**e**) Leaf symptoms

dark brown. Conidiophores arise directly from the main hyphae, mostly solitary, erect or slightly curved, septate, and pale brown to yellow-brown. Mature conidia are pale brown to yellow-brown, club-shaped, pear-shaped, ovoid, or nearly oval to round, with transverse, longitudinal, and oblique septa, measuring (6.46–14.89) μm × (17.67–46.98) μm (Fig. 21.8).

Occurrence Pattern

The pathogen overwinters as mycelia and conidial clusters on plant debris and spreads through conidia, which act as both primary and secondary inoculum, dispersed by wind and rain. Conditions that favor disease development include prolonged rainy or foggy weather with heavy dew, large diurnal temperature fluctuations, waterlogging, poor plant health, and excessive nitrogen fertilizer use, which leads to excessive plant growth and susceptibility.

Control Measures

1. Enhanced Cultivation Practices: Maintain optimal plant density to ensure adequate ventilation. Remove and destroy dead branches and diseased leaves promptly. Avoid continuous cropping; implement crop rotation.

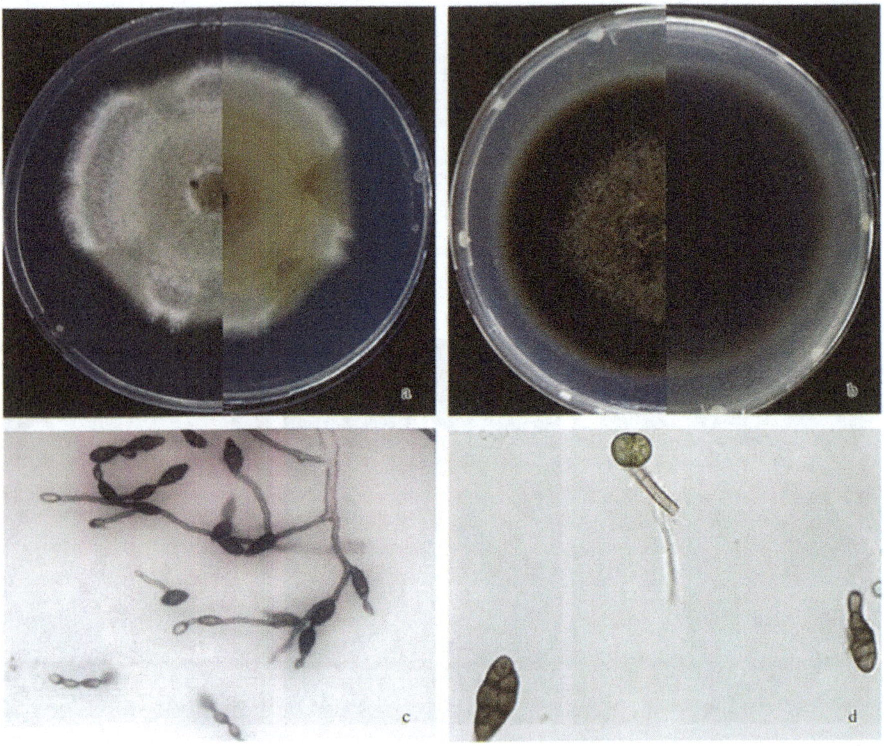

Fig. 21.8 Morphology of *Alternaria alternata*. (**a**) Colony morphology on PDA medium, (**b**) Colony morphology on PCA medium, (**c**) Chain-like conidia (400×), (**d**) Conidia (1000×)

2. Chemical Control: At disease onset, apply a 10% difenoconazole WG diluted 1,000-fold, spraying once every 10 days for a total of 3–5 applications. After symptoms appear, use a 3% procymidone ME diluted 500-fold, a 30% chlorothalonil AS diluted 500-fold, or an 80% garlic oil ME diluted 2,000-fold. Spray once weekly for a total of 3–4 applications, stopping before the flower buds begin to show color.

21.5 Viral Diseases of *C. morifolium*

Viral disease is widespread across *C. morifolium* cultivation areas, significantly impacting plants. The disease is especially severe in Zhejiang, with heavily affected fields experiencing nearly 100% infection rates and an average infection rate of 30%–35% across various regions.

Symptoms
Common symptoms of *C. morifolium* viral disease include mosaic patterns, vein clearing, chlorosis, necrotic spots, yellowing, leaf crinkling, and stunted growth.

The disease is caused by a combination of various viruses. Chrysanthemum B virus (CVB) typically leads to stunted growth, irregular chlorosis in young leaves, reduced leaf size, curling, and in severe cases, necrotic spots. Tomato aspermy virus (TAV) primarily causes stunting, leaf distortion, brown discoloration in vascular bundles of young stems, and floral deformities with reduced flower size. Tobacco mosaic virus (TMV) symptoms include translucent tissues in the minor veins and lateral veins of young leaves, with leaf tissues near the veins gradually turning pale green (see Fig. 21.9).

Fig. 21.9 Symptoms of *C. morifolium* viral disease. (**a**, **b**) Field symptoms, (**c**) Symptoms of Chrysanthemum B Virus, (**d–f**) Symptoms of Tomato Aspermy Virus

Pathogen

The pathogens that cause chrysanthemum viral disease include Chrysanthemum virus B (CVB), Tomato aspermy virus (TAV), and Tobacco mosaic virus (TMV). CVB, a positive-sense single-stranded RNA virus of the genus Carmovirus, forms rod-shaped particles approximately 685 nm × 12 nm in size, often surrounding chloroplasts. TAV, a positive-sense single-stranded RNA virus of the genus Cucumovirus, has spherical particles with icosahedral symmetry, approximately 22–28 nm in diameter. TMV consists of single-stranded RNA and coat proteins (CP), forming a rod-shaped structure. TMV particles are approximately 300 nm × 18 nm in size. It is highly stable, capable of persisting outside the host for years, and replicates rapidly within the host. TMV can be inactivated after 10 min at 90 °C, though it retains infectivity in diluted tobacco juice at concentrations as low as $10^{\wedge 6}$ (see Fig. 21.10).

Occurrence Pattern

C. morifolium viral disease is spread by vectors such as aphids, thrips, leafhoppers, and spider mites, and also through grafting and mechanical injury. The virus overwinters in mother plants retained for seed propagation and spreads through plant division and cuttings. High temperatures and humidity, combined with early aphid infestations, contribute to outbreaks.

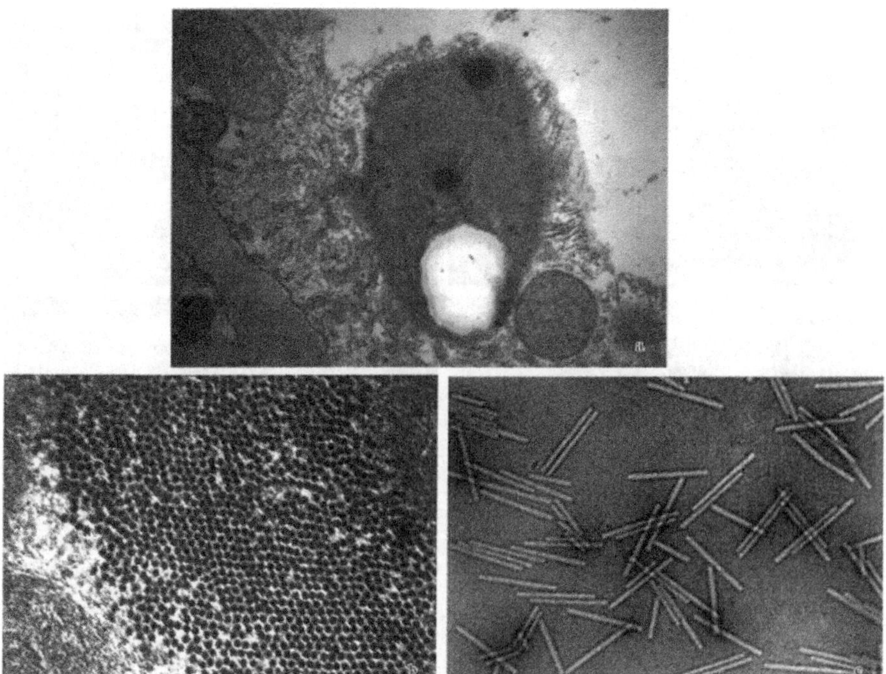

Fig. 21.10 *C. morifolium* Viral Pathogen Morphology. (**a**) CVB under Transmission Electron Microscopy, (**b**) TAV under Transmission Electron Microscopy, (**c**) TMV under Transmission Electron Microscopy

Control Measures

Strict Quarantine: Infected *C. morifolium* are virus carriers, so strict quarantine is required during introduction to prevent spreading to uninfected areas.

Use of Healthy Propagation Material: Select vigorous plants for seed retention, apply phosphorus and potassium fertilizers to enhance plant resistance, and use cuttings from virus-free plants.

Chemical Control: To control aphid vectors, apply 50% anti-aphid WP at 500 mg/L. At the onset of disease, spray each mu with 6% oligosaccharide chain of protein WP, or 5% JingDuQing (antiviral agent) WP at 2500 mg/L, or 20% morpholine hydrochloride and copper WP at 1667–2000 mg/L, or 20% morpholine guanidine SP at 2,000 mg/L. The drugs spread every 7–10 days for three consecutive treatments.

21.6 Leaf Blight Disease of *C. morifolium*

Leaf blight is a severe foliar disease common in major *C. morifolium* production areas such as Hubei, Anhui, Zhejiang, Jiangsu, and Henan. The disease has a high incidence rate, with infection rates exceeding 50% in many fields and, in severe cases, affecting entire plots.

Symptoms

The disease initially appears on lower leaves, gradually spreading upward. Early signs include necrosis at the edges and tips of leaves, which expand into irregular, light brown to dark brown large patches covered densely with small black spots. Eventually, the entire leaf succumbs to necrosis, curls, and dies, although dead leaves do not detach from the plant (Fig. 21.11).

Pathogen

The causal pathogen of *C. morifolium* leaf blight is *Phoma sorghina* (Sacc.) Boerema, Dorenb., et Kesteren, a deuteromycetous fungus in the Ascomycota. Initially, on PDA medium, the hyphae appear gray-white and later secrete a scarlet pigment, with the reverse side of the colony showing reddish-brown. Single-celled chlamydospores measure (7.91–32.23) μm × (12.03–38.42) μm, while multi-celled chlamydospores measure (6.32–25.10) μm × (21.75–100.05) μm. After 20–30 days on OA medium, black spherical pycnidia form, producing liquid-like orange-red conidial masses. Conidia are single-celled, colorless, round, and measure (4.10–5.93) μm × (1.98–3.18) μm (Fig. 21.12).

Occurrence Pattern

The pathogen overwinters within infected tissue by pycnidia. Under favorable conditions, conidia are produced and spread by wind and rain, entering plants through stomata or wounds. Continuous rainy weather in late spring, early summer, or autumn facilitates outbreaks.

Fig. 21.11 Symptoms of *C. morifolium* Leaf Blight Disease

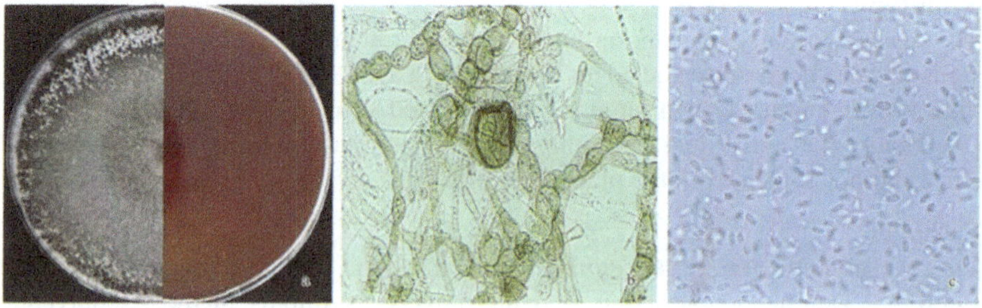

Fig. 21.12 Morphology of *Phoma sorghina*. (**a**) Colony morphology, (**b**) Chlamydospore morphology, (**c**) Conidial morphology

Control Measures

1. Enhanced Cultivation Practices: Adjust watering frequency according to weather, and control moisture during growth stages. Remove and destroy dead branches, leaves, and infected plant material. Avoid continuous cropping, and practice crop rotation.

2. Chemical Control: At the initial stage of disease, apply a 10% difenoconazole WG diluted 1,000-fold every 10 days for 3–5 applications. Once symptoms are evident, use a 3% oxathiapiprolin ME diluted 500-fold, a 50% carbendazim-thiram WP diluted 500-fold, or a 30% chlorothalonil AS diluted 500-fold. Apply weekly for 3–4 applications, discontinuing before flower buds start to show color.

21.7 Brown Spot Disease of *C. morifolium*

Brown spot disease is a severe foliar disease of *C. morifolium*, widely distributed across *C. morifolium*-growing regions such as Hubei, Zhejiang, Anhui, and Jiangsu. In these regions, infection rates can reach up to 100% in many fields, with an average incidence rate of around 30%.

Symptoms

Weak seedlings are particularly susceptible to brown spot disease. After infection, symptoms initially appear on the lower leaves and gradually spread upward. Early signs include small brown spots with conspicuous yellow halos on the leaves, which gradually expand into near-circular or irregular lesions with dark brown edges and slightly lighter brown centers that are prone to tearing. Tiny brown granules may appear in the central area of the lesions. Eventually, the entire leaf withers and dies, but remains attached to the plant (see Fig. 21.13).

Pathogen

The pathogen causing chrysanthemum brown spot disease is *Didyella americana* [(Morgan-Jones et J.F. White) Q. Chen et L. Cai], synonymously known as *Phoma americana* Morgan-Jones et J.F. White, which belongs to the Ascomycota division. On PDA medium, the mycelium appears sparse, varying from white to dark brown on the front and pale yellow on the reverse. On OA medium, the mycelium is even sparser, and after 7 days, black, spherical pycnidia form on the colony surface, producing numerous milky-white, liquid conidial masses after 10–15 days. The conidia are single-celled, colorless, and elongated-oval in shape, with dimensions of (4.14–7.01) μm × (8.58–16.56) μm (see Fig. 21.14).

Occurrence Pattern

The pathogen primarily overwinters as pycnidia within infected plant tissue. When conditions become favorable, it produces conidia that are disseminated by wind and rain, infecting host plants through stomata or wounds. This disease is most likely to develop during prolonged rainy periods in late spring, early summer, or autumn.

Fig. 21.13 Symptoms of *C. morifolium* Brown Spot Disease

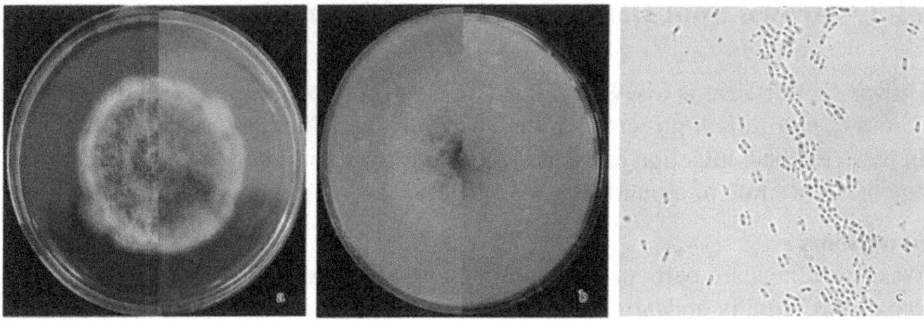

Fig. 21.14 Morphology of the *C. morifolium* Brown Spot Pathogen. (**a**) Colony morphology on PDA medium, (**b**) Colony morphology on OA medium, (**c**) Conidia

Control Measures

1. Enhanced Cultivation Management: Ensure appropriate plant spacing to improve ventilation; promptly remove and destroy dead branches, leaves, and infected foliage. Avoid continuous cropping and practice crop rotation.
2. Chemical Control: At the initial onset of the disease, apply a 10% benzothiazole WG diluted 500-fold, spraying once every 10 days for a total of 3–5 applications. Once symptoms appear, use a 50% benomyl WP diluted 500-fold, an 80% allicin oil-in-water emulsion (EW) diluted 2,000-fold, or a 3% thiabendazole ME diluted 500-fold, spraying once a week for 3–4 applications. Discontinuing before the flower buds show color.

21.8 Yellow Spot Disease of *C. morifolium*

Yellow spot disease has been observed only in Hubei Province, with an incidence rate of approximately 20%.

Symptoms

The disease initially affects the lower leaves, progressively moving upward. Early symptoms manifest as small, sunken brown spots on the leaf surface, which then expand into irregular or nearly circular brown to dark brown patches, slightly recessed with a distinct yellow halo around each lesion. In advanced stages, the spots merge into larger blotches, causing the entire leaf to wither; however, these dead leaves remain attached to the plant (Fig. 21.15).

Pathogen

The pathogen causing *C. morifolium* yellow spot disease is *Fusarium acuminatum*, a member of the genus *Fusarium* within the Ascomycota phylum. On PDA medium, the fungus grows rapidly, reaching the diameter of a 9 cm petri dish within approximately 7 days. The colony's surface is pale pink, while the reverse side releases pink to orange-red pigments. Macroconidia are spindle- or sickle-shaped, with thin walls

Fig. 21.15 Symptoms of *C. morifolium* Yellow Spot Disease

Fig. 21.16 Morphology of *Fusarium acuminatum*. (**a**) Colony morphology, (**b**) Conidia (400×), (**c**) Conidia (1000×)

and tapering ends, usually having three to four septa, occasionally five, measuring (28.64–46.08) μm × (4.00–7.96) μm. Microconidia, found on monophialides or short conidiophores, are kidney- to oval-shaped, mostly aseptate, occasionally with one to two septa, measuring (6.86–12.98) μm × (1.90–3.45) μm (Fig. 21.16).

Control Measures

1. Enhanced Cultivation Practices: Implement optimal plant spacing to ensure adequate ventilation; remove dead branches and leaves promptly and dispose of infected leaves. Avoid continuous cropping and practice crop rotation.
2. Chemical Control: At the disease's onset, apply a 10% difenoconazole WG diluted 500-fold, spraying every 10 days for 3–5 applications. If the disease progresses, use a 3% thifluzamide ME diluted 500-fold, a 50% carbendazim-thiram WP diluted 500-fold, or an 80% allium EW diluted 2,000-fold. Spray weekly for three to four applications, discontinuing treatment before the flower buds begin to show color.

21.9 Anthracnose of *C. morifolium*

Anthracnose is a prevalent disease in *C. morifolium*, widely distributed across China, with an average incidence rate of approximately 30%.

Symptoms

Anthracnose initially appears on the lower leaves of *C. morifolium* and gradually progresses upward. Early signs include small, grayish-white to yellow-brown spots, which then expand into irregular or nearly circular black-brown lesions. The lesion edges are dark brown, while the centers appear grayish-white or light brown. The spots are slightly sunken, with raised edges and sometimes concentric ring patterns. As the disease advances, these lesions merge into larger patches, causing the entire leaf to yellow and die, though it often remains attached to the plant (see Fig. 21.17).

Pathogen

The pathogens responsible for *C. morifolium* anthracnose are *Colletotrichum fructicola* Prihastuti and *Colletotrichum chrysanthemi* (Hori) Sawada, both from the *Colletotrichum* genus. When cultured on PDA medium for 7 days, *C. fructicola* appears as white or grayish-white colonies, with conidial acervuli forming after approximately 20 days. The conidiophores are typically colorless and generate short, oval, elliptical, or crescent-shaped conidia that are colorless and either non-septate or have a single septum. Elliptical conidia measure (3.29–5.68) μm × (11.93–17.66) μm, while crescent-shaped conidia measure (3.17–4.99) μm × (11.36–21.32) μm. *C. chrysanthemi* colonies have sparse white aerial mycelium, appearing dark olive underneath. Occasionally, orange conidial masses form near the mycelial mat, with conidia sizes of (7.9–13.7) μm × (2.5–4.1) μm. The optimal growth temperature for these fungi is 25–27.5 °C, though they can grow within a range of 10–35 °C (see Fig. 21.18).

Occurrence Pattern

The pathogen overwinters as mycelium or conidial acervuli on infected plant debris or seedlings. In spring, conidia are produced, serving as the primary infection

Fig. 21.17 Symptoms of *C. morifolium* Anthracnose

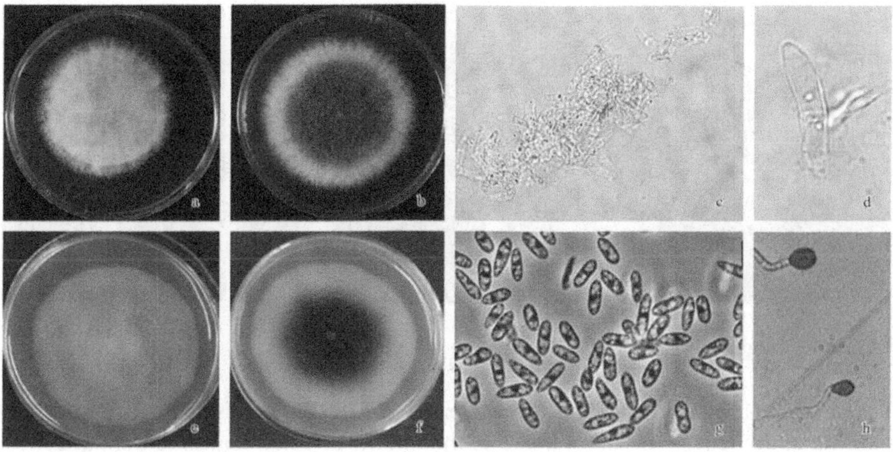

Fig. 21.18 Morphology of *C. morifolium* Anthracnose Pathogen. (**a**, **b**) *C. fructicola* colony morphology, (**c**, **d**) *C. fructicola* conidia, (**e**, **f**) *C. chrysanthemi* colony morphology, (**g**) *C. chrysanthemi* conidia, (**h**) *C. chrysanthemi* chlamydospores

source. Once infection is established, the disease generates large numbers of conidia, leading to further spread throughout the growing season. Newly infected leaves act as a continual source of inoculum, facilitating further pathogen spread. Conidia are dispersed by wind, rain, and insects, and upon landing on a leaf, germinate to form germ tubes, appressoria, and penetration pegs, enabling entry through stomata, wounds, or by direct penetration.

Control Measures

1. Enhanced Cultivation Management: During the seedling stage, avoid direct sunlight and adjust watering frequency according to weather conditions. During growth, manage humidity, practice crop rotation, and disinfect the soil.
2. Chemical Control: During disease onset, alternate sprays with 3% thiabendazole ME diluted 500-fold, 30% polyoxin AS diluted 500-fold, or 10% benzothiazole WG diluted 1,000-fold. Apply sprays every 7–10 days for several applications as needed.

21.10 Downy Mildew of *C. morifolium*

Downy mildew is a prevalent disease in *C. morifolium* cultivation areas of Anhui, Hubei, Zhejiang, and Jiangsu provinces, with severe outbreaks under moderate temperature and high-humidity conditions.

Symptoms

Downy mildew primarily affects leaves, petioles, flower stalks, buds, and young stems. The disease first appears on lower leaves, progressively spreading upward. Initial symptoms show as irregular gray-yellow patches on the leaf surface, with

Fig. 21.19 Symptoms of *C. morifolium* Downy Mildew

dense white fungal growth on the undersides. These patches expand, eventually causing entire leaves to wither and die, though the affected leaves remain attached. In spring, infections often lead to seedling wilting or weakened growth. In autumn, under cool and rainy weather, the disease reappears, covering leaves, flower stalks, and buds with a white fungal layer and ultimately causing the entire plant to turn brown and die (Fig. 21.19).

Pathogen

The causative agent of chrysanthemum downy mildew is *Peronospora radii* de Bary (synonym: P. danica Gäum), belonging to the genus *Peronospora* in the Oomycota phylum. Sporangiophores appear in clusters from stomata, forming white or grayish-white fungal colonies. The sporangiophores branch dichotomously, with terminal branches at right angles, straight or slightly curved. Sporangia are colorless, oval, ovoid, or round, without papillae, measuring (24.08–34.88) μm × (18.87–26.96) μm, and germinate via germ tubes (Fig. 21.20).

Occurrence Pattern

The pathogen overwinters as mycelium on plants reserved for seed or as oospores on plant debris. It spreads mainly through air currents, irrigation, farming activities, and insect vectors. Dense planting, excessive watering after transplanting, high soil moisture, and poor drainage conditions are conducive to disease spread. The disease typically proliferates during prolonged rainy periods in late spring, early summer, or autumn. In spring, disease development halts when maximum temperatures exceed 30 °C. After late September, the disease may recur under cool, rainy conditions.

Control Measures

1. Enhanced Cultivation Management: Implement appropriate spacing and timely drainage to control moisture levels; clear dead leaves and branches, promptly removing infected leaves for disposal.
2. Chemical Control: Before transplanting, soak seedlings in a 300-fold dilution of 40% dimethomorph WP. At early signs of infection, spray foliage with a 1,000-fold dilution of 66.8% propamocarb-mancozeb WP, 69% oxadixyl-mancozeb WP, 72% cymoxanil-mancozeb WP or an 800-fold dilution of 72.2% mefenoxam AS. Alternate chemical use to prevent resistance.

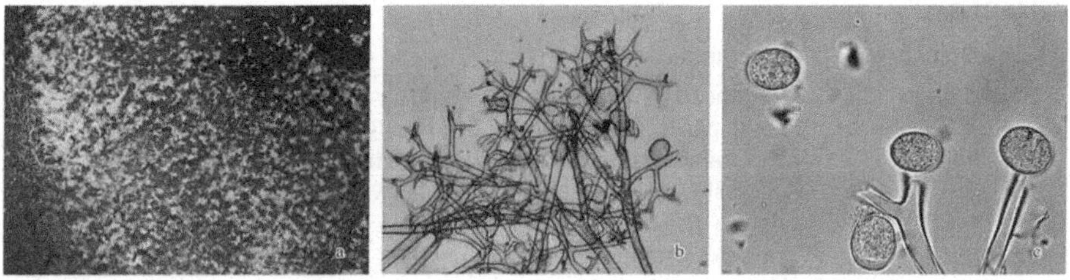

Fig. 21.20 Morphology of *Peronospora radii*. (**a**) White fungal colonies (10×), (**b**) Sporangiophore (400×), (**c**) Sporangia (1000×)

21.11 Occasional Diseases of *C. morifolium* (Table 21.1)

Table 21.1 Occasional diseases of *C. morifolium*

Disease (pathogen)	Symptoms	Occurrence pattern
Gray spot disease [*Albifimbria verrucaria* (synonym: *Myrothecium verrucaria*)]	Small silvery-gray spots initially appear on the upper surface of the leaves, with dark brown edges. The underside of these spots is brownish to dark brown. As the disease progresses, the spots gradually expand into nearly circular or irregular patches, becoming fragile	The disease can manifest throughout the entire growth period of *C. Morifolium*, with severity increasing under warm and humid conditions. The peak incidence typically occurs each year from July to august
Fusarium leaf blight [*Fusarium oxysporum*, *F. Equiseti*, *F. Armeniacum*, *F. Fujikuroi*]	*F. Oxysporum* and *F. Equiseti* cause irregular brown to reddish-brown spots along the leaf edges, which gradually spread toward the base of the leaf until the entire leaf turns brown or grayish-brown. *F. Armeniacum* causes sunken brown spots on the leaf surface that gradually expand into irregular or nearly circular brownish to dark brown patches, with a slight central depression surrounded by a prominent yellow halo. In advanced stages, these spots merge into larger lesions, causing the entire leaf to wither and die, though dead leaves do not detach. *F. Fujikuroi* leads to grayish-brown lesions along the leaf edges, sometimes accompanied by a yellow halo. These lesions tend to curl and are very brittle, eventually spreading toward the base until the entire leaf turns grayish-brown	July and august mark the peak infection period. After the rainy season, high temperatures, high humidity, and poor soil drainage conditions can lead to widespread outbreaks

(continued)

Table 21.1 (continued)

Disease (pathogen)	Symptoms	Occurrence pattern
Red spot disease [*Stagonosporopsis chrysanthemi*]	Red, nearly circular spots first appear on the upper leaf surface, with brown edges and sometimes a faint yellow halo. The underside of these spots appears brown. As the disease progresses, the spots enlarge, with the centers turning dark brown and the edges remaining dark brown	July and august are the peak months for outbreaks. Following the rainy season, conditions of high heat, humidity, and inadequate soil drainage contribute to extensive disease spread
Leaf spot disease [*Paraphoma chrysanthemicola*]	The disease initially affects the lower leaves, gradually spreading upward. Small brown spots first appear on the leaves, sometimes surrounded by a faint yellow halo, with the underside being light brown. Over time, these spots enlarge into near-circular or irregular patches with dark brown edges and lighter yellow--brown centers	July to august represents the high-risk period. After the rainy season, outbreaks can occur extensively under conditions of elevated temperature, humidity, and insufficient soil drainage
Rust disease [*Puccinia horiana*]	Primarily affecting young leaves and flower buds, the disease begins with small white spots on the underside of the leaves, with corresponding chlorotic (yellowed) spots on the upper surface. As the disease progresses, the white spots on the underside develop into small, pale yellow sticky masses, while the corresponding areas on the upper surface become slightly sunken. In advanced stages, the sticky masses on the underside expand into pale yellow, blister-like structures known as telia, which eventually turn white or grayish-white and produce large quantities of basidiospores. In severe cases, pale yellow telial clusters also appear on the upper leaf surface and may merge into larger areas, causing early yellowing, leaf drop, and ultimately, plant death	July to august is the peak outbreak season. Following the rainy season, high temperature, humidity, and inadequate drainage create ideal conditions for large-scale outbreaks

Chapter 22
The Diseases of *Andrographis paniculata*

Andrographis paniculata [*A. paniculata* (Burm. F.) Nees], also known as spring autumn willow, gold hook grass, bitter bile grass, and Indian herb, is an herbaceous plant in the Acanthaceae family. Widely cultivated in Hainan, Guangxi, Yunnan, Fujian, Guangdong provinces, and occasionally in Jiangsu and Shaanxi provinces, this plant is valued for its heat-clearing, detoxifying, blood-cooling, and swelling reducing properties. It is traditionally used to treat fever, sore throat, mouth ulcers, whooping cough, diarrhea, dysentery, painful urination, abscesses, and snake bites.

Primary diseases affecting *A. paniculata* include wilt, damping-off, stem blight, late blight, and viral disease.

22.1 Wilt Disease of *Andrographis paniculata*

Wilt disease in *A. paniculata* can lead to wilting, leaf yellowing, defoliation, and, in severe cases, plant death. This disease is prevalent in *A. paniculata* production areas in Guangxi and Guangdong provinces, with an incidence rate of 10–50%, reaching up to 80% in severely affected regions.

Symptoms

Wilt disease primarily affects the root and basal stem, causing rot, wilting, leaf yellowing, defoliation, and, ultimately, plant death. Key symptoms include elongated black lesions on the basal stem, decaying cortical tissue, and browning of vascular bundles. This disease occurs during both seedling and mature stages. In humid conditions, white, felt-like mycelium may appear around the basal stem and soil surface in seedlings. In mature plants, yellowing of apical leaves, stunted growth, blackening of the root and basal stem, and eventual plant death are common indicators (Fig. 22.1).

© China Agriculture Press Co., Ltd. 2026

T. Wang et al., *Atlas of Chinese Medicinal Plant Diseases*,
https://doi.org/10.1007/978-981-95-6326-5_22

Fig. 22.1 Symptoms of Wilt Disease of *Andrographis paniculata*

Pathogen

The pathogen responsible for *A. paniculata* wilt is *Fusarium oxysporum* (Snyder & Hansen), belonging to the genus *Fusarium*. On PDA medium, this pathogen produces abundant, cottony aerial mycelium, with colony undersides appearing pale purple to violet, occasionally white. It produces microconidia, macroconidia, and chlamydospores. Microconidia measure (6–14) μm × (2.7–3.6) μm, and macroconidia are typically three-septate, measuring (25–36) μm × 4.7 μm (Fig. 22.2).

Occurrence Pattern

Both seedlings and mature plants are susceptible, with the pathogen overwintering in soil. In Guangdong and Hainan provinces, disease onset typically occurs from August to October. The pathogen overwinters as mycelium, conidia, and chlamydospores in soil and plant residues, spreading through rain and irrigation water, infecting through wounds on the root or basal stem. High temperatures and humidity favor disease development.

Control Measures

Agricultural Control: Before transplanting, thoroughly remove diseased plant debris to reduce initial infection sources. Select drier planting sites and adopt high-ridge cultivation. Implement crop rotation, practice careful field management to avoid basal stem and root damage, ensure prompt ventilation to reduce humidity, cultivate soil early, and control nitrogen fertilizer while increasing phosphorus and potassium, which can enhance plant resistance.

Biological Control: *Trichoderma* spp. can be used as a preventative measure. Moisten seeds with water or an adhesive and mix them with Trichoderma wettable powder at a seed-to-powder ratio of 4:1 for uniform coating before sowing. Alternatively, mix Trichoderma powder with organic matter or low-concentration fertilizers in a 3:4 ratio for application. For field treatment, spray 200–300 g of Trichoderma WP per acre, diluted in 50–60 kg of water, every 5–7 days, for two to three consecutive treatments.

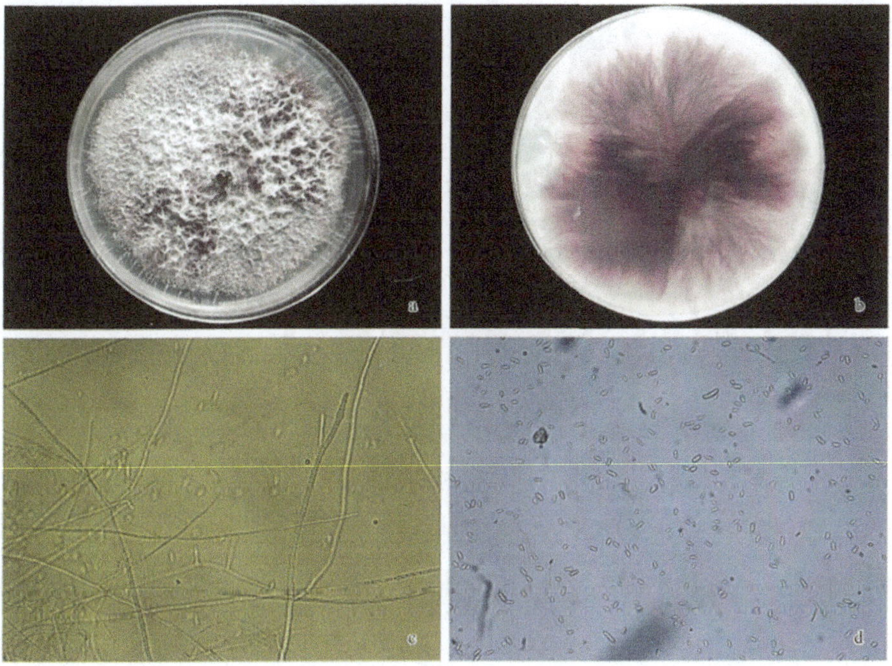

Fig. 22.2 Morphology of *Fusarium oxysporum*. (**a**) Front view of colony. (**b**) Back view of colony. (**c**) Mycelium. (**d**) Microconidia

Chemical Control: At the early stage of disease, drench or irrigate with 36% thiophanate-methyl SC diluted 600-fold or 50% benomyl WP diluted 800-fold. During the later stage of disease, use 75% mancozeb WP diluted 500-fold or 50% carbendazim WP diluted 500-fold at 10-day intervals, for a total of two applications.

22.2 Southern Blight of *Andrographis paniculata*

Southern blight is one of the primary diseases affecting *A. paniculata* seedlings, widely distributed across major Andrographis-producing regions in China. This disease, also known as "seedling wilt," is difficult to control and poses a significant threat to the growth and development of *A. paniculata*, severely impacting its yield.

Symptoms

Southern blight typically appears as yellowish-brown, water-soaked, elongated lesions on the basal stem near the ground of young *A. paniculata* seedlings. These lesions gradually expand around the stem, forming girdling bands. The affected area

Fig. 22.3 Symptoms of *Andrographis paniculata* Southern Blight. (**a**) Field symptoms. (**b**) Root symptoms. (**c**) Comparison between diseased roots (Left 1, 2) and healthy root (Right)

dehydrates, shrinks, and loses its ability to transport nutrients and water, causing seedlings to wilt, collapse, and die in clusters (Fig.. 22.3).

Pathogen

The pathogen causing *A. paniculata* southern blight is *Rhizoctonia solani* Kihn, an anamorphic fungus in the genus *Rhizoctonia*. On PDA medium, the colony initially appears colorless, later turning grayish-white, brown, gray-brown, or dark brown, eventually forming sclerotia. The mycelium is extensive, web-like, and robust, initially colorless and fine, about 5–6 μm wide, with near right-angle branching and a septum close to the branching point (Fig.. 22.4). Under staining, each cell typically contains 3–16 nuclei, most commonly 4–5. Mature hyphae often appear as a chain of barrel-shaped cells, yellow-brown, thick, and 8–12 μm wide, with similar right-angle branching at junctions. The sclerotia are irregularly shaped, light to dark brown, consisting of interwoven barrel-shaped hyphae connected by rope-like hyphae, loose in texture, and approximately 0.5–1.0 mm in diameter. Under artificial induction, basidia and basidiospores may form. The basidia are colorless, single-celled, cylindrical or oblong, with two to four short stalks at the top, each bearing one basidiospore; basidiospores are elliptical or ovoid, colorless, and single-celled.

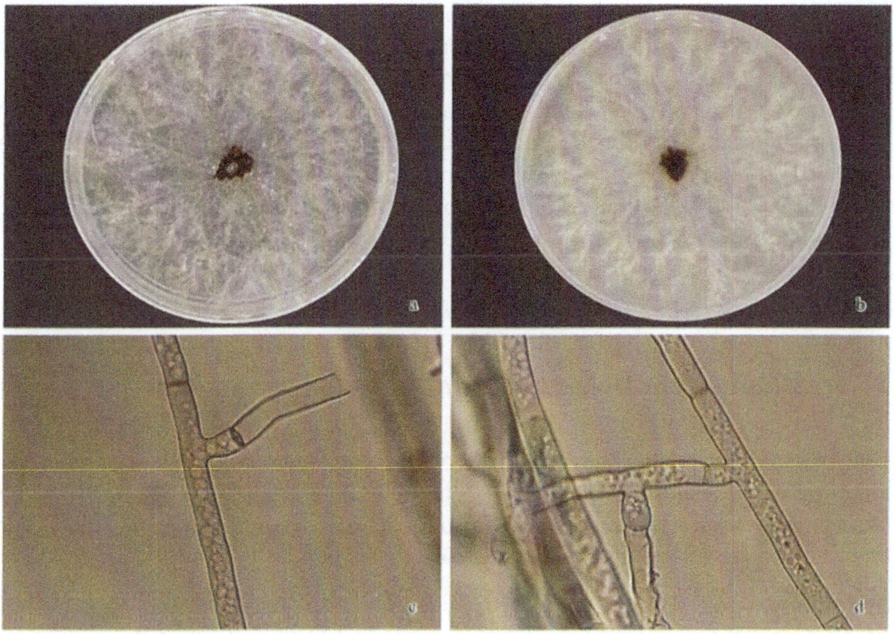

Fig. 22.4 Pathogen morphology of *Andrographis paniculata* Southern Blight. (**a**, **b**) Colony characteristics. (**c**, **d**) Hyphae

Occurrence Pattern

The pathogen overwinters in soil or crop residues as mycelium and sclerotia, surviving for 2–3 years due to its strong saprophytic capability. Uncomposted manure mixed with infected residues and overwintering mycelium and sclerotia on alternate host plants can serve as primary infection sources. The pathogen spreads through rainwater, runoff, contaminated farming tools, and infected compost, invading seedlings via wounds on the stem base or roots, and can also penetrate directly through the host's epidermis. The optimal growth temperature for the pathogen is 17–28 °C, with growth inhibited below 12 °C or above 30 °C. High nursery bed temperatures, excessive seedling growth, high soil moisture, heavy soil texture, poor drainage in low-lying areas, insufficient light, and poor photosynthesis can weaken plant resistance and promote disease.

Control Measures

Agricultural Control: Remove infected residues from the field to reduce disease incidence in seedlings. Regularly weed nursery beds and surrounding areas to minimize sources of infection. Apply fertilizers scientifically, avoiding the use of poorly decomposed manure mixed with infected plant residues. Increase potassium fertilizer application and spray micronutrient fertilizers to enhance plant resistance. Implement crop rotation and improve field management, ensuring timely ventilation, cooling, and moisture control.

Biological Control: Studies have shown that *Bacillus subtilis* strains GH18 and GH19 effectively control *R. solani* and increase defensive enzyme activity in plant leaves, making them suitable for managing Andrographis southern blight.

Chemical Control: Use 40% pentachloronitrobenzene dust formulation (DP) to treat soil, soak seeds, or drench infected areas. For soil treatment, apply 1–1.5 kg per mu (0.067 hectares), mixing evenly into the soil before sowing. For seed soaking, use a 500-fold dilution for 10 minutes. For soil drenching, apply a 200-fold dilution to wet the soil to a depth of 5 cm, or spray with an 800–1,000-fold dilution of 20% metalaxyl-mancozeb WP.

22.3 Black Stem Disease of *Andrographis paniculata*

Black stem disease, also known as bacterial wilt of *A. paniculata*, is a stem disease affecting *A. paniculata*. It occurs sporadically in Andrographis-producing areas of Guangxi province, with an infection rate of 5–20%. In some regions, the disease can be more severe, with an incidence rate exceeding 50%.

Symptoms

Black stem disease typically manifests as elongated black lesions on the stems near the soil surface, which gradually extend upward and downward. This results in the stems becoming constricted, thin, and weak, with leaves turning yellow-green, drooping, and curling inward along the edges. Cross-sectioning the stem reveals blackened internal tissues, and in severe cases, the entire plant wilts, yellows, and ultimately dies (Fig. 22.5).

Pathogen

The pathogens responsible for black stem disease in Andrographis are *Fusarium solani* (Mart.) Appel & Wollenw. ex Snyder & Hansen and *F. equiseti* (Corda) Sacc., both species within the genus *Fusarium*. On PDA medium, *F. solani* forms large, nearly circular, white, woolly colonies. The hyphae are septate, with each segment containing cytoplasm and one or more nuclei. The pathogen produces two types of conidia: macroconidia, which are fusiform to crescent-shaped, transparent, with two to four septa (typically three), measuring (22.5–37.5) μm × (3–4) μm; and microconidia, which are spindle-shaped or oval, with zero to one septum, measuring (4.5–24) μm × (2.5–4) μm. *F. equiseti* colonies are white, woolly, and light yellow-brown on the surface, with a yellow-brown reverse side on PDA medium. The macroconidia are sickle-shaped with a distinct foot cell base, thin and tapering at the tip, measuring (5.2–6.6) μm × (4.0–4.9) μm. Microconidia are rare, oval or elliptical. Chlamydospores are abundant, intercalary, occasionally terminal, colorless to light yellow-brown, smooth-surfaced, and spherical or elliptical, measuring (8.5–13) μm × (6.3–8.5) μm (Fig. 22.6).

Fig. 22.5 Symptoms of Black Stem Disease in *Andrographis paniculata*. (**a**) Field symptoms. (**b**) Root symptoms. (**c**) Comparison of diseased root (Left 1, 2) and Healthy root (Right)

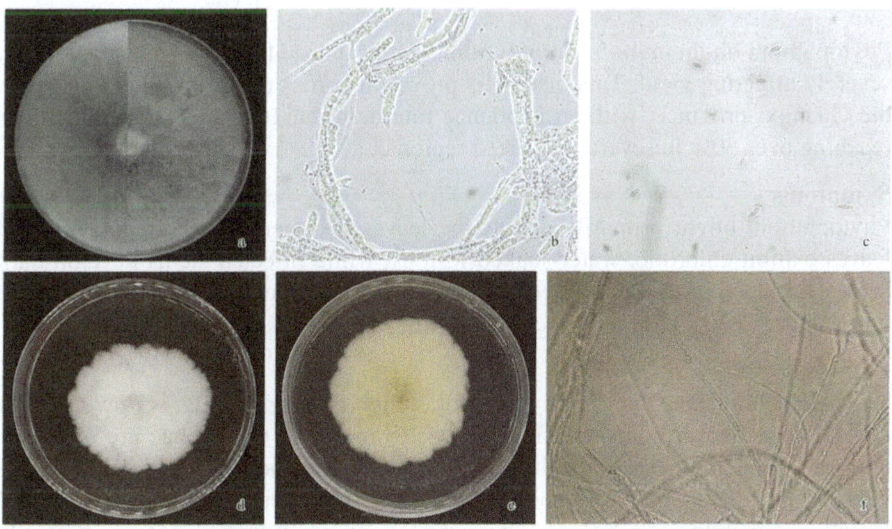

Fig. 22.6 Morphology of Black Stem Disease pathogen in *Andrographis paniculata*. (**a, b**) Colony characteristics of *Fusarium solani*. (**c**) Macroconidia of *F. solani*. (**d, e**). Colony characteristics of *F. equiseti*. (**f**) Hyphae of *F. equiseti*

Occurrence Pattern

The conidia of the pathogen can survive in soil and plant residues, spreading through wind and rain to cause reinfection. The disease is most severe in high-temperature, high-humidity seasons and in years with frequent rainfall. June marks the onset of the disease, while July and August are the peak infection periods.

Control Measures

Agricultural Control: Remove weeds and wild host plants around nursery beds to reduce sources of infection; practice appropriate crop rotation and avoid continuous cropping or rotation with susceptible host plants. Strengthen field management, apply fertilizers scientifically, avoid using poorly decomposed compost containing infected residues, and increase phosphorus and potassium fertilizer to boost plant resistance. After rainfall, drain standing water to lower soil moisture and reduce disease occurrence.

Biological Control: Research has shown that naturally occurring microorganisms, such as Trichoderma, Bacillus, and Actinomycetes, exhibit antagonistic effects against *F. solani*. These can be used to manage black stem disease of *A. paniculata* and help address environmental issues related to pesticide use.

Chemical Control: At the onset of disease, spray with an 800–1,000-fold dilution of 25% carbendazim WP, 50% thiophanate-methyl WP, or 75% chlorothalonil WP.

22.4 Phytophthora Blight of *Andrographis paniculata*

Phytophthora blight in *Andrographis paniculata* can lead to wilting and plant death, severely affecting yield. This disease is prevalent across major production areas in the Guangxi province, with an incidence rate between 10% and 50%, sometimes reaching over 80% in severely affected regions.

Symptoms

Phytophthora blight primarily affects the leaves and stems of the plant. Infected leaves exhibit chlorotic, water-soaked lesions, while infected stems develop dark brown streaks that eventually lead to constriction and decay, causing the plant to wilt and die (Fig. 22.7).

Pathogen

The pathogen responsible for Phytophthora blight in *A. paniculata* is *Phytophthora capsici* Leonian, an oomycete fungus. On V8 medium, the hyphae are relatively uniform in thickness, showing slight constrictions at branching points without vesicles or prominent chlamydospores. Sporangia show substantial morphological variation, appearing as ellipsoidal, pear-shaped, ovoid, or irregular forms, typically asymmetrical. They measure 25.5–63.0 μm in length and 20.0–32.5 μm in width, with a length-to-width ratio of about 1.6. Each sporangium has a single, indistinct papilla, approximately 3 μm thick. The sporangia detach readily, with sporangiophores varying in length, averaging between 1.0 and 50.0 μm. Sporangiophores are slender and may exhibit monopodial or irregular branching (Fig. 22.8).

Occurrence Pattern

Phytophthora blight generally occurs in July and August, particularly in hot, rainy conditions. The pathogen overwinters in soil as oospores within infected plant debris, serving as the primary infection source for the following year. It primarily

Fig. 22.7 Phytophthora Blight Symptoms of *Andrographis paniculata*. (**a**) Field symptoms. (**b–d**) Leaf symptoms

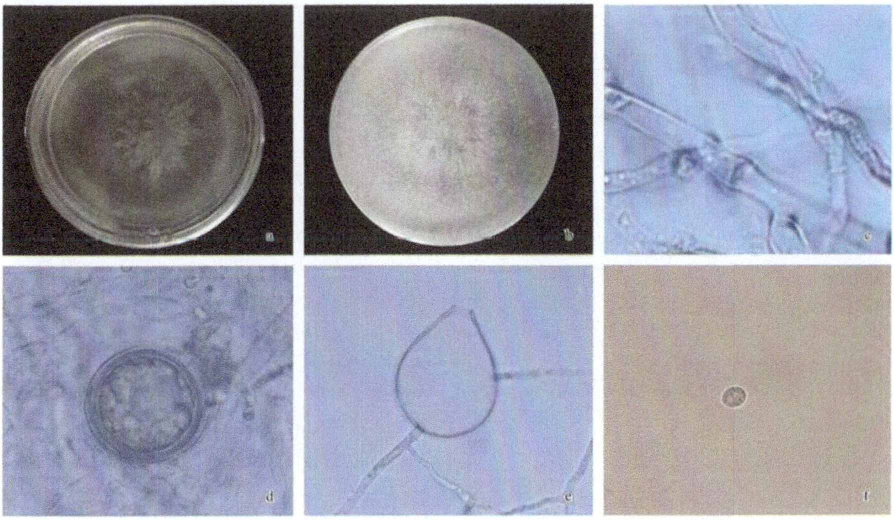

Fig. 22.8 Morphology of *Phytophthora capsici*. (**a**) Front view of colony. (**b**) Back view of colony. (**c**) Hyphae. (**d**) Oospore (**e**) Sporangium. (**f**) Zoospore

spreads through irrigation water, rain, farm activities, or air currents. High humidity, heavy rainfall, post-rain warmth, and trapped moisture can exacerbate the disease.

The pathogen grows within a 5–37 °C range, with an optimal growth temperature of 25–30 °C. Humidity is the most critical factor in disease outbreak within this favorable temperature range, with high soil moisture and elevated temperatures contributing to the rapid spread.

Control Measures

Agricultural Control: Promptly remove diseased plants and debris to reduce infection sources for the following season. Select elevated or well-drained land, or use ridge cultivation; water moderately to prevent flooding of the basal stem area, and ensure proper drainage. Apply organic manure, increase phosphorus and potassium fertilizers, limit nitrogen, and maintain moderate plant density and crop rotation to minimize disease incidence.

Physical Control: Disinfect seeds by soaking them in 52 °C water for 15 min.

Biological Control: *Trichoderma harzianum*, *Bacillus subtilis* LY-38 strain, and valinomycin have demonstrated inhibitory or lethal effects on the pathogen and are effective in managing Phytophthora blight.

Chemical Control: At the first signs of disease, apply a 500-fold dilution of 75% mancozeb WG or a 500-fold dilution of 50% carbendazim WP via foliar spray or root drench. Repeat applications every 10 days, adjusting the number of treatments (2–3) based on disease severity.

Chapter 23
The Diseases of Bletilla Species

The genus Bletilla (Bletilla Rchb.f.), part of the Orchidaceae family, includes several species commonly used in traditional Chinese medicine, such as *Bletlla formosana* (Hyata) Schltr., B. ochracea Schltr., *B. sinensis* (Rolf) Schltr., and *B. striata* (Thunb. ex Murray) Rchb.f. The dried tuber of Bletilla striata serves as the medicinal part of the plant. Bletilla is a perennial herb widely found in regions including southern Shaanxi, southeastern Gansu, Jiangsu, Anhui, Zhejiang, Jiangxi, Fujian, Hubei, Hunan, Guangdong, Guangxi, Sichuan, and Guizhou. The tuber is irregularly flat and round with a translucent, glossy appearance when cut, containing small, scattered vascular bundles. It has a sticky texture when chewed and is known for its hemostatic, anti-swelling, and wound-healing properties.

Common diseases affecting Bletilla include rust disease, anthracnose, leaf spot, and root rot.

23.1 Bletilla Rust Disease

Rust disease is widespread and rapidly spread in Bletilla, affecting production areas nationwide. It can cause premature leaf yellowing and death, significantly impairing the growth of Bletilla plants.

Symptoms
Bletilla rust disease mainly affects leaves; no infections have been observed on flowers or fruits. Infected leaves display numerous scattered, orange, nearly round pustules on the underside, arranged irregularly. In severe cases, the entire leaf underside is covered with orange spore clusters, leading to large necrotic patches and early leaf drop (Fig. 23.1).

Pathogen
The pathogen responsible for Bletilla rust disease is a fungus in the genus Coleosporium (*Coleosporium* sp.), belonging to the Basidiomycota division. The

© China Agriculture Press Co., Ltd. 2026

T. Wang et al., *Atlas of Chinese Medicinal Plant Diseases*,
https://doi.org/10.1007/978-981-95-6326-5_23

Fig. 23.1 Bletilla rust disease symptoms

summer spores are yellow, round or oval, with a net-like and warted surface. Winter spores are elongate, yellowish-brown, with sterile cells arranged in a single layer at the base (Fig. 23.2).

Occurrence Pattern

The optimal temperature range for Bletilla rust disease is between 26 °C and 28 °C. Disease spread is less favorable at average daily temperatures below 22 °C or above 33 °C. Initial symptoms appear on the upper leaf surface as small yellow spots in early June, with orange summer spore clusters forming on the underside. Under a microscope, these spore clusters are uncoated, raised, and scattered. By mid-October, reddish-brown winter spore clusters gradually form around the summer spore clusters, appearing either singly or in circular groups with a smooth, waxy surface. High humidity and frequent rain increase the disease incidence. Altitude also influences infection rates: areas above 800 m and between 200 m and

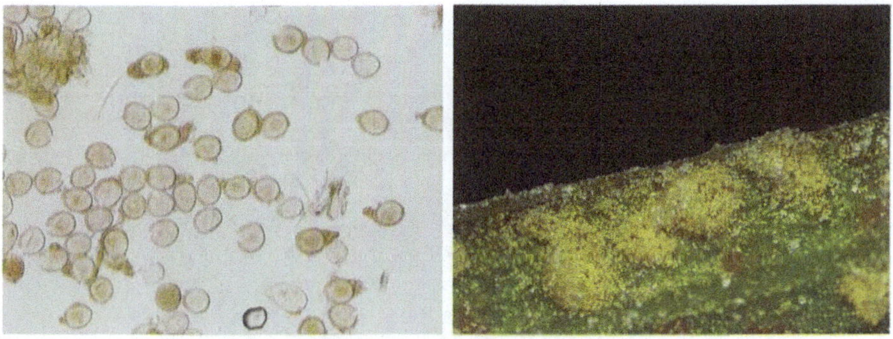

Fig. 23.2 The morphology of summer spores of Bletilla rust pathogen

480 m show lower incidence, while areas at 500–750 m show higher infection rates. Soil type impacts infection rates as well, with clay soils experiencing higher incidence and severity of Bletilla rust disease.

Control Measures

1. Enhanced Field Management: Improve fertilization and water management, use balanced fertilization techniques, and apply fully decomposed organic fertilizers. Increase phosphorus and potassium fertilizers to enhance plant resistance.
2. Chemical Control: At the early stages of infection, spray a mixture of 70% mancozeb WP diluted 1,000-fold and 15% triadimefon WP diluted 2,000-fold every 10–15 days for 2–3 consecutive treatments.

23.2 Anthracnose of Bletilla

Anthracnose is a common foliar disease in Bletilla, widely distributed across Bletilla cultivation areas with a high incidence rate, posing a serious threat to the growth and development of Bletilla plants.

Symptoms

Bletilla anthracnose manifests as nearly circular, dark brown lesions on infected leaves, with a grayish center and black edges, without any halo. In severe cases, lesions may cause perforations in the leaves (Fig. 23.3).

Pathogen

The pathogen causing Bletilla anthracnose is a fungus of the Colletotrichum genus (*Colletotrichum* sp.), an ascomycete anamorph. On PDA medium, the colony initially appears white, then turns grayish-white to gray, and finally becomes dark gray. The aerial mycelium is dense and woolly, with well-developed hyphae. Around the spore clusters on the culture, a black-brown ring of granular structures, known as acervuli, is formed. The acervuli develop subepidermally and are either colorless or brown, with densely packed, unbranched conidiophores (Fig. 23.4).

Fig. 23.3 Symptoms of Bletilla anthracnose

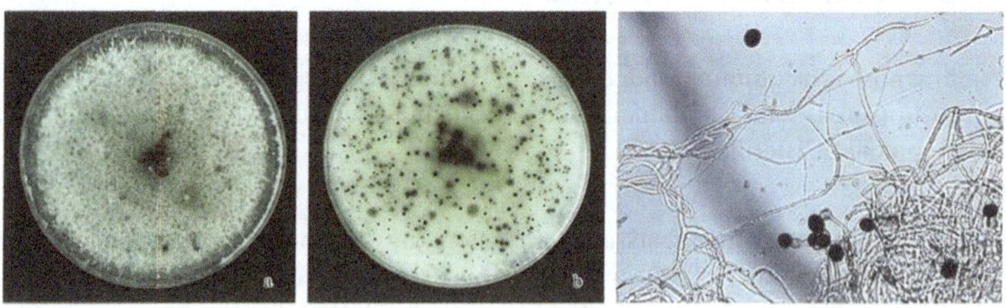

Fig. 23.4 Morphology of Bletilla anthracnose pathogen. (**a, b**) Colony characteristics. (**c**) Hyphae

Occurrence Pattern

The pathogen mainly overwinters as conidia or acervuli on infected plant debris. Under favorable conditions in the following year, it germinates and spreads through wind and rain. High temperatures and humidity encourage the disease, with typical onset in May and June and a peak in July and August. Dense planting and excessive nitrogen fertilization also increase the likelihood of disease occurrence.

Control Measures

1. Field Management: Implement proper plant spacing and promptly remove infected plant debris.
2. Chemical Control: At the early onset of the disease, spray with an 800-fold dilution of 80% thiram foranthracnose control WP or a 500–800-fold dilution of 50% carbendazim WP to effectively control the disease.

23.3 Bletilla Leaf Spot Disease

Bletilla leaf spot disease is widespread across production areas in China, with incidence rates generally between 10% and 30%, posing a significant threat to Bletilla cultivation.

Symptoms

The disease primarily affects Bletilla leaves, where lesions appear as circular, oval, or irregular brown or light brown spots. There is a clear boundary between diseased and healthy tissue, often surrounded by a yellow halo. In severe cases, the entire leaf becomes discolored and dies. Under high humidity, grayish-black mold may develop on the surface of the lesions (Fig. 23.5).

Pathogen

The causal organism of Bletilla leaf spot disease is an *Alternaria* spp. fungus, an asexual form of an Ascomycete. The mycelium is dark gray or blackish-brown, with short, single or clustered conidiophores, most of which are unbranched. Conidia are club-shaped or spindle-shaped, with septa, and may appear in chains (Fig. 23.6).

Occurrence Pattern

The pathogen overwinters as mycelium and conidia in crop residues or in soil, initiating primary and secondary infections in the following year. The disease typically begins in early May, with peak infection occurring after the rainy season in June and July as temperatures rise.

Control Measures

1. Field Management: Remove crop residues to reduce pathogen levels in the field.
2. Chemical Control: At the onset of disease, spray with an 800-fold dilution of 80% thiram foranthracnose control WP or a 500–800-fold dilution of 50% carbendazim WP.

23.4 Bletilla Root Rot Disease

Root rot is a common and serious root disease in Bletilla, with high incidence and wide distribution, making it challenging to control.

Fig. 23.5 Bletilla Leaf Spot Disease symptoms

Fig. 23.6: Bletilla Leaf Spot pathogen morphology. (**a**, **b**) Culture characteristics. (**c**) Conidia

Fig. 23.7 Bletilla Root Rot Disease symptoms

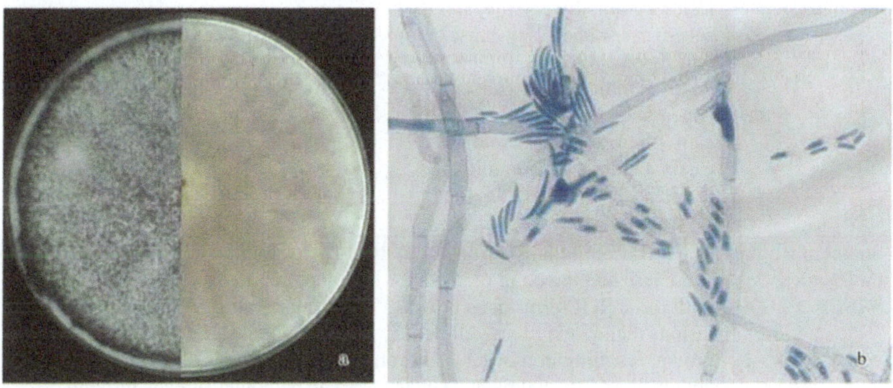

Fig. 23.8 Bletilla Root Rot pathogen morphology. (**a**) Culture characteristics. (**b**) Conidia

Symptoms

Root rot primarily infects the root system. Affected plants show yellowing leaves, with wilting starting from the leaf tips. Infected roots exhibit brown, fibrous rot. In high-humidity conditions, abundant white aerial mycelium may be observed (Fig. 23.7).

Pathogen

The causative agent of Bletilla root rot is *Fusarium oxysporum* Schltdl. ex Snyder et Hansen, an asexual form of an Ascomycete. Large conidia with three septa measure (25–36) μm × 4.7 μm, while those with four septa measure (32–41) μm × (3.9–4.7) μm. Small conidia are blunt at both ends, either long-oval or kidney-shaped, single- or two-celled, measuring 9.4–19.7 μm in diameter (Fig. 23.8).

Control Measures

1. Cultivation Management: Avoid continuous cropping; rotate with Poaceae crops; ensure adequate drainage during the rainy season; disinfect seedlings; and control soil-borne pests.
2. Chemical Control: In early stages of infection, irrigate affected areas with a 3000-fold dilution of 70% meiling WP.

23.5 Bletilla Opportunistic Diseases (Table 23.1)

Table 23.1 Characteristics of Bletilla opportunistic diseases

Disease (Pathogen)	Symptoms	Occurrence pattern
Blight (Tobacco blight) (Phytophthora nicotianae)	Initially appears as small spots on the leaves, followed by brownish-gray lesions with irregular shapes. The lesions gradually enlarge, and in severe cases, they may cause complete wilting of old leaves. Infected leaf edges exhibit a water-soaked appearance. Severe infections may cause the leaves to rot and fall off easily. Under high humidity, the lesions exhibit a white layer of mycelium	The pathogen prefers humid environments, with an optimal temperature of 28 °C. It is most severe during the rainy seasons from April to May
Southern Blight (*Sclerotium rolfsii*)	The infection initially manifests at the stem base and the point of contact with the soil. It forms deep brown lesions that quickly turn yellow--brown, resulting in root rot, stem rot, and death of affected plants. Under warm, humid conditions, infected plants die rapidly. White, silky mycelium and small, brownish-yellow sclerotia are visible around the roots and the infected area	The pathogen overwinters in soil and plant debris. Favorable conditions promote rapid growth, and the pathogen invades Bletilla roots and stems, leading to wilting and death. The pathogen spreads through irrigation water and rain splashes, with peak incidence occurring from early June to mid-July, especially after periods of heavy rainfall

Chapter 24
The Diseases of *Dendrobium*

Dendrobium, a member of the Orchidaceae family, is also known as "Immortal Orchid," "Everlasting Herb," "Purple Orchid," and "Hanging Orchid." Notable species include *Dendrobium nobile* Lindl., *D. huoshanense* C.Z. Tang & S.J. Cheng, *D. chrysotoxum* Lindl., and *D. fimbriatum* Hook., all of which are listed in the Pharmacopoeia of the People's Republic of China. There are records of over 30 species of *Dendrobium* with edible uses. Known as the "First of the Nine Immortal Herbs" and "Gold of the Herb World." *Dendrobium* is a perennial herb with roots that are used medicinally. It is primarily distributed in southern Anhui, Taiwan, southern Hubei, Hong Kong, Hainan, western to northeastern Guangxi, southern Sichuan, southwestern to northern Guizhou, southeastern to northwestern Yunnan, and southeastern Tibet regions. After drying, the root is firm and easily broken, with a smooth, gray-white to gray-green cross-section. It has a mild aroma, slightly bitter taste, and is used to promote stomach health, and clear heat.

Dendrobium is susceptible to several diseases, including white silk disease, leaf spot disease, viral disease, blight, and rust. Among these, white silk disease, leaf spot disease, and viral disease are more common, severely affecting Dendrobium's quality and yield, necessitating timely control measures.

24.1 White Silk Disease of Dendrobium

White silk disease is one of the most common diseases affecting Dendrobium, occurring in major production areas such as Anhui, Hubei, Sichuan, and Guangxi provinces. It primarily damages the leaves and stems of Dendrobium. In severe cases, it can cause the entire plant to wither and die, with an infection rate typically between 10% and 30%, greatly impacting yield.

© China Agriculture Press Co., Ltd. 2026
T. Wang et al., *Atlas of Chinese Medicinal Plant Diseases*,
https://doi.org/10.1007/978-981-95-6326-5_24

Symptoms

White fungal threads form at the stem base or on the cultivation substrate, then spread to the stem and leaves, causing necrosis in the leaves and eventually leading to stem rot, bending, and death. In later stages, fungal sclerotia form on dead leaves and stems, and rot symptoms appear at the stem base and roots (Fig. 24.1).

Pathogen

The pathogen responsible for white silk disease in Dendrobium is classified within the genus *Sclerotium*, specifically *S. rolfsii* Sacc. and *S. delphinii* Welch. On PDA medium, the fungus grows in a cottony or silky white radiating pattern, initially forming milky-white spherical clusters. After 5 days, the colony turns pale yellow to yellow-brown. After 2 weeks, the colony darkens to a deep brown, resembling rapeseed, with a smooth, glossy, spherical or nearly spherical shape, 1–5 mm in diameter (Fig. 24.2). Most sclerotia measure around 1–2 mm, with fewer at 3 mm $\times$ 5 mm. Sclerotia are scattered within the culture dish, more concentrated around the edges than in the center, sometimes clustering at the inoculation point.

Fig. 24.1 Symptoms of White Silk Disease in Dendrobium

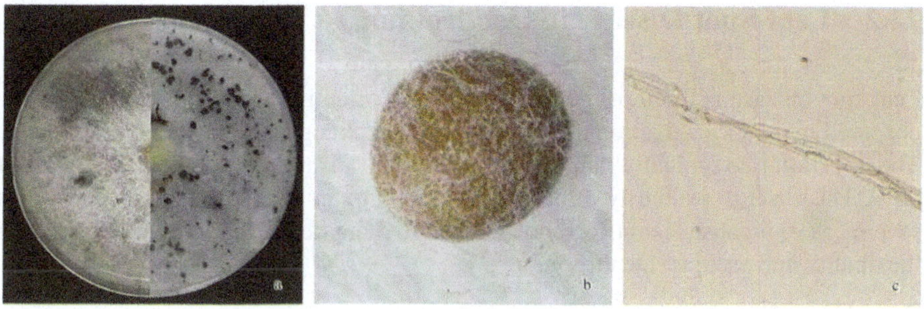

Fig. 24.2 Pathogen morphology of White Silk Disease in Dendrobium. (**a**) Colony characteristics. (**b**) Sclerotia. (**c**) Mycelium

Occurrence Pattern

White silk disease typically occurs from May to October, with a peak from June to September. The disease slows at temperatures above 35 °C, with the optimal growth temperature at (25 ± 2)°C. Growth stops below 8 °C or above 40 °C. The pathogen overwinters as mycelium and sclerotia in cultivation substrates or plant debris. It spreads over long distances via water, soil, cultivation substrates, or through tools and personnel. There are two primary sources of the pathogen: local sources and external introductions. It primarily invades through natural openings or wounds, often from insect damage, on *Dendrobium* plants. The sclerotia are highly resistant to extreme temperatures, surviving 5–6 years under natural conditions.

Control Measures

1. Selection of Resistant Varieties

   Plant high-yield, disease-resistant varieties to effectively control and reduce disease occurrence, such as the highly resistant "Senshan No. 1."

2. Agricultural Management

   Remove infected plants immediately, and take them out of the cultivation area for deep burial or destruction. Disinfect the affected site and replace the substrate with new material, sterilized with 3% lime water.

3. Biological Control

   Trichoderma harzianum exhibits strong antagonism against the pathogen. Use *T. harzianum* T-22 WP (3 × 10^6 CFU/g) diluted 1000–1500 times for spraying or root drenching. *Bacillus subtilis* is also highly antagonistic to the pathogen and can be applied as a wettable powder diluted 1500–2000 times.

4. Chemical Control

   At the onset of disease, apply 80% carbendazim WP diluted to 700–800-fold as a spray.

24.2 Leaf Spot Disease in Dendrobium

Leaf spot disease is the most common affliction affecting Dendrobium, commonly observed in major production areas such as Anhui, Jiangsu, and Guangxi provinces. The infection rate typically ranges from 10% to 50%, and in some fields can exceed 80%. This disease primarily affects the leaves but can also spread to the stems. In severe cases, it causes leaf drop and can lead to plant death, significantly impacting the quality and yield of Dendrobium.

Symptoms
Infected areas on the leaves initially turn yellow and then develop into light brown or dark brown spots with distinct edges, often surrounded by a yellow halo. Over time, the lesions enlarge in tandem with plant growth, with the surrounding tissue turning yellow or light green and forming a sunken appearance (Fig. 24.3).

Pathogen
The pathogen responsible for Dendrobium leaf spot disease is *Phoma cucurbitacearum* (Fr.: Fr.) Sacc. (synonym: *Phyllosticta citrullina* Chester), an asexual form of the ascomycete fungus in the genus *Phoma*, with its sexual form classified as *Didymella bryoniae* (Auersw.) Rehm. On PDA medium, the colony appears raised and fluffy with black mycelium. The hyphae are septate, and most conidia are oval or nearly cylindrical (Fig. 24.4).

Occurrence Pattern
The timing and severity of leaf spot disease vary across regions and cultivation sites. It typically emerges from late spring to early summer, with peak occurrence from May to September. The pathogen overwinters as conidia within infected tissue, producing conidia under suitable conditions that spread via wind and rain, entering

Fig. 24.3 Symptoms of leaf spot disease in dendrobium

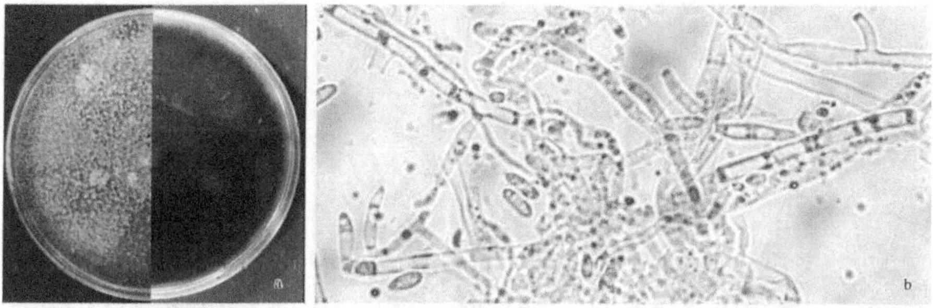

Fig. 24.4 Morphology of the Pathogen Causing Dendrobium Leaf Spot Disease. (**a**) Culture characteristics. (**b**) Hyphae and Conidia

plants through stomata or wounds. Conditions such as continuous cropping, overcrowding, poor ventilation, and high humidity facilitate disease development.

Control Measures

1. Agricultural Control: After harvest, promptly remove plant debris, perform precise pruning in early spring to eliminate infected branches and leaves, and immediately destroy them to effectively reduce sources of infection. During the growing season, remove infected plants promptly, dispose of them outside the cultivation area by deep burial or incineration, and disinfect affected areas. Replace the substrate of infected plants, ensure proper ventilation and ample light, strictly control watering during outbreaks, and avoid overhead watering.
2. Biological Control: Apply a 1000–1500-fold dilution of *Trichoderma harzianum* strain T-22 WP (3×10^{8} CFU/g) by spraying onto plants or as a root drench.
3. Chemical Control: During the disease period, apply a 0.5% Bordeaux mixture every 15 days, totaling two to three applications. Alternating applications of mancozeb and copper-based products can also be effective.

24.3 Viral Disease of Dendrobium

Dendrobium is primarily propagated through asexual methods such as cuttings and division, leading to viral accumulation over time. This buildup results in genetic degradation, as viruses are retained in the propagation material. Dendrobium viral disease, known as the "cancer" of orchids, is incurable and significantly affects both the quality and yield of Dendrobium. Viral diseases are common across all major production areas, though the infection rate generally remains low.

Symptoms

Infected leaves exhibit yellowing, chlorosis, mosaic patterns, and deformities. In severe cases, diseased leaves undergo necrosis and fall off (Fig. 24.5).

Fig. 24.5 Symptoms of viral disease in Dendrobium

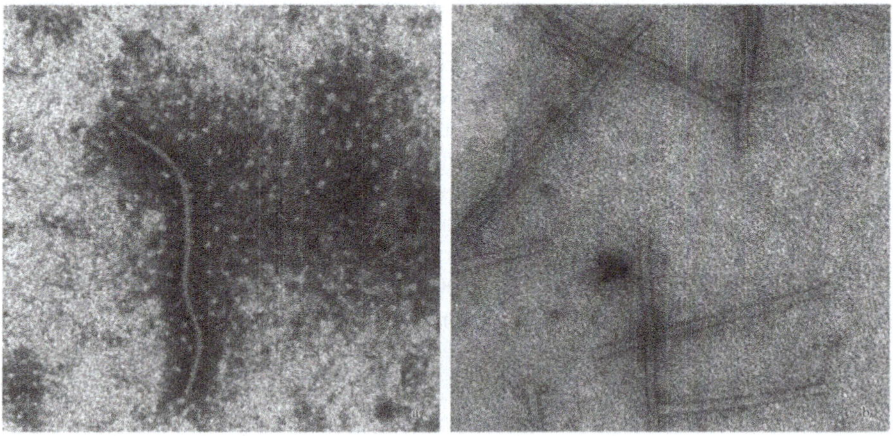

Fig. 24.6 Morphology of Dendrobium viral pathogens. (**a**) CymMV particles. (**b**) ORSV particles

Pathogen

Reported viruses affecting Dendrobium include cucumber mosaic virus (CMV), cymbidium mosaic virus (CymMV), odontoglossum ringspot virus (ORSV), and orchid fleck virus (OFV) (Fig. 24.6).

Occurrence Pattern

CMV spreads through a variety of means, primarily transmitted by aphids through nonpersistent transmission. The primary insect vectors are *Aphis gossypii* and *Myzus persicae*. CMV can also spread through mechanical contact with infected sap and via seeds, with a typical seed transmission rate of 4–8%. CymMV and ORSV, however, enter plants only through mechanical wounds, so any action during tissue culture or greenhouse cultivation that creates surface wounds-such as cutting flower stems, pruning, or even leaf-to-leaf friction-may allow viral infection. OFV is primarily transmitted through sap and seedlings.

Control Measures

1. Agricultural Control

   Promptly remove and destroy infected plants, thoroughly disinfect pruning tools, and avoid using infected plants for propagation. Select virus-free seedlings, clear surrounding weeds to reduce infection sources, and apply foliar fertilizers to enhance plant resistance to viruses.
2. Physical Control

   Place reflective silver sheets in the field to repel aphid vectors, such as *A. gossypii* and *M. persicae.*
3. Chemical Control

   Use insecticides to control aphids and other pests with piercing-sucking mouthparts. Suitable options include 1.8% abamectin EC at a dilution of 3,000–5,000-fold, 10% imidacloprid WP diluted 2,000-fold, or 50% pymetrozine WP at a 1,500–2,000-fold dilution. Alternatively, 30% dimethoate WP at a 500-fold dilution can be applied for preventive measures. Before planting, apply 150–250 g per mu of 20% moroxydine hydeochloride WP, with a spray interval of 2–3 days, followed by an additional application 10–15 days after planting.

24.4 Occasional Diseases in Dendrobium (Table 24.1)

Table 24.1 Occasional diseases in Dendrobium

Disease (pathogen)	Symptoms	Occurrence pattern
Blight (*Phytophthora nicotianae*)	Primarily affects the roots of young, transplanted dendrobium plants, often leading to plant death. In the early stages, dark brown, water-soaked lesions appear at the stem base, which gradually turn black. As the disease progresses, it causes root and stem rot, resulting in leaf yellowing, defoliation, and wilting. Under humid and cloudy conditions, the disease spreads rapidly to surrounding plants	The pathogen spreads through irrigation and splashing water. The peak infection period is from April to august, during the rainy season, when the disease becomes most severe. In waterlogged conditions, the pathogen produces a large number of zoospores that spread easily to the dendrobium root base via water flow, resulting in extensive outbreaks
Rust (*Coleosporium zanthoxyli*)	Primarily infects the leaves. In the initial stages, small, round, yellow-green spots appear on the upper side of the leaves, gradually expanding. After about 7 days, orange or brick-red summer spore clusters form on the lesions and then disperse. In severe cases, photosynthesis is hindered, weakening the plant and reducing yield the following year. Symptoms typically begin in September, with visible rust by June the following year. Winter spore clusters form at the site of the summer spore clusters	The pathogen overwinters as mycelium and spores within infected plant debris. The disease period is generally from may to November. As temperatures rise and humidity drops, the disease progression slows. However, from July to September, conditions are favorable for severe infections, often causing premature leaf drop

Chapter 25
The Diseases of Asarum

Asarum sieboldii Mig., commonly known as Asarum, is a perennial herbaceous plant in the Aristolochiaceae family. In traditional Chinese medicine, Asarum is valued for its effects in dispelling wind, dispersing cold, promoting water movement, and opening the orifices. It is used to treat ailments such as cold-induced headaches, nasal congestion, toothache, phlegm induced cough, and rheumatic pain. There are two main varieties: Liaoning Asarum (also known as Northern Asarum), primarily produced in Jilin and Liaoning, and Huaxin Asarum, mainly produced in Shaanxi, Sichuan, and Hubei. Asarum from northeast China is generally considered to be of the highest quality, while Huaxin Asarum from Huayin County, Shaanxi, is also well-regarded. Asarum is also cultivated in Japan and Korea.

The primary diseases affecting Asarum include leaf blight, sclerotinia rot, blight, and rust. Of these, leaf blight and sclerotinia rot are particularly damaging.

25.1 Leaf Blight in Asarum

Leaf blight is the most severe disease affecting Asarum production and occurs across all major cultivation areas, including the main production regions of Liaoning such as Qingyuan, Xinbin, and Huanren Fengcheng. In affected fields, the incidence of disease can reach up to 100%, with yield losses typically ranging from 30% to 50%. In severe cases, the entire above-ground portion of the crop may die, resulting in total crop loss.

Symptoms
Leaf blight affects Asarum plants at all growth stages, primarily targeting the leaves but also infecting petioles, flowers, fruits, and buds, though it rarely affects the roots. Early symptoms on leaves appear as small brown spots that gradually enlarge into near-circular lesions, 5–18 mm in diameter, with 6–8 concentric rings and a yellow-brown or reddish-brown halo. In severe cases, the lesions merge, leading to

© China Agriculture Press Co., Ltd. 2026

T. Wang et al., *Atlas of Chinese Medicinal Plant Diseases*,
https://doi.org/10.1007/978-981-95-6326-5_25

perforations and eventual leaf death. Lesions on petioles are spindle-shaped, dark brown, and progressively enlarge and deepen, cutting off vascular tissue and causing leaf wilting. Infected flowers and fruits develop blackish-brown, sunken lesions, 3–6 mm in diameter, leading to petal darkening and premature fruit drop. Bud infection causes rotting, and in severe cases, root rot may follow. Under high humidity, all infected areas can produce a brown mold, which represents the pathogen's conidiophores and conidia (Fig. 25.1).

Pathogen

The pathogen responsible for Asarum leaf blight is *Mycocentrospora acerina* (R. Hartig) Deighton, an asexual form in the genus Mycocentrospora. Conidiophores are light brown and slightly geniculate. Conidia are clavate or club-shaped, colorless to light brown, with 3–11 septa, a flat base, and a tapered apex forming a long beak, either straight or curved. The conidia may or may not have basal appendages

Fig. 25.1 Symptoms of Leaf Blight in Asarum

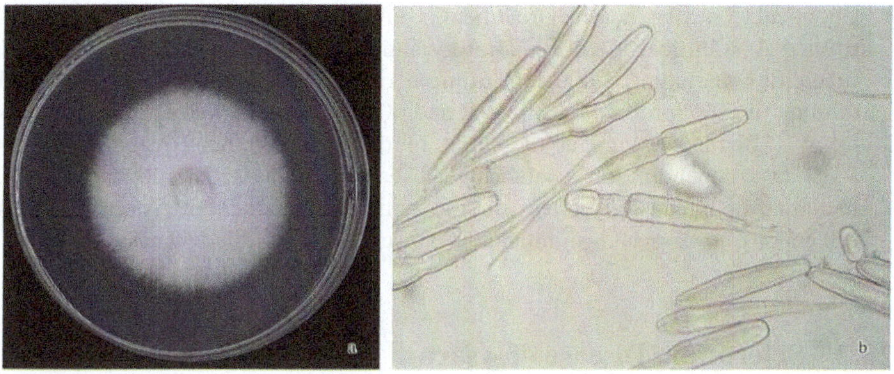

Fig. 25.2 Morphology of Asarum Leaf Blight pathogen. (**a**) Culture characteristics. (**b**) Conidia

and measure (15–30) μm × (15–20) μm. Cultured colonies appear red, although they rarely produce spores on PDA medium (Fig. 25.2).

Occurrence Pattern

The pathogen overwinters as conidia and mycelium on plant debris and infected buds in the field. It can be transmitted through infected seedlings. Conidia spread primarily within fields through wind and rain splash. Low temperatures, high humidity, and rainy weather favor disease development and spread. The optimal temperature for infection is 15–20 °C, while temperatures above 25 °C inhibit pathogen growth. Shaded cultivation reduces disease incidence compared to full--sun exposure. Symptoms typically begin in early May, with peak incidence from June to July. Low temperatures during the leaf emergence phase reduce the resistance of Asarum stems and leaves, while cool, rainy conditions in the mid-growing season also promote disease outbreaks. Loose, well-draining humus or sandy soil has a lower disease incidence, while low-lying, poorly drained clay soils are more susceptible. Areas with dense planting, poor ventilation, and excessive weed growth, especially if infected Asarum debris is not removed, create conditions favorable for pathogen persistence and reinfection, significantly increasing disease incidence.

Control Measures

1. Seedling Disinfection: Before planting, soak Asarum seedlings in an 800-fold dilution of 50% procymidone WP for 4 h to eliminate pathogens on the seedlings and effectively prevent seedling-transmitted infection.
2. Field Sanitation: After natural wilting in the fall, promptly remove and destroy plant debris by burning or deep burial outside the field. Before Asarum emerges in spring, disinfect the field surface by spraying a 400-fold dilution of 50% ammonium thiosulfate solution to reduce overwintering inoculum.
3. Shaded Cultivation: Shaded cultivation significantly reduces disease incidence compared to full-sun exposure. Utilize forested areas or shading screens to reduce disease severity.

4. Chemical Control: Chemical treatment is currently a necessary measure for controlling Asarum leaf blight. Laboratory and field tests have identified effective fungicides, including 1,000-fold dilution of 50% procymidone WP, 800-fold dilution of 50% iprodione SC or WP, and 600-fold dilution of 50% chlorothalonil WP. Begin spraying at the onset of disease symptoms, repeating every 7–10 days depending on weather and disease conditions. Thorough coverage, including both sides of the leaves, is essential, especially after Asarum reaches full canopy cover, which significantly improves control effectiveness.

25.2 Sclerotinia Disease of Asarum

Sclerotinia disease is one of the major threats to Asarum cultivation, prevalent across all Asarum-growing regions. Provinces currently known to be affected by this disease include Liaoning and Jilin. From 1979 to 1982, the disease spread extensively in the primary Asarum-producing areas of Xinbin, Fengcheng, and Kuandian in Liaoning Province, with typical infection rates between 15% and 30%. In certain fields, widespread plant death occurred, leading to complete crop destruction. The disease spreads rapidly and poses a significant threat to crops, particularly in older infested areas, where an infection center can expand over 2–3 years, ultimately causing total field devastation.

Symptoms

Sclerotinia disease in Asarum primarily affects the root system but can also infect stems, leaves, flowers, and fruits. Typically, infection starts below ground and gradually spreads to aboveground parts. Lesions appear brown or pinkish, with a granular surface, eventually forming sclerotia. These sclerotia are oval or irregular, smooth on the surface, dark brown on the exterior, and white internally. Root-based sclerotia are larger, measuring 6–20 mm in diameter, while those on leaves and flowers are smaller, measuring 0.4–1.6 mm. In severe cases, underground roots decay, leaving only the outer skin, with sclerotia found both inside and outside. Affected plants develop yellowish-brown leaves that gradually wilt and die. Early-stage infected plants appear nearly identical to healthy plants, making it challenging to identify and easy to miss the optimal time for treatment (Fig. 25.3).

Pathogen

The pathogen responsible for sclerotinia disease in Asarum is *Sclerotinia asari* (*S. asari* Wu et Wang), a fungus from the Sclerotinia genus in the Ascomycota class. On PDA medium, the mycelium grows along the substrate, forming a thin, nearly colorless to light-white colony that produces white sclerotia after 5–8 days, later darkening to black. Sclerotia germinate in spring, forming one to nine apothecia that produce ascospores, which facilitate further infection. *Sclerotinia asari* thrives at temperatures ranging from 0 to 27 °C, with an optimal range of 7–15 °C, making it a low-temperature fungus. Sclerotia can germinate at 2–23 °C, usually taking

Fig. 25.3 Sclerotinia Disease symptoms in Asarum

5–15 days to start germinating posttreatment. The pathogen primarily infects Asarum and has not been observed infecting other plants.

Occurrence Pattern

The pathogen overwinters as mycelium and sclerotia within plant debris, soil, and contaminated seedlings. Initial infection mainly arises from mycelium, which invades roots, stems, and leaves. Under natural conditions, sclerotia primarily produce apothecia upon germination. In northeastern regions, Asarum emerges in mid-to-late April; when soil temperatures reach 1–4 °C, sclerotia start to germinate. Apothecia emerge in early May and naturally decay by late May. Ascospores mostly enter through wounds rather than directly invading plant tissues. The disease onset occurs in early-to-mid May, peaks in late May, and gradually declines after mid-June. Sclerotinia is a low-temperature disease, with infection beginning as low as 2–4 °C and spreading most rapidly at 6–10 °C soil temperatures, ceasing above 15 °C. Conditions of low temperature, high humidity, poor drainage, and dense planting intensify the disease's severity.

Control Measures

1. Selection of Disease-Free Seedlings: Ensure thorough disinfection of seedlings by soaking them in an 800-fold dilution of 50% procymidone wettable powder for 4 h.
2. Enhanced Cultivation Management: In early spring, promptly drain fields before and after Asarum emergence to reduce soil moisture. Timely weeding and soil loosening help to raise soil temperature, which significantly reduces sclerotinia spread. In plots with sparse weeds, leaf cover, and good water retention, imple-

ment no-till cultivation to prevent soil-borne pathogen spread. Carefully inspect fields for infected plants before weeding, and prevent soil-borne pathogen spread by avoiding contaminated tools. Remove severely infected plants, disinfect the surrounding soil with lime, and apply fungicides like procymidone or carbendazim to eradicate soil-borne pathogens.

3. Chemical Control: Use fungicides as a drench at the early stage of disease. Recommended options include an 800-fold dilution of 50% procymidone WP, a 1,000–1,300-fold dilution of 50% iprodione WP, or a combination of 200-fold dilution of a carbendazim WP with 800-fold ammonium fungicide solution. Use 2–8 kg of diluted solution per square meter, ensuring thorough saturation of the cultivated soil layer. Apply every 7–10 days, repeating three to four times as necessary.

25.3 Phytophthora Blight in Asarum

Phytophthora blight is one of the primary diseases affecting Asarum production, with occurrences reported in all major cultivation areas in China, particularly severe in Liaoning and Jilin.

Symptoms
Phytophthora blight primarily infects the base of the petioles and leaves. Lesions on the petioles appear as elongated, dark green, water-soaked spots that are prone to soft rot. On leaves, lesions are larger, circular, dark green, and water-soaked. During high humidity and rainy seasons, a white mold consisting of the pathogen's mycelium and sporangia develops on the lesions. Under conditions of heavy rain and high humidity, the disease progresses rapidly, causing petioles to soften and collapse and leaves to rot and droop, often leading to the death of entire patches of Asarum plants (Fig. 25.4).

Pathogen
The pathogen responsible for Phytophthora blight in Asarum is *Phytophthora cactorum* (Lebert & Cohn) J. Schröt., a species in the Oomycota phylum. The pathogen's aerial mycelium is white, cottony, and non-septate. Sporangia are borne on slender sporangiophores, either terminally or laterally, and are ovoid with a nipple-like projection at the apex. Upon germination, sporangia release zoospores. The oospores are spherical, yellow-brown, and single-celled.

Occurrence Pattern
The pathogen overwinters as mycelium or oospores in plant debris or soil. In the following spring, when conditions are favorable, it infects the base of Asarum stems and other aerial parts. Zoospores produced at infection sites spread through wind and rain, leading to secondary infections and disease spread. Warm temperatures, heavy rainfall, and high humidity are conducive to disease outbreaks.

Fig. 25.4 Symptoms of Phytophthora blight in Asarum. (**a**) Early-stage symptoms (**b**) Late-stage symptoms

Control Measures

1. Agricultural Control: Select well-draining, sandy loam soils for Asarum cultivation. Promptly remove infected plants to eliminate sources of infection. Disinfect the soil at infection sites with quicklime or a 0.5–1% potassium permanganate dilution to reduce the pathogen population in the soil.
2. Chemical Control: Before the onset of the rainy season, apply a 1:1:200 Bordeaux mixture, or spray with a 300-fold dilution of 40% fosetyl-aluminum WP, a 600-fold dilution of 25% metalaxyl WP, or a 500–600-fold dilution of 58% metalaxyl–mancozeb WP (Ridomil-Mancozeb). These fungicides are applied preventively to protect against oomycete diseases during periods of high humidity and rainfall.

25.4 Rust Disease in Asarum

Rust disease is a significant ailment affecting Asarum production, found across all major cultivation areas in China. This disease was first identified in an Asarum garden in Huanren County, Liaoning Province, by Professor Fu Junfan from Shenyang Agricultural University in 1993, where some plants exhibited leaf dieback due to infection. Today, rust disease has become a common problem in prominent Asarum gardens.

Symptoms

Rust disease primarily infects the leaves, though it can also spread to flowers and fruits. Telia (winter spore clusters) form on both sides of the leaves and on the petioles, appearing circular or elliptical. Initially, they develop beneath the host's epidermis, forming raised, mound-like structures that later rupture, releasing a yellow-brown to chestnut-brown powder. These clusters can merge to form circular patterns on the leaf surface, more prominent on the upper side than the underside, with a diameter of 4–7 mm. On petioles, telia appear as elliptical or elongated lesions, reaching lengths of up to 7–50 mm, sometimes encircling the petiole and causing it to swell. In severe cases, the entire leaf may die (Fig. 25.5).

Pathogen

The pathogen responsible for Asarum rust disease is *Puccinia asarina* Kunze, a basidiomycete fungus in the genus Puccinia. Neither uredospores nor aeciospores have been observed. The teliospores are two-celled, elliptical, elongated, spindle-shaped, or irregular, measuring (30–51) μm × (16–25) μm, and range in color from yellow-brown to dark brown. The ends of the spores are either rounded or slightly tapered, with a slight constriction at the septum. The wall thickness is uniform, measuring 1.5–2 (up to 2.5) μm. Each cell has one germ pore with a transparent papilla. The germ pore is located terminally on the upper cell and near the septum on the lower cell. The spore stalk is colorless, slender, easily broken, and can extend up to 45 μm (Fig. 25.6).

Occurrence Pattern

The overwintering location and mechanism of the pathogen are currently unknown. In northeastern regions, the disease typically begins in early May, with peak incidence in July and August. Infected plants are often concentrated in shaded areas, such as under trees, where high humidity, frequent rain, and dew favor disease development. Teliospores spread via air currents and rain splash.

Fig. 25.5 Symptoms of Rust Disease in Asarum. (**a**) Symptoms on the upper side of the leaf. (**b**) Symptoms on the lower side of the leaf

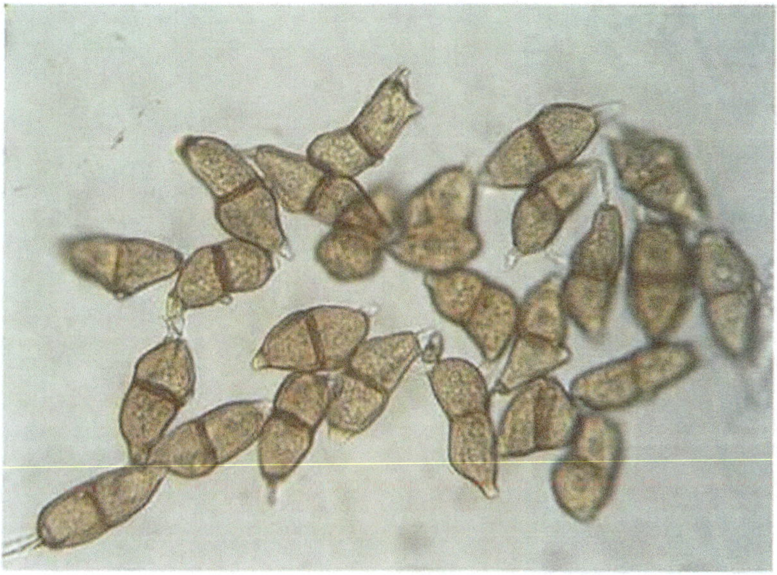

Fig. 25.6 Morphology of Teliospores of the rust pathogen in Asarum

Control Measures

1. Improved Cultivation Management: Promote robust plant growth to enhance resistance to disease. During the rainy season, promptly drain any excess water from the fields. In autumn, thoroughly remove and destroy diseased plant debris outside the field. Regularly remove heavily infected leaves to reduce the pathogen load within the field.
2. Chemical Control: At the onset of infection, spray a 1,000–1,500-fold dilution of 25% triadimefon WP, a 600-fold dilution of 62.25% cymoxanil-mancozeb WP, or a 300-fold dilution of 95% sodium selenite WP. Apply sprays every 7–10 days, repeating 2–3 times as needed.

Chapter 26
The Diseases of Aconitum

Aconitum, a member of the Ranunculaceae family, is an important medicinal plant genus in China. The Chinese Pharmacopoeia includes various Aconitum-based herbal medicines, such as Chuanwu, Zhichuanwu, Fuzi, Caowu, and Caowu leaves. Chuanwu and Fuzi refer to the dried mother and daughter roots of *Aconitum carmichaelii* Debeaux, respectively, while Caowu and Caowu leaves refer to the dried tuber and leaves of *A. kusnezoffii* Reichb. Aconitum species are perennial herbaceous plants used in traditional medicine, primarily distributed in Yunnan, Sichuan, Hubei, Guizhou, Hunan, Guangxi, Anhui, Shaanxi, and Henan, with major cultivation centered in Sichuan. *A. kusnezoffii*, commonly known as Caowu, is mainly found in Shanxi, Hebei, Inner Mongolia, Liaoning, Jilin, and Heilongjiang. These plants contain various alkaloids, including aconitine, mesaconitine, and hypaconitine, known for their efficacy in dispelling wind, eliminating dampness, warming the channels, and relieving pain.

The primary diseases affecting medicinal Aconitum plants include damping-off, root rot, leaf spot, downy mildew, blight, southern blight, anthracnose, sclerotinia rot, and viral diseases. Among these, root rot, southern blight, and downy mildew have the greatest impact on yield, while viral diseases are also widespread.

26.1 Root Rot in *A. kusnezoffii*

Root rot is the most common soil-borne fungal disease affecting *A. kusnezoffii*, with widespread occurrence and severe impacts due to complex causes. It primarily affects the tubers and daughter roots, causing decay in the underground parts, leading to wilting and death of the entire plant. The general incidence rate is 5–10%, but in severe cases, it can exceed 80%, making it a destructive disease.

© China Agriculture Press Co., Ltd. 2026

T. Wang et al., *Atlas of Chinese Medicinal Plant Diseases*,
https://doi.org/10.1007/978-981-95-6326-5_26

Symptoms
Root rot primarily affects the tubers and daughter roots. Initial symptoms appear in the aerial parts of the plant, with wilting indicating water deficiency. As the disease progresses, the lower leaves begin to die, and the symptoms gradually spread upward, leading to the death of the entire plant. The primary and lateral roots, as well as the daughter roots, can become infected, often starting from the mother root, with brown or dark brown patches of varying sizes that spread outward. Early decay of the mother root prevents the formation of daughter roots; late infection results in undersized or decayed daughter roots. The disease gradually spreads to surrounding plants, forming visible infection centers in the field. Disease severity increases in hot, rainy seasons (Fig. 26.1).

Pathogen
The pathogens causing root rot in *A. kusnezoffii* are *Fusarium solani* (Mart.) Appel et Wollenw. ex Snyder et Hansen and *F. equiseti* (Corda) Sacc., both from the genus Fusarium. On PDA medium, *F. solani* colonies initially appear yellowish-white, with sparse aerial mycelium, dense hyphae, and regular edges, remaining yellowish--white as they mature. Macroconidia are sickle-shaped, measuring (14.79–25.83) μm × (2.71–5.82) μm, while microconidia are round, single-celled, measuring (4.52–9.95) μm × (1.94–3.51) μm. *F. equiseti* colonies initially appear white on

Fig. 26.1 Symptoms of root rot in *A. kusnezoffii*. (**a**) Field symptoms. (**b–d**) Root symptoms

PDA medium, with abundant, cottony aerial mycelium that is dense and irregular-edged. Macroconidia are sickle-shaped with two to five septa, measuring (34.16–43.14) μm × (1.81–4.75) μm; microconidia are round and single-celled (Fig. 26.2).

Occurrence Pattern

The pathogen overwinters as mycelium and conidia in soil or plant debris, serving as the primary infection source in the following year. The pathogen enters the plant through wounds on the root or rhizome, spreading via rainwater or irrigation. Low-lying areas, poor drainage, waterlogged fields, continuous cropping, root injury from pests or agricultural practices, and dense planting increase disease severity. In northeastern regions, symptoms usually appear in late June, with peak incidence in July and August during rainy periods. High temperature, high humidity, and rainy years exacerbate the disease. Older plants and fields with dense planting show higher infection rates due to accumulated inoculum.

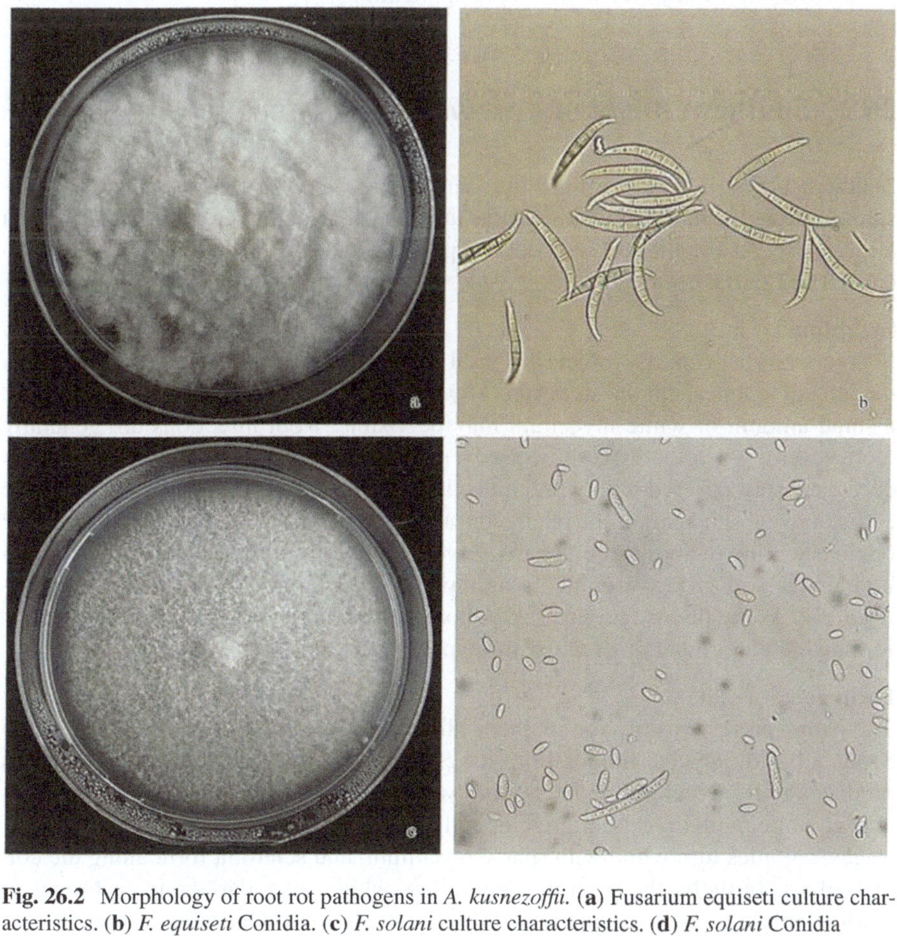

Fig. 26.2 Morphology of root rot pathogens in *A. kusnezoffii*. (**a**) Fusarium equiseti culture characteristics. (**b**) *F. equiseti* Conidia. (**c**) *F. solani* culture characteristics. (**d**) *F. solani* Conidia

Control Measures

1. Scientific Site Selection: Choose well-drained sites with slightly sandy soil and high organic matter for transplanting. Avoid continuous cropping in the same field; instead, rotate with crops from different plant families.
2. Enhanced Field Management to Improve Resistance: Base fertilization primarily on organic compost, supplemented with moderate amounts of phosphorus and potassium. Ensure timely drainage in the field to reduce soil moisture. Employ appropriate planting density to maintain good ventilation and light penetration in the field.
3. Sanitation: Remove and destroy infected plants immediately upon detection to prevent spread, disinfect affected areas, and regularly clear plant debris from the field. In autumn, thoroughly remove above-ground stems, leaves, and other debris from the field and dispose of them to maintain cleanliness.
4. Chemical Control: Apply a 500-fold dilution of 50% carbendazim WP or a 500–1,000-fold dilution of 50% thiophanate-methyl WP as a root drench, repeating every 7–10 days for two to three applications.

26.2 Southern Blight of *Aconitum kusnezoffii*

Southern blight is a severe root and stem disease affecting *A. kusnezoffii*, widespread across the main cultivation areas in China. The pathogen's resilient sclerotia allow it to persist in the soil for extended periods, making the disease challenging to control and quick to spread.

Symptoms

Southern blight primarily infects the basal stems and roots of *A. kusnezoffii*. Initially, the infection appears at the stem base as water-soaked brown lesions. In mild cases, a small amount of white mycelium may be visible. Root infections are generally more severe; when the disease is discovered, the roots typically exhibit significant decay, densely covered with white mycelium. When the disease develops early, the entire aboveground part of the plant wilts and dies. In later stages, symptoms include obvious wilting of aboveground parts. Severe infections produce dense white mycelium on the soil surface around diseased plants, which rapidly spreads to neighboring areas. As the disease progresses, yellow, yellow-brown, or dark brown sclerotia form, which are round or irregular in shape (Fig. 26.3).

Pathogen

The pathogen responsible for southern blight in *A. kusnezoffii* is *Sclerotium rolfsii* Sacc., a basidiomycete fungus from the Sclerotium genus. In its asexual phase, it produces only mycelium and sclerotia, without spores. On PDA media, colonies appear white, with mycelium growing radially outward and a sparse center. In later stages, colonies turn white with sparse mycelium, and sclerotia form along the colony edges. Initially, the sclerotia are yellow-white, later darkening to brown or reddish-brown and resembling mustard seeds in size and shape (Fig. 26.4).

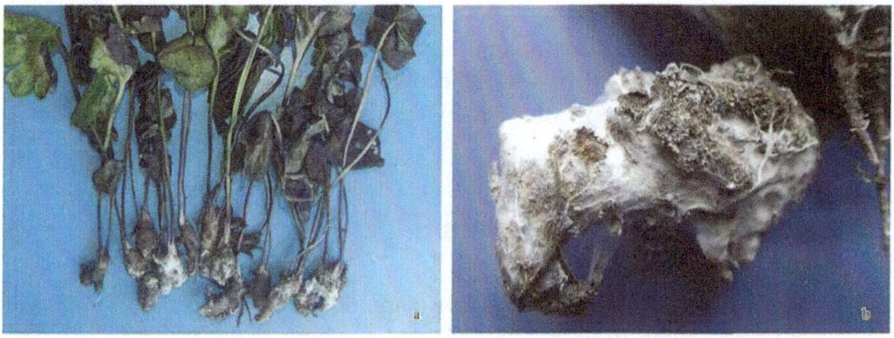

Fig. 26.3 Southern Blight symptoms of *Aconitum kusnezoffii*. (**a**) Whole-plant symptoms. (**b**) Root symptoms

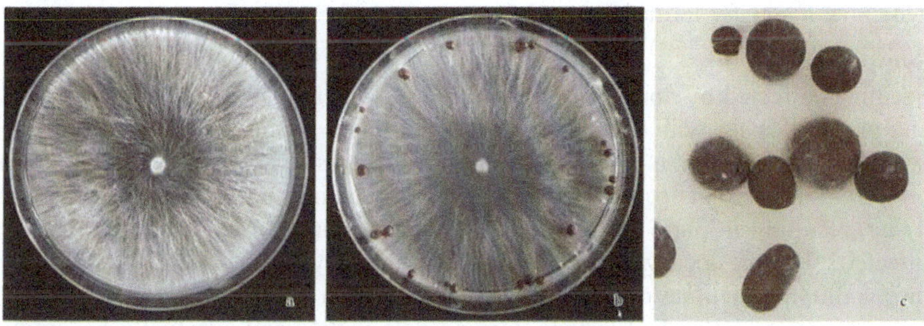

Fig. 26.4 Pathogenic characteristics of *Sclerotium rolfs*ii. (**a**) Five-day culture. (**b**) Fourteen-day culture. (**c**) Sclerotia

Occurrence Pattern

The pathogen overwinters as mycelium and sclerotia in plant debris and soil. Under favorable conditions in the following year, overwintered mycelium and sclerotia germinate, producing mycelium that invades plants directly through the basal stem, roots, or wounds. Mycelium spreads through contact and further infects the area. Infected seedlings can facilitate long-distance transmission. The disease thrives in high temperatures, high humidity, and rainy conditions, with greater incidence in densely planted, poorly drained, and wound-prone areas.

Control Measures

1. Enhanced Cultivation Management

   Avoid overcrowding, reduce plant wounding, and apply organic fertilizers to strengthen plants and improve resistance. Drain excess water promptly to reduce field moisture levels. Diseased plants should be promptly removed, and lime should be applied to disinfect the surrounding soil.
2. Crop Rotation

   Southern blight is a typical soil-borne disease, and its pathogen can survive in the soil for extended periods. Implementing a 3–5-year rotation with cereal crops can significantly reduce disease incidence.

3. Chemical Control

   At the onset of the disease, apply a 1,500-fold dilution of 40% fludioxonil WP, a 3,000-fold dilution of 43% tebuconazole SC, or a 1,500-fold dilution of 50% iprodione WP. Spray every 7 to 10 days, repeating 2–3 times.

26.3 Downy Mildew in Aconitum

Downy mildew in Aconitum is particularly common during the seedling stage, with incidence rates ranging from 3% to 20%, and reaching as high as 30% in severe cases. Known for its high infection rates and strong contagiousness, heavily infected seedlings eventually turn brown and die, leading to plant gaps that severely impact the yield of Fuzi, an essential medicinal part of Aconitum, and cause significant economic losses for farmers.

Symptoms

Seedlings affected by downy mildew show gray-green, curled, narrow, thickened, and upright leaves, with a grayish-purple mold layer forming on the underside—commonly referred to as "gray seedlings." As the disease progresses, leaves turn gray from the bottom up, with severely affected seedlings ultimately browning and dying, resulting in plant gaps. In mature plants, young leaves at the top may be infected, showing localized chlorosis and turning a yellow-white color, with the discoloration sometimes extending over 20 cm, known as "white tips." Initially, leaf spots appear as oil-soaked lesions, turning pale yellow and eventually becoming purplish-red. These spots are confined by leaf veins, causing leaves to twist, with the midveins turning brown. The underside of the leaves also develops a purplish--brown downy layer (Fig. 26.5).

Pathogen

The pathogen responsible for Aconitum downy mildew is *Peronospora aconiti* Yu, a member of the genus Peronospora in the phylum Oomycota. The pathogen's mycelium forms a dense, gray or grayish-purple layer on the underside of leaves. Sporangia emerge from the stomata, either singly or in clusters (1–5), colorless, with the main axis occupying 35–75% of the total length. The basal part often swells, and the tips branch three to six times, typically at acute angles, appearing slightly curved or straight. Sporangia are elliptical, lemon-shaped, oval, or nearly spherical, ranging from colorless to light brown, with some showing nipple-like projections. Oogonia form within infected leaf tissue and are subglobose, oval, or irregular with smooth walls. Oospores are spherical, light yellow-brown, smooth, and thick-walled (Fig. 26.6).

Occurrence Pattern

Infested soil, contaminated seed roots, and residual infected plant tissues containing oospores or mycelium are typical sources of initial infection. Symptoms usually appear in early March, as the infection begins in seedlings. When the average

Fig. 26.5 Symptoms of Downy Mildew in Aconitum. (**a**) Field symptoms. (**b–f**) Leaf symptoms

temperature reaches 16 °C and relative humidity is around 77% in April, sporangia spread via wind and rain, resulting in secondary infections in the field. When temperatures exceed 19 °C and humidity levels reach over 80% in early to mid-May, secondary infections peak. As temperatures rise above 24 °C, leaf aging reduces the likelihood of pathogen germination and infection. Consequently, downy mildew develops most easily in the cool, humid, and alternating wet-dry conditions of April and May, or during cool, humid, and rainy seasons, leading to rapid and severe disease progression.

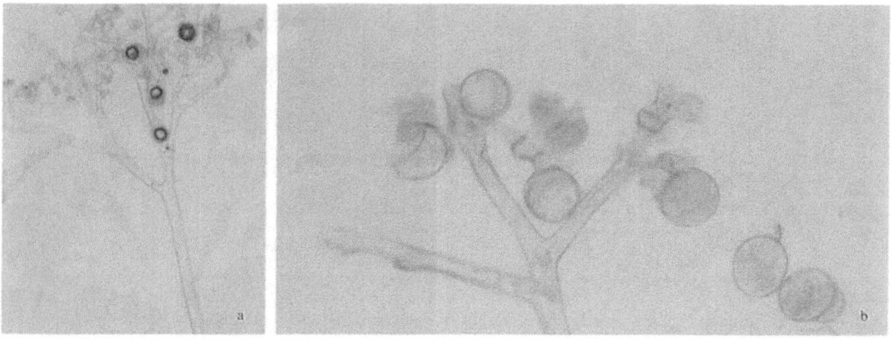

Fig. 26.6 Morphology of Downy Mildew pathogen in Aconitum. (**a**) Sporangiophores. (**b**) Oospores

Control Measures

1. Enhanced Cultivation Management: Employ appropriate planting density and ensure timely drainage to control humidity. Remove plant debris and diseased leaves promptly, disposing of them away from the cultivation area to reduce sources of infection.
2. Chemical Control: At the onset of infection, apply a foliar spray of a 1,000-fold dilution of 66.8% propineb-cymoxanil WP, a 1,000-fold dilution of 80% dimethomorph WG, a 1,000-fold dilution of 72% metalaxyl-mancozeb WP, or an 800-fold dilution of 72.2% mefenoxam AS.

26.4 Damping-Off Disease of *Aconitum kusnezoffii*

Damping-off disease is a significant fungal threat to *Aconitum kusnezoffii* seedlings, prevalent across the main cultivation areas in Northeast China. Infected fields exhibit clear centers of infection, with the disease spreading rapidly and often causing large patches of seedling death. The typical infection rate ranges from 10% to 15%, with severely affected areas seeing rates over 40%, thus greatly impacting the availability of healthy, high-quality seedlings.

Symptoms

Damping-off disease primarily infects the basal stem area of seedlings. Early infection is indicated by individual seedlings showing dark brown, irregular lesions at the stem base. As the disease progresses, the affected area becomes sunken and constricted, expanding and leading to the death of the basal stem. The aboveground parts of the plant then wilt, with leaves turning yellow, and the entire plant either remains standing or collapses and withers. Infection centers in the field are often highly visible, with the disease rapidly spreading outward, causing clusters of seedling death and resulting in noticeable gaps in the rows. Under high humidity conditions, white filamentous mycelium can be observed on the basal stems. Infected plants, when uprooted, reveal significantly reduced fibrous roots compared to healthy plants; in severe cases, both the main and fibrous roots are extensively decayed (Fig. 26.7).

Pathogen

The pathogen causing damping-off disease is *Rhizoctonia solani* Kühn, an asexual fungus of the genus Rhizoctonia within the Basidiomycota division. On PDA medium, colonies are initially white or pale yellow, gradually turning yellow as radiating mycelial growth occurs. Mycelial aggregations in the center produce yellow-brown sclerotia. The mycelium is thick, with diameters ranging from 6.9 to 10.4 μm, multinucleated, and branches at nearly right angles, showing slight constrictions at the base and visible septa. In older mycelia, constrictions become more pronounced, with some cells swelling into bead-like shapes (Fig. 26.8).

Occurrence Pattern

The pathogen overwinters as mycelium and sclerotia in plant debris or soil, serving as the primary infection source in the following year. It can survive in soil for 2–3 years or longer. The pathogen penetrates seedlings through the basal stem or root wounds, and can also directly invade host epidermal tissue, spreading through rain, irrigation, and farming activities. Disease incidence is highest when soil

Fig. 26.7 Damping-Off Disease symptoms in *Aconitum kusnezoffii*. (**a**) Early stage. (**b**) Advanced stage. (**c**) Infection center. (**d**) Typical symptoms

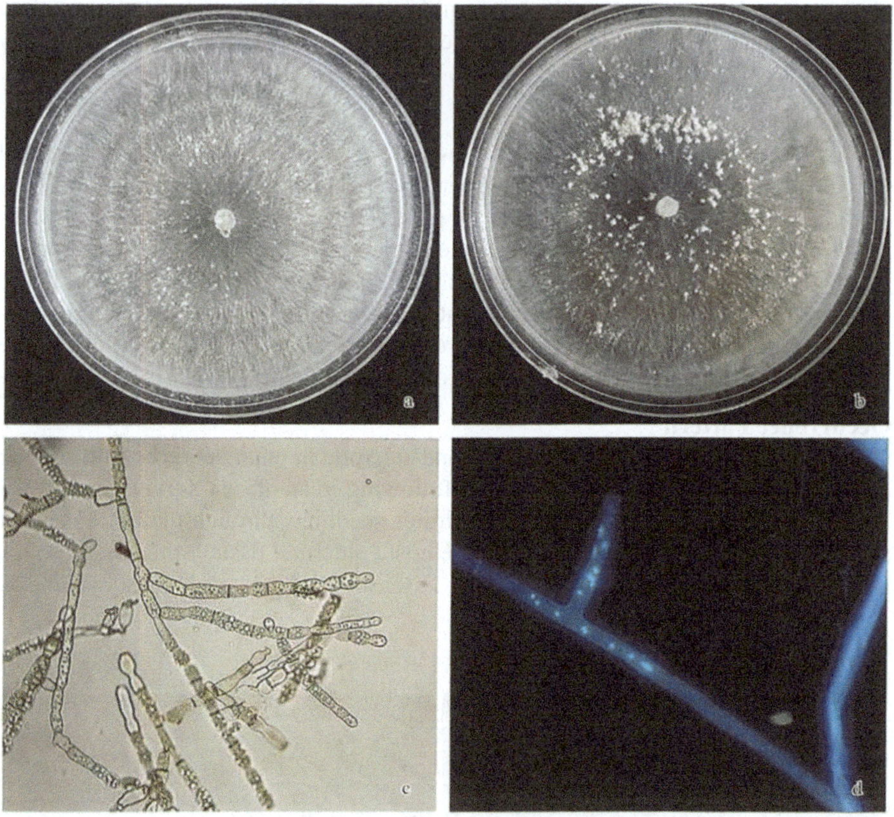

Fig. 26.8 Pathogen characteristics of *Rhizoctonia solani*. (**a**) Four-day culture. (**b**) Fourteen-day culture. (**c**) Mycelium. (**d**) Nuclear staining

temperatures exceed 15 °C, with favorable humidity levels. Low-lying, compacted, and poorly drained soils contribute to disease onset and spread.

Control Measures

1. Scientific Site Selection

   Choose well-drained, fertile, aerated soils, ideally sandy loam, for seedbeds. Raised beds help prevent waterlogging, with proper drainage needed during the rainy season.

2. Seed Treatment with Fungicides

   Before sowing, treat seeds with 0.1–0.3% of seed weight using a 50% WP of carbendazim, thiram, or procymidone.

3. Soil Disinfection

   Before bed preparation, disinfect soil by mixing 10–15 g/m^2 of a 50% WP of carbendazim, 75% chlorothalonil, 50% thiram, or 50% procymidone into the top 5 cm of soil.

4. Enhanced Seedling Management

   Regularly loosen the soil after seedlings emerge to increase soil temperature and promote aeration. Use clean rice straw or pine needles as mulch, avoiding excessively thick layers. Ensure any organic fertilizers are well-decomposed to prevent root rot or stem damage.
5. Chemical Control

   At the initial onset of symptoms, apply a 50% WP of carbendazim, 200–300-fold dilution of 30% copper sulfate SC, or 800–1,000-fold dilution of 30% metalaxyl AS to the bed surface, ensuring the solution penetrates 3–5 cm into the soil. Increase application around infection centers.

26.5 Leaf Spot Disease in *A. kusnezoffii*

Leaf spot disease in *A. kusnezoffii* has recently become widespread in Liaoning Province, affecting large areas and causing significant damage to the leaves. Both young and mature leaves are susceptible, with an average infection rate of 10–30%, reaching up to 70% in severe cases, which significantly impairs the normal growth of *A. kusnezoffii*.

Symptoms

Leaf spot disease primarily infects the leaves of *A. kusnezoffii*. During the seedling stage, brown circular or elliptical spots appear on the leaves. In mature plants, the disease initially manifests as small black spots on the leaves, which expand along the leaf veins into circular, elliptical, or elongated spindle-shaped lesions as the infection progresses. The spots are often confined by the leaf veins, appearing gray-brown with black edges and surrounded by yellow halos. The lesions necrotize from the center outward, eventually leading to perforations, causing the entire leaf to decay and wilt. Severe infections result in the dieback of the entire above-ground portion of the plant (Fig. 26.9).

Pathogen

The pathogen responsible for leaf spot disease in *A. kusnezoffii* is a fungus of the genus Phoma, an ascomycete. On PDA medium, it forms circular white colonies with irregular edges, where the central mycelium is denser than that at the edges. Pycnidia initially develop beneath the epidermis and, upon maturation, emerge as small blackish-brown spots that are nearly spherical or flattened-spherical. The conidia are single-celled, colorless, spindle-shaped, or elliptical, containing an internal oil droplet, and measure (15.5–21.1) μm × (6.3–7.2) μm (Fig. 26.10).

Occurrence Pattern

The pathogen overwinters as mycelium and pycnidia in soil or plant debris, acting as the primary infection source for the following year. During the growing season, when temperature and humidity are optimal, conidia are spread by wind and rain, leading to secondary infections that intensify the damage. The spread of leaf spot

Fig. 26.9 Symptoms of Leaf Spot Disease in *A. kusnezoffii*. (**a**) Leaf symptoms in seedling stage. (**b**) Leaf symptoms in mature plants

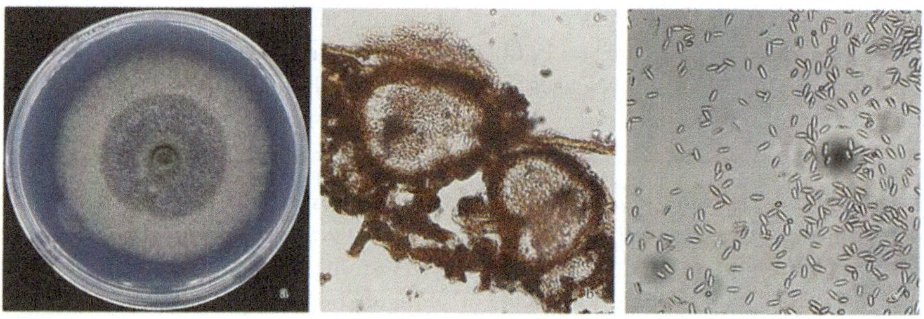

Fig. 26.10 Morphology of *A. kusnezoffii* leaf spot pathogen. (**a**) Culture characteristics. (**b**) Pycnidia. (**c**) Conidia

disease is more likely in conditions with high rainfall, frequent rain events, and suitable temperatures. Dense plantings with poor ventilation and light penetration also promote infection. In Liaoning Province, the disease typically appears in late June and persists throughout the growing season, with peak incidence in July and August.

Control Measures

1. Enhanced Field Management: Maintain field sanitation by removing plant debris in late winter or early spring for disposal or deep burial, thus reducing the initial infection source for the following year. Ensure proper ventilation and light exposure, apply fertilizer as needed, maintain balanced irrigation, and promptly drain excess water after rainfall. Use fully decomposed organic fertilizer and increase phosphorus and potassium applications, while avoiding excessive nitrogen to strengthen plant resistance.
2. Chemical Control: At the early stages of infection, apply fungicides such as a 600-fold dilution of 50% mancozeb WP, an 800–1,000-fold dilution of 70%

thiophanate-methyl WP, a 1,500-fold dilution of 50% iprodione WP, or a 1,000-fold dilution of 50% benomyl WP. Spray at intervals of 7–10 days, repeating two to three times as needed.

26.6 Leaf Blight Disease of *Aconitum kusnezoffii*

Leaf blight disease is a significant foliar disease affecting *A. kusnezoffii*, common across major cultivation areas in Liaoning Province, with an average infection rate of 20% to 30%. In severe cases, the infection rate can exceed 90%, which greatly impacts the growth and development of *A. kusnezoffii*.

Symptoms
Leaf blight disease primarily affects the leaves, producing lesions on both surfaces. In the early stages, numerous scattered, small, round or nearly round, dark brown spots appear on both sides of the leaves. As the disease progresses, these spots enlarge into distinct, dark brown lesions with irregular or polygonal shapes, often confined by the leaf veins. The lesion centers become grayish-white, with some merging into larger patches. Lesion sizes average around 3.4 mm × 6.2 mm, bordered by brown or dark brown edges and surrounded by a faint yellow halo. Small black dots—pycnidia of the pathogen—are visible on the lesions. Infection at the leaf tips and edges may cause leaf curling. In advanced stages, lesions become dense and coalesce, leading to leaf necrosis and premature defoliation (Fig. 26.11).

Pathogen
The pathogen responsible for leaf blight in *Aconitum kusnezoffii* belongs to the genus Septoria, an asexual ascomycete fungus. On PDA medium, colonies appear white or grayish-white on the surface and brown to reddish-brown on the reverse. Optimal growth occurs at 20–25 °C, although conidia production is infrequent. Pycnidia are embedded or partially embedded in the leaf tissue, typically scattered or in sparse clusters. They are spherical to nearly spherical and dark brown, with sizes ranging from (86.3–89.9) μm × (126.8–146.4) μm and wall thickness of 4.2–13.4 μm. Conidia are needle-shaped, blunt at the base, colorless, and transparent, with straight or slightly curved forms and pointed tips. They are septate and measure (23.8–59.2) μm × (1.3–2.6) μm (Fig. 26.12).

Occurrence Pattern
The pathogen overwinters as mycelium in infected plant debris. In the following spring, conidia are dispersed by air currents, initiating primary infections. Conidia continue to form throughout the growing season, spreading via wind and rain, thus causing secondary infections and expanding disease spread. Heavy nitrogen fertilization, dense planting, and wet, rainy conditions favor the disease's development. In Northeast China, initial symptoms generally appear in late June, with peak infection from July to August.

Fig. 26.11 Leaf Blight Disease symptoms of *Aconitum kusnezoffii*: (**a**) Early-stage symptoms; (**b**) Advanced-stage symptoms

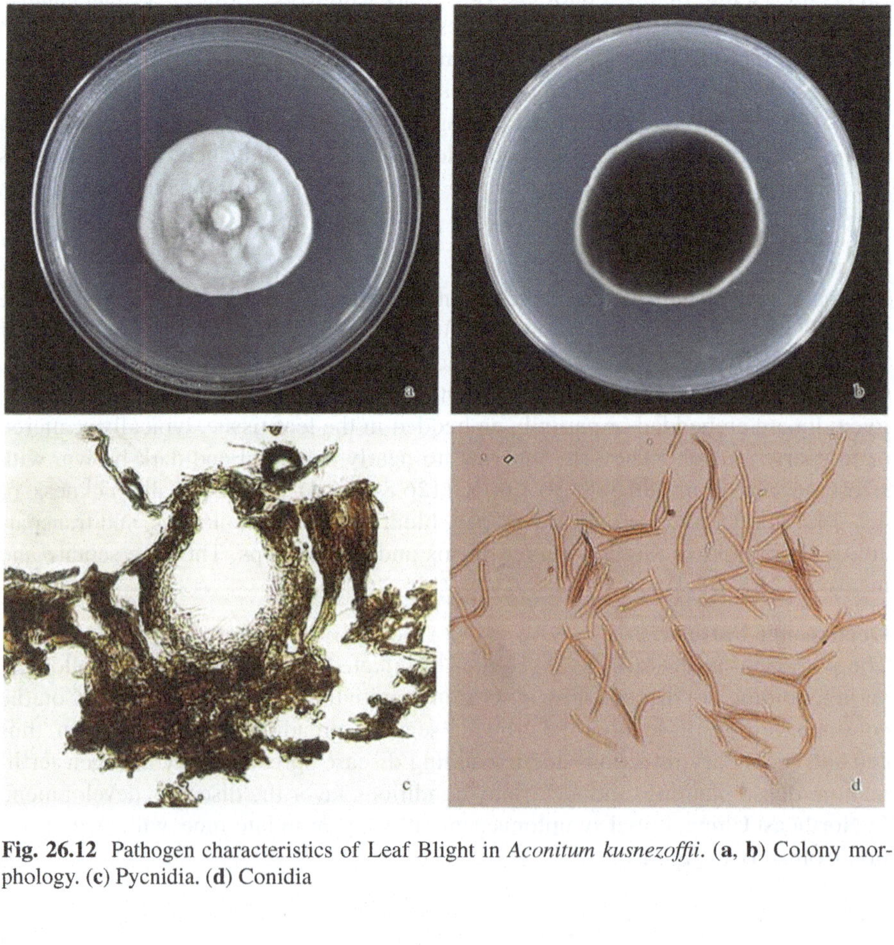

Fig. 26.12 Pathogen characteristics of Leaf Blight in *Aconitum kusnezoffii*. (**a, b**) Colony morphology. (**c**) Pycnidia. (**d**) Conidia

Control Measures

1. Field Management

   Since *Aconitum kusnezoffii* is a perennial plant, thorough field sanitation should be conducted before plants emerge each year. Applying soil disinfectants at appropriate concentrations helps reduce the quantity of overwintering pathogens in the soil and plant debris. During rainy periods, drainage channels should be created to reduce field humidity. Vigorous seedling cultivation, controlled sowing density, and regular weeding support healthy plant growth and enhance disease resistance. After harvest, all infected plants and leaves should be promptly removed from the field and destroyed to prevent reinfection.
2. Chemical Control

   At the onset of symptoms, apply chemical treatments. Using a 500–800-fold dilution of a 50% mancozeb WP or a 45% ammonium metiram AS. Spray at 7-day intervals for a total of three applications.

26.7 Soft Rot Disease in Aconitum

Soft rot disease is a newly identified bacterial infection affecting Aconitum, particularly prevalent in Mianyang, Sichuan Province. The disease incidence is around 20% and can reach up to 80% in severe cases. As a systemic disease, soft rot causes extensive plant decay, leading to wilting, complete plant death, and significant reductions in Aconitum yields.

Symptoms

Soft rot affects the roots, stems, and leaves of Aconitum. Early symptoms appear as water-soaked lesions on the leaves, gradually spreading to cover the entire leaf. As a systemic disease, it causes vascular decay, resulting in the softening and rotting of the Aconitum stem, which emits a foul odor. In severe cases, the stem breaks, causing the entire plant to collapse, displaying a water-soaked appearance before dying (Fig. 26.13).

Pathogen

The pathogen responsible for soft rot in Aconitum is *Pectobacterium brasiliense*, a bacterium from the genus Pectobacterium within the family Pectobacteriaceae, class Gammaproteobacteria. On NA medium, the pathogen forms milky-white, round colonies with raised centers and irregular edges. It is Gram-negative and non-spore-forming. Under electron microscopy, the bacterium appears as a short rod with peritrichous flagella, measuring approximately (0.5–1.0) μm × (1.0–3.0) μm. The optimal growth temperature for the pathogen is 27–30 °C, with an ideal pH of 7.1 (Fig. 26.14).

Occurrence Pattern

The pathogen overwinters in soil and plant debris, acting as the primary infection source for the following year. Under favorable environmental conditions, it infects

Fig. 26.13 Symptoms of Soft Rot Disease in Aconitum. (a, b) Plant symptoms. (c) Stem symptoms

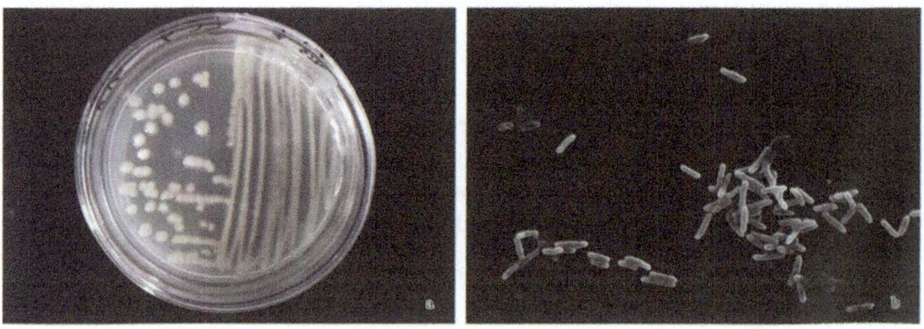

Fig. 26.14 Morphology of Soft Rot pathogen in Aconitum. (a) Culture characteristics. (b) Bacterial cell morphology

Aconitum through wounds or natural openings. Rainwater and insects are the main vectors for disease spread, with high temperature and humidity accelerating onset. Infection generally begins in early June, peaking in mid to late July as spread intensifies. Exposure to sunlight and dry conditions can inhibit disease progression.

Control Measures

1. Enhanced Field Management: Maintain field sanitation by promptly removing diseased plant debris to reduce initial infection sources. Ensure adequate ventilation and light exposure, balanced irrigation, and promptly drain excess water after rainfall to lower field humidity.

2. Selection of Healthy Seedlings: Use only healthy seedlings, disinfecting them before planting.
3. Chemical Control: During the growing season, antibiotics and copper-based treatments can be applied for disease control. Recommended treatments include spraying with a 0.3% kasugamycin AS or a 3% thiodiazole-copper WP.

26.8 Virus Disease of Aconitum

Also known as Aconitum mosaic disease, this virus disease occurs widely across Aconitum cultivation areas. In severely affected years, infection rates in fields can reach 70–80% in the Jiangyou region of Sichuan.

Symptoms
Infected plants display ring spots and mosaic patterns. In the early stages, leaves exhibit patches of dark green, light green, or pale yellow. In severe cases, leaves become distorted, wrinkled, curled, and ultimately wither. Affected plants experience stunted growth, with tubers that are malformed, undersized, and of poor quality (Fig. 26.15).

Pathogen
The causative virus of Aconitum virus disease is the Cucumber Mosaic Virus (CMV), classified in the family Bromoviridae and genus Cucumovirus. CMV is a tripartite, positive-sense, single-stranded RNA virus, consisting of three genomic RNAs: RNA1, RNA2, and RNA3, known for its high variability and adaptability. RNA1 encodes the 1a protein, while RNA2 encodes the 2a protein—both involved in viral replication. The 2b protein, encoded by the subgenomic RNA4A, is a multifunctional protein that facilitates long-distance movement of the virus and relies on RNA2. RNA3 encodes the 3a movement protein (MP), essential for viral mobility, and the coat protein (CP), which determines virion movement and aphid transmissibility (Fig. 26.16).

Occurrence Pattern
CMV is transmitted through various pathways, predominantly by nonpersistent aphid vectors, including cotton aphids and green peach aphids. It can also spread via sap contact, mechanical transmission, and seed transmission, with seed transmission rates typically between 4% and 8%. Symptoms generally appear when seedlings reach the four- to five-leaf stage, with peak infection occurring from May to June. Aphids are the primary agents of secondary transmission; higher aphid populations significantly elevate infection rates.

Control Measures
1. Selection of Disease-Resistant Varieties
   Use disease-tolerant varieties where available.

Fig. 26.15 Aconitum virus disease symptoms. (a) Field symptoms. (b–e) Leaf symptoms

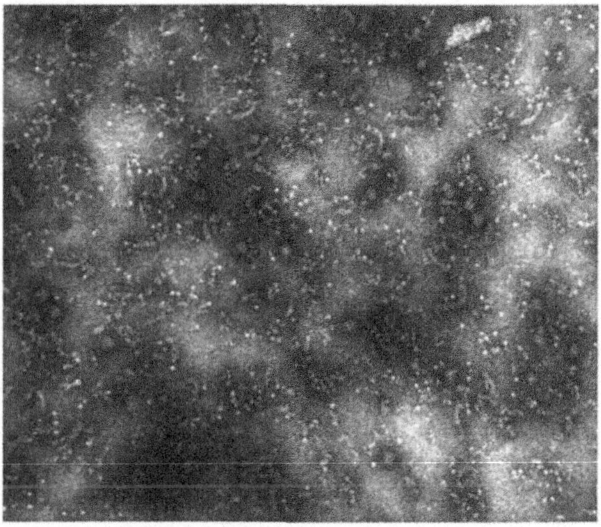

Fig. 26.16 Cucumber Mosaic virus structure

2. Enhanced Cultivation Practices

   Cultivate robust seedlings and plant them at the optimal time, typically right after the region's last frost. In protected environments, planting may be done slightly earlier. Use insect-proof nets to prevent aphid transmission.

3. Balanced Fertilization

   Apply organic active fertilizers or seaweed-based fertilizers, employing a balanced fertilization approach to support overall plant health.

Chapter 27
The Diseases of Coptis (Chinese Goldthread)

The primary medicinal sources of Coptis in traditional Chinese medicine are the plants *Coptis chinensis* (Chinese Goldthread), *Coptis deltoidea* (Triangle-leaf Goldthread), and *Coptis teeta* (Yunnan Goldthread) from the Ranunculaceae family. Known commercially as Wei Lian, Ya Lian, and Yun Lian, these medicinal products are derived from the dried rhizomes. Coptis is primarily cultivated in eastern Sichuan, western Hubei, as well as in Shaanxi, Hunan, Guizhou, and Gansu provinces. It has properties for clearing heat, drying dampness, and detoxifying, and is used in traditional treatments for digestive ailments, infections, inflammation, and skin conditions. Root diseases, particularly root rot, significantly impact yield and quality, posing a major threat to the Coptis industry.

27.1 Root Rot Disease in Coptis

Root diseases pose a significant threat to Coptis cultivation and limit the potential for large-scale production. The incidence of root rot disease is about 40% annually, reaching 80–90% in severely affected fields, which can lead to complete crop loss. This disease drastically affects both the yield and quality of Coptis as a medicinal herb, causing substantial economic losses.

Symptoms
In the early stages of infection, fibrous roots, storage roots, and rhizomes develop blackish-brown, dry rot symptoms, which eventually darken until the roots die. Leaf symptoms begin as irregular purple-red lesions at the leaf tips and edges, with the leaf undersides turning from yellow-green to purple-red. The leaves progressively dry out, yellow, wilt, and finally turn dark purple-red and necrotic (Fig. 27.1).

Pathogen
The root rot disease in Coptis is primarily caused by fungi of the *Fusarium* genus, which may infect plants individually or in combination. The main pathogenic

© China Agriculture Press Co., Ltd. 2026

T. Wang et al., *Atlas of Chinese Medicinal Plant Diseases*,
https://doi.org/10.1007/978-981-95-6326-5_27

Fig. 27.1 Symptoms of root rot in Coptis

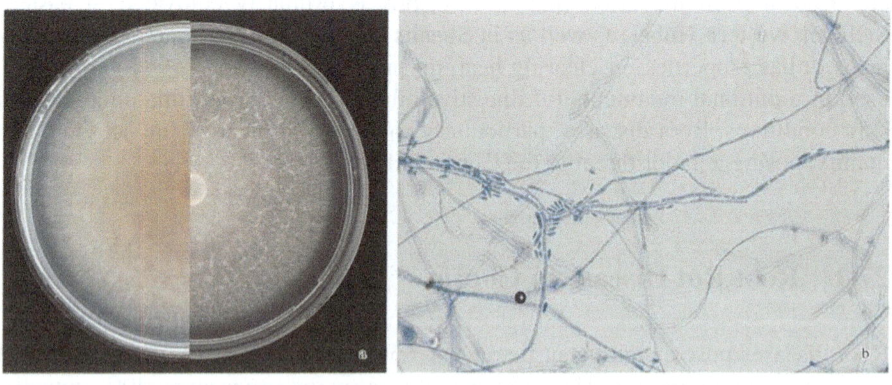

Fig. 27.2 Morphology of *Fusarium oxysporum*. (**a**) Culture characteristics. (**b**) Mycelium

species include *Fusarium oxysporum* (Schltdl. et Snyder et Hansen), *F. solani* [(Mart.) Appel et Wollenw. ex Snyder et Hansen], *F. tricinctum* [(Corda) Sacc.], and *F. carminascens* (Lombard, Crous et Lampr.). Among these, *F. oxysporum* is the dominant pathogen isolated, making it the principal causative agent (Fig. 27.2). *F. oxysporum* forms colonies with a fluffy texture and dense mycelium, initially white and later turning a pale purple-red. Microconidia are ovoid or kidney-shaped, measuring (6–23) μm × (3–4) μm, while macroconidia are sickle-shaped, slightly curved or straight, measuring (16–40) μm × (3–6) μm. Chlamydospores are abundant, terminal or intercalary, with diameters of 6–14 μm. *F. solani* colonies appear as thin gray wool with sparse aerial mycelium that does not alter the color of the medium.

The hyphae are septate, with kidney-shaped microconidia containing zero to one septa, measuring (6.0–8.0) μm × (2.5–4.0) μm. Macroconidia are nearly sickle--shaped, with blunt ends and slightly curved tips, typically containing two to three septa, measuring (10.0–20.0) μm × (2.5–4.5) μm. F. tricinctum colonies exhibit abundant aerial mycelium, with the medium base turning deep purple-red. The colony center is yellow-brown, fading outward to pink and then white. Microconidia are long-elliptical or obovate, measuring (6–13) μm × (3–7) μm, with zero to one septa. Macroconidia are slender and sickle-shaped, with pointed ends and indistinct basal cells, containing three to five septa and measuring (17–39) μm × (3–7.5) μm. *F. carminascens* colonies grow rapidly on PDA medium under light at 28 °C for 7 days, reaching up to 8 cm in diameter. Aerial mycelium is white and thinly woolly, while the colony underside turns purple-red as it matures. Macroconidia are sickle--shaped, with three to five septa, measuring (23.0–41.3) μm × (3.1–5.4) μm. Microconidia vary in shape, appearing kidney-shaped or ovoid with zero to one septa, measuring (6.21–17.9) μm × (2.7–4.3) μm (Fig. 27.2).

Occurrence Pattern

Root rot disease typically emerges in the second or third year after planting. Symptoms appear from April to May, peaking in July and August, and gradually decreasing after September. Infected plants show blackish-brown, dry rot symptoms in the fibrous roots, while affected leaves wilt and turn dark purple-red. In the fourth year, the disease may cause severe deterioration, with large sections of the field affected.

Control Measures

1. Scientific Site Selection: Choose deep, fertile soil with good drainage and rich organic content for cultivating Coptis. The ideal location is forested or partially shaded areas at elevations between 1200–1800 meters, with slopes of 15–25° and minimal direct sunlight.
2. Soil Disinfection: Before planting, use soil solarization to raise soil temperatures for disinfection or apply a 96% thiram solution diluted to 800 times. Applying lime in infected areas can effectively prevent the rapid spread of the disease.
3. Enhanced Cultivation Management: Avoid continuous planting of Coptis on the same land. Implement crop rotation with unrelated plant species to reduce pathogen buildup. Plowing exposes plant debris to sunlight, helping reduce soil pathogens. Timely planting helps avoid peak periods of pest and disease activity. Weed the field four to five times during the first and second years, three to four times in the third and fourth years, and once in the fifth year. After winter, promptly remove and dispose of infected plant debris to prevent reinfection.
4. Chemical Control: After planting, apply a 300-fold dilution of 30% oxine-copper AS as a preventive measure. For active root rot outbreaks, use a 1000-fold dilution of 25% propiconazole EC.

27.2 White Rot Disease of Coptis (Huanglian)

White rot disease has emerged as one of the more serious diseases affecting Coptis in recent years, with an infection rate ranging from 5% to 20% across various production regions.

Symptoms
The pathogen initially infects the root collar of the plant, causing the area to turn brown, and white, silk-like mycelium forms near the soil surface. In advanced stages, the mycelium spreads to the leaves, with the petioles and entire leaves turning purple-brown or orange-yellow. Severe cases result in water-soaked rot in the root collar and petioles, leading to wilting, wrinkling, and eventually the death of the entire plant (Fig. 27.3).

Pathogen
The causal pathogen of Coptis white rot disease is *Athelia rolfsii* (Curzi) C.C.Tuet Kimbr., from the Athelia genus in the Basidiomycota phylum. Its asexual form is *Sclerotium rolfsii* Sacc., in the Sclerotium genus. On PDA medium, the pathogen's mycelium appears white and silky, radiating outward. After 7 days of incubation at 28 °C, abundant sclerotia form. These sclerotia transition from milky white to yellowish, and eventually to tea-brown, taking on a spherical or irregular shape (Fig. 27.4). Microscopically, the hyphae are septate with a diameter of approximately 6.45 μm. Optimal growth and sclerotia germination temperatures range from 20 °C to 30 °C, with 30 °C being the most favorable. Mycelium is killed at 60 °C, while sclerotia die at 65 °C. The pathogen thrives across a pH range of 2.0 to 13.0, with pH 6.0 supporting optimal growth and maximum sclerotia formation. It can utilize a variety of carbon sources, including monosaccharides, disaccharides, and

Fig. 27.3 Coptis White Rot Disease symptoms

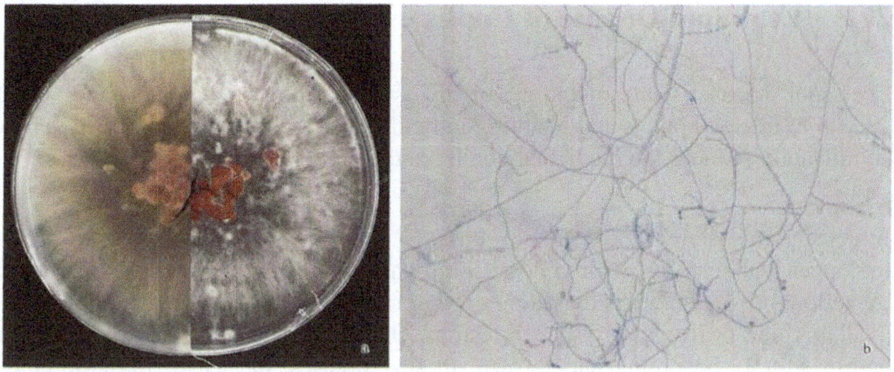

Fig. 27.4 Pathogen morphology of Coptis White Rot Disease: (**a**) Cultivation characteristics. (**b**) Mycelium

polysaccharides, and shows high nitrogen utilization with ammonium chloride and ammonium sulfate as sources.

Occurrence Pattern

Coptis white rot disease occurrence is closely related to climatic conditions; temperatures above 25 °C, coupled with rainy, high-humidity conditions, lead to high incidence rates. The pathogen's mycelium and sclerotia overwinter in crop residues and soil, acting as the initial infection source in the following growing season. Under warm and humid conditions, the sclerotia germinate readily, spreading through water and infecting plants upon contact. Sclerotia have strong resilience and can survive in soil for 5–6 years or more. The disease typically appears in late May, with a peak in August, affecting approximately 30% of Coptis plants in the field. Infection rates vary with field age, with lower rates in 1- to 2-year-old fields and higher rates in 3- to 5-year-old fields. In fields with similar cultivation conditions, low-lying, waterlogged areas experience more severe infections.

Control Measures

1. Soil Disinfection: Treat field soil with 50% lime water or irrigate with a 500-fold dilution of 50% carbendazim WP for disinfection.
2. Enhanced Cultivation Management: Avoid continuous cropping by implementing rotation with crops like grasses or *Achyranthes bidentata*. Avoid rotating with other susceptible crops such as Scrophularia and Paeonia. Remove diseased plants along with surrounding soil, then bury or destroy them outside the cultivation area. Disinfect the affected area with lime powder.
3. Chemical Control: At early onset, irrigate with 50% lime water. For more severe infections, apply 40% fluazinam EC at 1,000-fold dilution, 96% hymexazol concentrate at 3,000-fold dilution, or 70% mancozeb WP at 1,000-fold dilution. These treatments can achieve over 90% control effectiveness.

27.3 Leaf Spot Disease of Coptis

Leaf spot disease is one of the major diseases affecting Coptis, causing stunted growth in infected plants and resulting in field infection rates exceeding 30%. This significantly reduces the medicinal quality and commercial value of Coptis.

Symptoms

Infected plants show stunted growth. Initially, small, circular black spots appear on the upper leaf surfaces, with yellowish-brown lesions on the undersides, bordered by yellow-brown halos. As the disease progresses, these spots enlarge into irregular oval, circular, or irregularly shaped lesions, changing from black to yellow-brown and eventually becoming dark brown. In severe cases, the leaves may die completely (Fig. 27.5).

Pathogen

The pathogen causing Coptis leaf spot disease is *Phoma aquilegiicola* M. Petrov, an asexual ascomycete fungus of the genus Phoma. On MA medium, it forms dense colonies with abundant hyphal growth, reaching an average diameter of 6.72 cm after 5 days. The pathogen grows best at 20 °C, a temperature that aligns well with the optimal growth conditions for Coptis. It shows minimal sensitivity to light variation and can grow across a pH range of 4.0–10.0, with optimal growth at pH 6.0. Among carbon sources, maltose is most effective, and among nitrogen sources, peptone supports the highest growth, followed by glycine (Fig. 27.6).

Fig. 27.5 Symptoms of coptis leaf spot disease

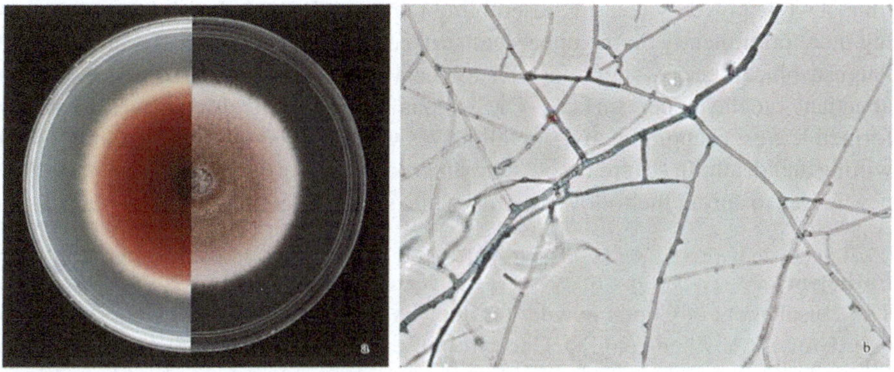

Fig. 27.6 Morphology of Coptis Leaf Spot Disease pathogen. (**a**) Colony morphology. (**b**) Mycelium

Occurrence Pattern

Coptis leaf spot disease generally occurs during the hot and humid months of July and August. The pathogen overwinters as pycnidia on infected plant debris, spreading more readily in rainy, humid, and misty conditions.

Control Measures

1. Scientific Site Selection: Choose broadleaf forest areas with low-lying, sunny, fertile, and well-drained soil for cultivating Coptis.
2. Enhanced Cultivation Management: Practice crop rotation; after transplanting, set up shade structures to control light exposure, gradually increasing light levels as needed. Ensure proper drainage to prevent moisture buildup.
3. Chemical Control: During the seedling stage, spraying with 70% thiophanate-methyl WP at 1,000–1,500-fold dilution or 40% carbendazim SC at 500–1,000-fold dilution can effectively reduce seedling mortality due to the disease. When the disease is present, apply a 10% difenoconazole WG at 6,000–7,000-fold dilution, 30% epoxiconazole-propiconazole EC at 1,000–1,500-fold dilution, or a fluopyram-tebuconazole spray at 206.7 g/L. Apply sprays every 10 days, with at least two consecutive treatments.

27.4 Powdery Mildew of Coptis (Huanglian)

Powdery mildew, also known as "winter melon disease," is a significant disease affecting Coptis, commonly found in production areas. It can cause severe plant thinning and seedling death, often reducing yields by over 50%.

Symptoms

In the early stages, small yellow-brown spots measuring 2–25 mm in diameter appear on the underside of the leaves, gradually expanding into larger lesions. Over time, a white powdery coating, resembling winter melon powder, develops, appearing more prominently on the upper leaf surface than the underside. By late July and August, black specks—granular cleistothecia—can be seen on the lesions, more abundant on the upper surface. The lesions spread from older leaves to newly formed leaves, eventually covering all leaves with powdery mildew, leading to leaf wilting and death. In severe cases, stem and root decay ensue; mild cases may allow new leaves to sprout the following year, while severe infections result in plant death (Fig. 27.7).

Pathogen

The causal agent of Coptis powdery mildew is *Erysiphe aquilegiae* DC. var. ranunculi (Grev.) R.Y. Zheng et G.Q. Chen from the Ascomycota phylum. The mycelium primarily grows on the upper side of the leaf, with some on the underside and petioles. Most conidia are cylindrical, with a minority barrel-shaped, measuring (25.4–38.1) μm × (11.4–17.5) μm. The cleistothecia are scattered or clustered, dark brown, flattened, spherical, and measure (73.2–83.5) μm × (112.0–125.6) μm. The cell walls are irregularly polygonal, 6.3–18.5 μm in diameter. Appressoria number between five to thirteen, typically unbranched, with most curved and a few nearly straight, often displaying pronounced bends or undulations, occasionally appearing nodular. Length ranges from 50.5 to 350.3 μm, with a uniform thickness or slight variations, and a diameter of 3.8–7.5 μm. Appressoria have five to eleven septa, brownish at the base, fading toward the apex, and sometimes nearly colorless. Asci contain two to eight ellipsoid to nearly spherical ascospores, some with short stalks, measuring (26.5–35.0) μm × (40.6–53.3) μm. Ascospores are ellipsoid or long-ellipsoid, yellow, and measure (16.3–18.8) μm × (6.8–10.2) μm.

Occurrence Pattern

The pathogen overwinters as mycelium and cleistothecia on infected plant debris or old Coptis leaves, serving as the primary source of infection for the next season. By

Fig. 27.7 Coptis powdery mildew symptoms

mid-May, as daily temperatures exceed 16 °C, the overwintered mycelium germinates and releases conidia, initiating infection. The disease spreads rapidly once daily temperatures exceed 24 °C, reaching peak infection rates and severity during July and August when temperatures are between 24 °C and 26 °C. Drought conditions exacerbate disease severity.

Control Measures

1. Scientific Site Selection: Choose broad-leaved woodland with low-lying, sunny areas with fertile, well-drained soil for Coptis cultivation.
2. Enhanced Cultivation Management: Implement crop rotation; regulate shading to increase light exposure and ensure adequate drainage. In winter, remove plant debris and fallen leaves from the medicinal garden and destroy them. Apply organic fertilizers during early disease onset to promote seedling regrowth and improve disease resistance.
3. Chemical Control: Apply 30–50% lime sulfur carefully as a spray at the initial stage of infection. Spray every 5–7 days for two to three applications. Apply once at disease onset and two to three times during peak infection. Additionally, 70% thiophanate-methyl WP at 1,000–1,500-fold dilution can be applied two to three times.

27.5 Coptis Anthracnose Disease

Coptis anthracnose is a major disease affecting Coptis growth, characterized by high pathogenicity, a short incubation period, rapid progression, and high mortality rates. The pathogen primarily infects young leaves, especially newly emerged ones, with field infection rates ranging from 10% to 100% and resulting in yield losses of 10–30%.

Symptoms

In the early stages of infection, oil-soaked lesions appear on leaves, which gradually expand into large spots with grayish-white centers and dark red margins. These lesions, ranging from 3–20 mm in size, often develop small, raised black dots on the leaf surface (representing the pathogen's conidia). In later stages, leaves may become perforated. Infected petioles may exhibit purple-brown lesions that penetrate inward, leading to leaf drop. Severe infections cause widespread leaf death, significantly reducing both the yield and quality of Coptis (Fig. 27.8).

Pathogen

The causal agent of Coptis anthracnose is a fungus of the Colletotrichum genus (Ascomycota, asexual stage). On PDA medium, the pathogen initially forms a thin, dense white colony that later develops into a fluffy, white outer ring, with dark, dense aerial mycelium in the inner layer and concentric radial growth. After 12 days, conidia clusters emerge, surrounded by dark brown setae. The fungus thrives at temperatures between 10 °C and 30 °C, with an optimal range of 20–24 °C. Acervuli

Fig. 27.8 Coptis anthracnose disease symptoms

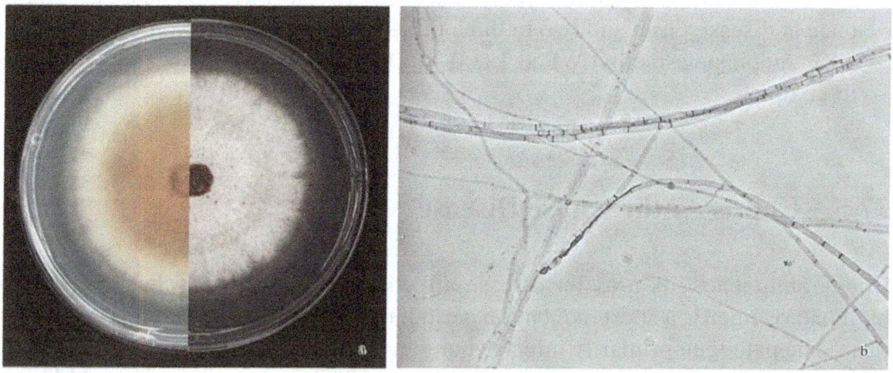

Fig. 27.9 Morphology of *Coptis Anthracnose* pathogen. (**a**) Colony morphology. (**b**) Mycelium

are black and lack setae, with hyaline, unbranched conidiophores bearingone to three septa, measuring 15.0–20.0 μm × 3.0–4.0 μm. The conidia are hyaline, single-celled, cylindrical to elongated, measuring 15.0–18.0 μm × 3.5–6.0 μm (Fig. 27.9).

Occurrence Pattern

Coptis anthracnose commonly appears from late March through July in major Coptis production areas and has a pattern of repeated infections. The pathogen can directly invade plant tissues, with an incubation period of 3–4 days at temperatures between 20 °C and 30 °C, and shorter periods at higher temperatures. Conidia are primarily spread by rain, enabling repeated infection cycles. While air currents may aid in spore dispersion, rain breaks up the spore clusters, facilitating distant transmission. In dry conditions, spores tend to clump together, reducing wind dispersion, making drought conditions less conducive to disease spread.

Control Measures

1. Seed Treatment: Prior to sowing, coat seeds with a protective film to retain moisture, enhance water absorption, and improve germination rates, leading to stronger seedlings.
2. Field Management: Maintain field sanitation by removing diseased plant debris and weeds in winter. Apply a protective film on the soil surface to prevent pathogen entry and eliminate overwintering sources.
3. Chemical Control: At the onset of infection, spray with a 1:1:100 Bordeaux mixture combined with a protective film, applying every 7 days for three to four applications. Upon infection, promptly remove diseased leaves and apply an 800–1,000-fold dilution of 50% sulfur - carbendazim WP every 7 days for two to three applications.

Chapter 28
The Diseases of *Paeonia* x *suffruticosa* Andr.

The peony (*Paeonia* x *suffruticosa* Andr.) is a deciduous shrub belonging to the family Ranunculaceae. It can grow up to 2 meters tall, featuring short and thick branches. Peonies prefer warm and humid climates, are relatively tolerant to cold and drought, but are sensitive to waterlogging, high temperatures, and intense sunlight. They thrive in deep, well-drained, fertile, and loose sandy loam or silty loam soils; saline-alkali and heavy clay soils are unsuitable for cultivation.

Peony bark, known as mudanpi in traditional Chinese medicine, is the dried root bark of peonies grown in Anhui, Shaanxi, Sichuan, Shandong, Henan, and other regions. It has a bitter and pungent taste, is slightly cold in nature, and is associated with the heart, liver, and kidney meridians. Mudanpi has the effects of clearing heat and cooling blood, promote blood circulation and removing blood stasis, and reducing deficient heat. The "Fengdan" peony bark from Fenghuang Mountain in Tongling, Anhui Province, is considered top quality.

Currently, more than 30 types of peony diseases have been reported domestically. Fungal diseases are the most prevalent, and in recent years, they have been worsening year by year, especially during the later growth stages. These diseases have become the most significant limiting factor in peony production, directly affecting the yield and quality of peony bark and diminishing the ornamental value of the plants. The most severe fungal diseases affecting peonies include leaf mold, anthracnose, black spot, and root rot.

Peony viral diseases are also a critical factor restricting peony production. To date, five viral diseases of peonies have been reported: yellows, mosaic, ringspot, leaf curl, and a peony virus caused by tobacco rattle virus. In China, peony virus diseases associated with yellowing and mosaic are the most common.

The main nematodes harming peonies are root-knot nematodes. Morphological identification indicates that the northern root-knot nematode (*Meloidogyne hapla*) is predominant, along with the southern root-knot nematode (*M. incognita*), peanut root-knot nematode (*M. arenaria*), and *Ditylenchus triformis*. These nematodes primarily damage the fibrous roots of peonies. After the fibrous roots are damaged, many nearly spherical galls with the size of mung beans are produced, leading to

© China Agriculture Press Co., Ltd. 2026

T. Wang et al., *Atlas of Chinese Medicinal Plant Diseases*,
https://doi.org/10.1007/978-981-95-6326-5_28

weakened growth of the above-ground parts. New leaves become wrinkled and yellow, eventually withering and falling prematurely. In severe cases, the entire plant may die. Reports on bacterial diseases of peonies are scarce both domestically and internationally, and the extent of their harm remains unclear.

Among noninfectious diseases, damage from intense sunlight—also known as peony sunburn—is quite common. It often causes severe wilting of leaves or entire plants, significantly affecting plant growth and yield of peony bark. Peony iron-deficiency chlorosis is another common physiological disorder. Initially, the leaf mesophyll turns yellow while the veins remain green. In severe cases, the yellowed portions become necrotic, branches fail to mature properly, and flowering becomes difficult or results in smaller flowers. Additionally, there are reports of fluoride poisoning in peonies, mainly caused by fluoride pollution leading to withering of leaf tips. Resistance tests among different peony varieties have shown variations in tolerance to atmospheric fluoride pollution. Moreover, because peony flowers are sensitive to climatic factors such as low temperatures and cold fronts, they often suffer varying degrees of damage, sometimes even severe outbreaks.

28.1 Peony Leaf Mold Disease

Peony leaf mold disease, also known as peony red spot disease, is a primary disease affecting peonies. In recent years, it has become widespread across peony-growing regions, causing severe damage that often leads to early leaf wilting. This significantly impacts plant growth and flower bud differentiation, making it a pressing issue in peony cultivation. With the increase in peony cultivation areas and varieties, the incidence of leaf mold disease is rising. The disease incidence varies by variety, with most gardens experiencing approximately a 50% disease incidence, and some severely affected areas reporting incidence exceeding 90%. According to Yuan et al. (2010), a survey in Heze revealed that in heavily affected gardens, disease incidence exceeded 92%, with most gardens reporting incidence between 70% and 80%.

Symptoms

The disease primarily affects leaves but can also damage stems, petioles, sepals, petals, fruits, and seeds. Initial symptoms on leaves appear as faded green, light brown, pin-sized, sunken spots. These spots develop from the leaf underside and gradually protrude, expanding into circular or irregular shapes. On the leaf surface, lesions are reddish-brown to dark brown (with notable variations among different varieties), with purple-brown or yellow edges. The abaxial surface of the leaf is light brown and has distinct concentric ring patterns, with a diameter of 3–30 mm. Under moist conditions, a dark green mold layer appears on both sides of the lesions, which are conidiophores and conidia of the pathogen. In severe cases, multiple spots merge with each other, causing leaf scorching, premature leaf fall, and occasional perforation (Fig. 28.1).

Fig. 28.1 Peony leaf mold symptoms

On stems, early symptoms present as small, raised, dark purplish-red, oblong spots, which slowly expand to 3-5 mm. The middle of the lesions often cracks and sinks. In advanced stages, spots may coalesce, and there is often visible dark green mold layer on the lesions. This causes plant stunting or branch wilting and death. They cannot sprout branches and bloom, and even the whole plant may die, diminishing aesthetic value and reducing the yield of peony bark. On petioles and fruits, lesions are oblong or nearly circular, brown to dark brown, and measure 2-18 mm in diameter, with a dark green mold layer that becomes particularly obvious under humid conditions.

Pathogen

The causative pathogen of peony leaf mold is *Cladosporium paeoniae* Pass., an asexual form of ascomycete belonging to the genus *Cladosporium*. On PDA culture medium, it grows slowly, forming a dense dark green mold layer in the center, representing conidiophores and conidia. The colony is compact, with uniform hyphal growth at the edges, often showing a star-shaped crack on the reverse side. Conidiophores are distinct, dark brown, occurring singly or in clusters, linear or slightly curved, occasionally branching with knobby swellings on one side, usually with two to seven septa and arranged in clusters of three to seven. Conidia are single or in chains, dark brown, ovoid, fusiform, or spindle-shaped, either unicellular or multicellular, measuring (6.6–27.1) μm × (5.5–6.7) μm (Fig. 28.2).

Occurrence Pattern

The pathogen overwinters as mycelium in plant debris within fields. In spring, it produces conidia, which are spread by wind, rain, or human activities, and directly invade host tissues through natural wounds. The disease can occur from post-bloom in spring through autumn, initially affecting lower leaves and intensifying by late April. In Heze and Jinan, symptoms first appear on young stems and petioles in late March; by early April, small green spots emerge on newly formed leaves. As the disease progresses, the spots gradually expand and even connect into patches. The peak incidence period is from mid-June to early August, with leaves wilting and

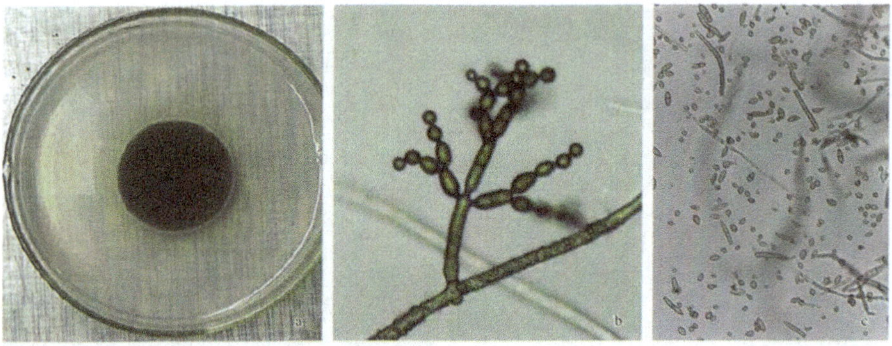

Fig. 28.2 Peony leaf mold pathogen morphology. (**a**) Culture characteristics. (**b**) Conidiophore and Conidium arrangement. (**c**) Conidium

scorching due to disease, and incidence in severely affected areas reaches over 90%. By early to mid-September, most leaves have withered and scorched, and even fallen off, directly affecting the differentiation of buds. In mid-November, the pathogen enters the overwintering period. In Nanjing, the disease onset is in late March, peaking from June to July; in Beijing, it starts in late April to early May, peaking from July to August. Low temperatures, heavy rainfall, high humidity, standing water, dense weeds, and poor management intensify infection. Nutrient deficiencies or imbalanced nutrients also increase susceptibility.

Control Measures

1. Utilize Disease-Resistant Varieties

   Developing and using disease-resistant peony varieties is the most economical and effective way to manage leaf mold. Although no immune variety has been found, obvious resistance differences exist among varieties. Planting more resistant varieties can substantially reduce disease impact. Resistant germplasm can also be used to breed new varieties, making it essential for disease control.
2. Maintain Field Cleanliness

   After the growing season, promptly remove diseased leaves and branches, destroying or burying them 15 cm below the soil to reduce primary inoculum sources for the following spring.
3. Enhance Cultivation Management

   Ensure appropriate spacing during planting to improve ventilation and sunlight exposure. Apply fertilizers, water, and weed regularly to prevent increased susceptibility due to nutrient deficiency or excessive humidity. During blooming, provide sufficient water, and promptly remove withered flowers post-bloom to prevent the reduction of stress resistance and decrease nutrient consumption. In winter, remove above-ground branches without damaging root buds to improve airflow and light transmission conditions, ensure growth vigor, and reduce disease incidence.
4. Biological Control

At early disease onset, apply wettable powder formulations of *Trichoderma harzianum* (3 × 10^8 CFU/g) or 2% kasugamycin for effective control. A 0.5% matrine solution can also be used.

5. Chemical Control

   Start fungicide application from mid-March to mid-July with four sprays to achieve 90% control effectiveness and reduce disease incidence. Effective fungicides for leaf mold include mancozeb, thiophanate-methyl, and carbendazim. Chlorothalonil and lime sulfur are also effective against peony leaf mold.

28.2 Peony Black Spot Disease

Peony black spot disease is relatively common, with field incidence generally reaching around 40%. This disease can cause peony leaves to wither and fall off, and in severe cases, lead to entire plant death, seriously affecting the yield and quality of peony bark.

Symptoms

Peony black spot disease mainly affects leaves but can also damage petioles, stems, and fruits. In the initial infection stage, small, circular lesions of 1.1–3.2 mm in diameter appear on the leaves, which are grayish-black or dark brown with a slightly lighter center. These lesions gradually enlarge to 6.2–32.5 mm in diameter, taking a circular or irregular shape with a dark brown color and producing a blackish moldy surface. When touched, the lesions feel rough, and they sometimes fall off, often forming perforations. In severely infected plants, all leaves eventually wither and fall off (Fig. 28.3).

Pathogen

The primary pathogen causing peony black spot disease is *Alternaria alternata* (Fr.) Keissl., a fungus of the Alternaria genus. Additionally, *Alternaria tenuissima* (Nees: Fr.) Wiltshire, *A. suffruticosicola* M. Zhang, Z.S. Chen, and T.Y. Zhang, and *Alternaria sufuticosae* M. Zhang, Z.S. Chen, and T.Y. Zhang can also infect peonies, resulting in black spot disease. *Alternaria* spp. grows rapidly on PDA medium, and can cover the entire plate within approximately 7 days. Initially, the mycelium is white, then gradually turns gray-green or gray-black in the center with white edges, eventually developing into a gray-green or gray-black colony with sparse aerial hyphae. Within 2 days, abundant conidia form on the medium surface. Conidiophores are light brown to dark brown, produced from the mycelium, and occur singly or in clusters of 3–4, or on the stroma. They are mostly geniculate, and a few are straight, branched or unbranched, with evident enlarged scar cells at the base and containing 2–8 septa. Conidia are borne at the tips of the conidiophores, light to dark brown, oval to obclavate, with transverse and vertical septa, often borne in chains. The conidial surface may be smooth or warty, with apical cells frequently containing beak-like structures. Conidia measure (13.1–68.2) μm × (7.3–13.1) μm, with one to seven transverse septa (usually four to five) and zero to three

Fig. 28.3 Peony Black Spot Disease symptoms. (**a**) Field symptoms. (**b**) Early symptoms. (**c**) Advanced symptoms

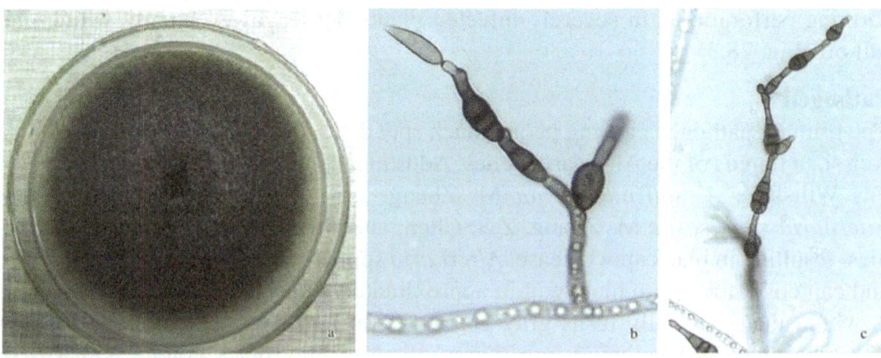

Fig. 28.4 Alternaria spp. Morphology. (**a**) Culture characteristics. (**b**, **c**). Conidiophores and Conidia

longitudinal septa. The beak-like cells measure (0–20.8) μm × (0–5.2) μm with zero to one septum (Fig. 28.4).

Occurrence Pattern

The pathogen primarily overwinters as mycelium or conidia in infected plant debris in the field. In the following spring, mycelium germinates, producing conidia that spread via wind and rain. Once germinated, the spores enter host leaves through wounds. Peony black spot disease primarily occurs during summer and autumn, with peak infection from August to September. High field humidity, susceptible varieties, and densely planted fields increase disease severity. Poor field management, insufficient or untimely fertilization, also exacerbate disease severity.

Control Measures

Planting disease-resistant peony varieties and enhancing field management with proper fertilization, irrigation, and weeding to boost plant resistance are key to preventing and controlling black spot disease in peonies.

1. Appropriate Spacing and Scientific Fertilization
   Avoid excessive planting density for medicinal peonies, and promote the use of well-decomposed organic fertilizer to encourage root growth and strengthen disease resistance. During the rainy season, prevent prolonged humidity and avoid overhead irrigation. Thoroughly remove and destroy diseased debris at the end of autumn to reduce initial source for the following year.
2. Enhanced Field Management
   At the onset of sporadic infections, remove infected leaves promptly as part of routine field management to reduce inoculum sources and improve ventilation and light transmission conditions.
3. Biological Control
   At early infection stages, spray with 3% wettable powder of polyoxin complex (900-fold dilution), 4% pyrimidine-based antibiotics aqueous solution (600–800-fold dilution), or other suitable biological agents for effective control.
4. Chemical Control
   During peak infection periods, supplement enhanced field management with effective chemical control. At the early stage, 50% iprodione wettable powder (1000-fold dilution), 40% chlorothalonil suspension (600-fold dilution), or 70% propineb wettable powder (600-fold dilution) can be sprayed. In more widespread cases, apply 29% lime sulfur (100–200-fold dilution), 20% thiediazole copper suspension (500–800-fold dilution), 25% azoxystrobin suspension (1250–2500-fold dilution), or 70% mancozeb wettable powder (400–600-fold dilution). Bordeaux mixture (1% standard solution) or 75% chlorothalonil wettable powder (800-fold dilution) can also reduce pathogen pressure and protect plants.

28.3 Root Rot Disease of Peony

Root rot, commonly referred to as "rotting root disease," is a primary disease affecting medicinal peony, prevalent across cultivation regions. It is a major limiting factor in peony bark production. In recent years, with expanded cultivation areas and continuous cropping practices, peony root rot has shown an increasing trend in severity year by year.

Fig. 28.5 Symptoms of Peony Root Rot. (**a**) Field symptoms. (**b**) Plant symptoms. (**c, d**) Root symptoms

Symptoms

Peony root rot primarily affects the plant's root collar and root system, including the main root, lateral roots, and fibrous roots, with older roots being especially vulnerable. In the initial stage, irregular black spots appear on the root skin of the main root, which then expand, causing most of the root to blacken and spread into the woody tissue. The disease progresses from the wood to the pith, leading to decay in the fleshy roots, tissue rot, and distinct lesions with clear demarcation between diseased and healthy areas. In severe cases, the entire root system rots, causing aboveground parts to wilt and eventually die. When lateral and fibrous roots are infected, they turn black and rot, potentially spreading to the main root. Affected plants display slowed growth, weakened development, smaller yellowed leaves, and increased wilting due to root damage, especially during high evaporation periods. Severely affected plants eventually die (Fig. 28.5).

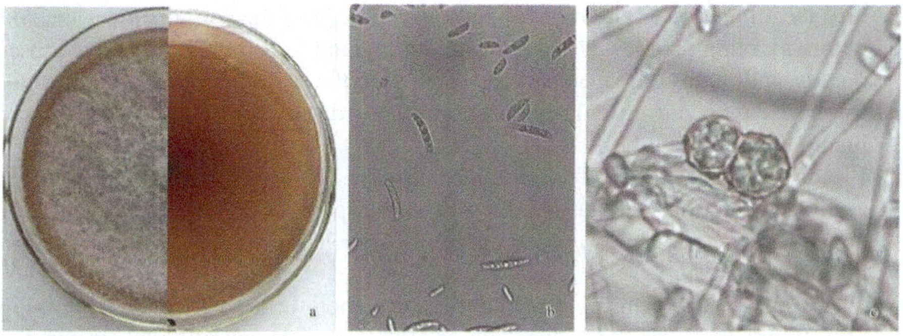

Fig. 28.6 Morphology of *Fusarium solani.* (**a**) Colony characteristics. (**b**) Conidia. (**c**) Chlamydospores

Pathogen

The primary causal agent of peony root rot is *Fusarium solani* (Mart.) Appel & Wollenw., an ascomycetous fungus in the Fusarium genus. Other species such as *F. oxysporum* Schltdl. ex Snyder et Hansen, *F. verticillioides* (Sacc.) Nirenberg, and *F. equiseti* (Corda) Sacc. also cause root rot in peonies. On PDA medium, *F. solani* forms circular, fuzzy colonies with even edges. Colony colors vary between strains, appearing as orange, green-black, or wine-red. Some strains produce mucilaginous spore masses. *F. solani* produces two types of conidia: macroconidia, which are sickle- or crescent-shaped with a pointed tip and multiple septa (usually 3–5 transverse septa), and microconidia, which are oval or elongated and usually single celled. The fungus also produces numerous chlamydospores that are terminal or intercalary, spherical, and thick-walled (Fig. 28.6).

Occurrence Pattern

The pathogen primarily overwinters as chlamydospores on diseased roots, in the soil, or in manure. Conidia and mycelium can also overwinter on diseased plant residues in the field, with chlamydospores remaining viable for many years. The pathogen spreads through soil, rainwater, and irrigation water, entering plants via wounds from pests, mechanical injuries, or nematodes. Peony, which thrives on fertilizers, is particularly susceptible when nutrients are insufficiently or improperly applied, leading to severe root rot. Continuous cropping promotes root rot, as longer cropping periods exacerbate the disease, alter soil properties, and reduce soil fertility. Peonies in alkaline soils are especially vulnerable to this disease.

Additionally, as peony plants age, root rot tends to worsen. Older plants, especially those over 20 years old, are more susceptible due to declining growth and reduced resistance, making root rot particularly severe.

Control Measures

1. Enhanced Cultivation Management: Select well-drained planting sites to minimize field water accumulation in the field. Implement crop rotation when feasible to reduce the pathogen's long-term accumulation of pathogens.

2. Soil Disinfection: As the root rot pathogens reside in the soil, disinfect the soil before planting peonies. Use organic and inorganic fertilizer blends, mixed thoroughly with finely screened soil and applied to the planting area to significantly lower disease occurrence. Promptly remove any diseased plants and disinfect the planting hole with lime.
3. Biological Control: For effective control, irrigate with 4% pyrimidine nucleoside antibiotic solution, 4% kasugamycin wettable powder, 1% shenqinmycin suspension, or *Paenibacillus polymyxa* KN-03 suspension at a concentration of five billion CFU/g.
4. Chemical Control: During transplanting, dip roots in a paste made by mixing 36% thiophanate-methyl suspension or 10% carbendazim colloidal suspension (700× dilution) with appropriate micronutrients and soil before planting.

28.4 Peony Gray Mold Disease

Peony gray mold disease, prevalent in all cultivation areas, is particularly severe in humid regions along rivers and lakes. This disease is an important disease in peony production, frequently causing rot of peony leaves and flowers, as well as stem collapse, which severely impacts the harvest of peony bark medicinal materials and the ornamental value of peonies.

Symptoms
Gray mold disease primarily affects the leaves and stems of peonies, with the lower leaves being most susceptible. In the early stages, near-circular or irregular water-soaked lesions appear, commonly at leaf tips and edges. Later, lesions expand to over 1 cm in size, and the lesions are brown to purplish-brown, sometimes with ring patterns. Under high humidity, a grayish mold layer forms on both the upper and lower leaf surfaces. Stem lesions appear brown and often soften, causing the plant to break or collapse entirely. Infected flowers turn brown and rot, producing a gray mold layer, with occasional black sclerotia appearing on the diseased parts (see Fig. 28.7).

Pathogen
The causative pathogen of peony gray mold is *Botrytis paeoniae* Oudem., which is an asexual morph of ascomycete belonging to the genus *Botrytis*. On PDA medium, *Botrytis paeoniae* initially forms a white, woolly colony with fast-growing mycelium. Later, the colony gradually turns gray-white, and many tiny gray-white sclerotia grow on it in the early stages. In the later stage, the sclerotia become black and granular, and appear like mouse droppings. Conidiophores are slender with hemispherical tips, and produce 2–7 small branches where conidia form. Conidia are nearly round or oval, unicellular, and measure (6.8–13.6) μm × (6.8–10.2) μm (Fig. 28.8).

Fig. 28.7 Peony Gray Mold Disease symptoms. (**a**) Initial symptoms. (**b**) Later symptoms. (**c**) Leaf symptoms. (**d**) Stem symptoms

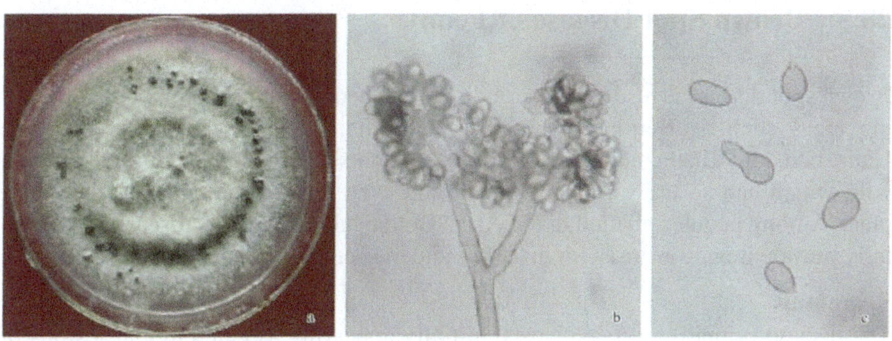

Fig. 28.8 Morphology of Peony Gray Mold pathogen. (**a**) Colony characteristics. (**b**) Conidiophores and Conidia arrangement. (**c**) Conidia

Occurrence Pattern
The pathogen overwinters as sclerotia in soil or plant debris. In the following spring, the sclerotia germinate to produce mycelium, conidiophores, and conidia, which are spread by wind and rain to the leaves and other susceptible parts of the host and invade through wounds. High temperatures and rainfall are conducive to the large-scale formation and spread of conidia. Overcrowding, high humidity in the field, and insufficient light weaken the growth of peonies, increasing susceptibility to infection. Additionally, continuous cropping, excessive nitrogen fertilization, or nutrient imbalance result in severe outbreaks. When peonies are forced to flower in greenhouses, they are also extremely prone to the disease, often causing flower buds to shrink and rot. In central plain regions, infections typically begin around April 20, and the affected plants wither and become scorched before mid-June and reaching peak damage levels in July and August.

Control Measures
1. Agricultural Control: Implement crop rotation, maintain appropriate planting density, and ensure ventilation and adequate sunlight. Apply fertilizers scientifically, and increase phosphorus and potassium levels to boost disease resistance. Water appropriately and drain fields promptly after rainfall to prevent excessive moisture.
2. Biological Control: Spray with 0.3% eugenol solution or 1% carvacrol squeous solution for effective control.
3. Chemical Control: At the early stage of disease onset, spray with 50% procymidone WP at 1,000–1,500-fold dilution, 65% iprodione WP at 1,000-fold dilution, or 50% carbendazim WP at 900-fold dilution. In regions where pathogens are resistant to these fungicides, use 65% metalaxyl-mancozeb WP at 1,000-fold dilution, rotating different fungicides to prevent resistance.

28.5 Brown Spot Disease of Peony

Brown spot disease is also a common disease of peonies. It occurs in peony - growing areas such as Beijing, Shanghai, Changsha, Nanjing, Chengdu, Guiyang, Hangzhou, and Zhengzhou. Heavily infected plants may experience premature leaf wilt, which can severely impact growth and even lead to the death of the entire plant. Among peonies planted in the Xi'an Botanical Garden, this disease is particularly prevalent and severe, with disease incidence reaching as high as 89.3%.

Symptoms
The disease primarily affects leaves, initially appearing as pale circular lesions of varying sizes. As the disease progresses, the lesions turn brown and become more defined, typically round or near-round, with a diameter of 2–20 mm, displaying a characteristic concentric ring pattern. Individual lesions are densely scattered across a leaf and eventually merge into large, irregular blotches. A gray mold layer often

Fig. 28.9 Symptoms of brown spot disease on peony

forms over the lesions, particularly under high humidity. Severe cases result in leaf necrosis (Fig. 28.9).

Pathogen

The main pathogen causing peony brown spot disease is *Cercospora paeoniae* Tehon et Dan. In addition, *C. varicolor* Wint. can also infect peonies. Both fungi are in the Cercospora genus (asexual fungi). *C. paeoniae* grows slowly on PDA medium, producing sparse spores in later stages. The mycelium is white and fuzzy, with a slightly sunken colony center and red-brown specks visible underneath. The colony shows distinct concentric rings, with a darker center. After 10 days, it can cover the entire petri dish, and reddish-brown spots (growing under the culture medium) can be seen on the back. The colony shows distinct concentric rings, with a darker center. Mycelium is typically endophytic, emerging from stomata on the lower surface of leaves, yellow-green, septate, and about 2.0–3.0 μm wide. The sporodochia grow under stomata, spherical to oblong, dark brown to nearly black, measuring 30.0–75.0 μm in diameter, or up to 40.0–75.0 μm long and 30.0–60.0 μm wide. Conidiophores cluster densely on sporodochia or may grow individually as lateral branches on the mycelium. They are yellowish-brown to light yellow-brown, with paler, narrower tips. They are generally upright or slightly curved, unbranched or occasionally branched, conical at the top, with zero to one geniculate bend, 0 - 2 septa, mostly without septa, with dimensions of (4.0–40.0) μm x (3.0–4.0) μm. Conidia are linear or tail-shaped, multi-septate, and measure (9.8–20.7) μm × (2.6–3.3) μm (Fig. 28.10).

Occurrence Pattern

The pathogen overwinters as mycelium and sporodochium within infected tissue or plant debris, acting as the primary infection source for the following year. Conidia spread via wind, rain, or insect activity, infecting leaves, and often resulting in

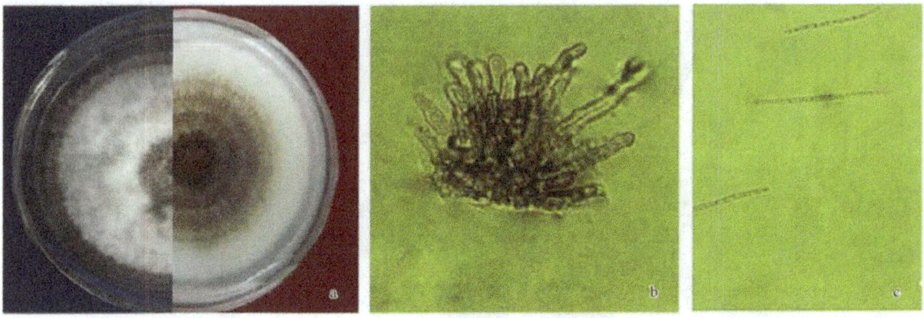

Fig. 28.10 Morphology of *Cercospora paeoniae.*(**a**) Colony characteristics. (**b**) Conidiophores and Conidiophore stalks. (**c**) Conidia

intensified outbreaks after windy and rainy weather. Initial infections typically appear on lower leaves, creating early lesions. As lesions expand and produce new conidia, secondary infections spread the disease upward. After August, lesions multiply, and with the onset of the rainy season, disease severity peaks. Diseased leaves start to fall off in late August. High autumn temperatures, frequent rainfall, dense planting, and poor ventilation are the main factors contributing to late-season disease severity.

Control Measures

1. Enhanced Cultivation Management: After harvest, thoroughly remove infected plants and fallen leaves, gathering and destroying them to reduce the following year's initial infection source. Implement balanced fertilization, with increased phosphorus and potassium, to enhance disease resistance. Irrigate as needed and promptly drain excess water after rainfall to avoid high humidity levels.
2. Biological Control: At the onset of symptoms, spray an 800-fold dilution of a 2% Wuyiencin solution, which has shown to be effective. Alternatively, 0.5% matrine aqueous solution or 3% polyoxin wettable powder may be used as a biological control measure.
3. Chemical Control: To prevent infection, apply a 600–800-fold dilution of 75% chlorothalonil wettable powder. Once disease symptoms appear or in early stages, apply a 400–600-fold dilution of 50% mancozeb wettable powder, a 600–800-fold dilution of 25% prochloraz emulsifiable concentrate, or an 800-fold dilution of 80% carbendazim wettable powder.

28.6 Peony Powdery Mildew

Symptoms

This disease primarily affects peony leaves but can also affect fruits, petioles, and stems. In the initial stages, small white powdery spots appear on the upper surface of leaves, which gradually expand, sometimes merging to cover the entire leaf. In

Fig. 28.11 Peony Powdery Mildew symptoms. (**a**) Infected plant. (**b**) Leaf symptoms

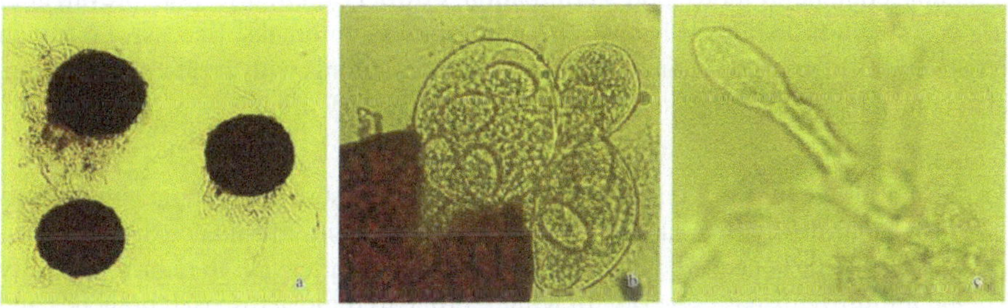

Fig. 28.12 Morphology of *Erysiphe paeoniae*. (**a**) Cleistothecia. (**b**) Asci and Ascospores. (**c**) Conidia

severe cases, both sides of the leaves, petioles, and stems develop a dirty white powdery layer scattered with small black dots, representing cleistothecia of the pathogen (see Fig. 28.11). Affected leaves grow slowly, age gradually, and may drop early. When the infection is severe, the plants wither early.

Pathogen

The pathogen responsible for peony powdery mildew is *Erysiphe paeoniae* Zheng & Chen, an ascomycete fungus. The cleistothecia of *E. paeoniae* are dispersed, dark brown, spherical or oblate. The cell of the cleistothecial wall are polygonal, and there are appendages on the cleistothecial wall. Filamentous appendages occasionally branch. There are 5–8 asci insdie the cleistothecia. The asci are ovoid or elliptical with a short stalk, measuring (54.4–61.2) μm × (28.9–34.0) μm and containing four to five ascospores. The ascospores are ovoid, measuring (17.0–23.8) μm × (10.2–13.6) μm. Conidia are produced singly, oblong, smooth, colorless, and grow in chains on unbranched conidiophores, measuring (25.5–27.2) μm × (10.2–11.9) μm (see Fig. 28.12).

Occurrence Pattern

The pathogen overwinters as mycelium and spores in infected buds. In spring, when the infected buds germinate, the pathogen also starts to grow, producing conidia that are spread via air or rain, infecting leaves and other susceptible tissues directly. In greenhouses or under plastic covers, the pathogen can continuously reproduce and cause infections. The pathogen is highly cold-resistant, retaining viability at 0 °C, making overwintered conidia on plant residues an essential initial infection source, capable of causing both primary and secondary infections, though ascospores generally play a minor role. The optimal growth temperature is 21 °C, with a maximum of 33 °C and a minimum of 3–5 °C. Conidia germinate best at relative humidity (RHs) levels of 97–99%, and a few conidia can still germinate at 23% RH, but the germination rate is very low in water droplets. Under ideal conditions, the pathogen completes its asexual cycle within 72 h, while it takes 7–10 days in the field. Powdery mildew tends to peak in May–June and September–October in open fields, while greenhouse infections persist year-round. Overcrowding, poor ventilation, and insufficient sunlight increase susceptibility, while young leaves with high β-alanine content promote spore germination. Powdery mildew is more prevalent in dry years with little rain, as heavy rains can suppress outbreaks. Excessive nitrogen fertilization, lush growth, or overly tender leaves, along with prolonged shading, also contribute to the large-scale occurrence of powdery mildew.

Control Measures

1. Agricultural Control: Maintain proper spacing to ensure ventilation; apply fertilizers scientifically; increase phosphorus and potassium to enhance disease resistance; irrigate as needed, and drain fields promptly after rain to reduce humidity and minimize disease severity. In winter, prune to remove diseased branches and buds, and remove infected leaves as soon as they appear.
2. Biological Control: At the early stages, apply a 300–400-fold dilution of 4% nucleoside antibiotic or use biopesticides like Wuyiencin and amino-oligosaccharides, both of which offer effective control.
3. Chemical Control: At the onset of symptoms, spray with fungicides like 25% pyraclostrobin suspension, 50% azoxystrobin water-dispersible granules, or 8% flusilazole microemulsion. Other options include 70% thiophanate-methyl WP, 20% triadimefon EC, 25% propiconazole EC, or 0.2–0.3 Baume sulfur-lime solution. Apply sprays once every 7–10 days, repeating two to three times to effectively control powdery mildew in peonies.

28.7 Occasional Parasitic Diseases of Peony (Table 28.1)

Table 28.1 Characteristics of occasional parasitic diseases in peony

Disease (Pathogen)	Symptoms	Occurrence pattern
Rust Disease (*Cronartium flaccidum*)	Initially, leaf lesions appear as nearly round or irregular spots, yellow-green in color. On the undersides of the leaves, there are granular aecium clusters. When the epidermis ruptures, yellowish-brown powdery aeciospores scatter. In the later stage, on the grayish-brown lesions on the undersides of the leaves, clusters of dark brown, bristle-like telia grow. In severe cases, leaves may wilt and die prematurely	The pathogen is a heteroecious fungus. It overwinters as teliospores on peonies. In the next spring, it produces pycniospores and aeciospores, which are spread by wind and rain to infect *Paeonia lactiflora*. In the later growth stage on *Paeonia lactiflora*, teliospores are produced. The teliospores germinate to produce basidiospores, which infect peonies and overwinter on them
Violet Root Rot (*Helicobasidium mompa*)	Primarily affects the root collar and roots. Infected areas are covered with a cottony purple fungal mat. Initially, lesions are yellowish-brown, then turn black. The disease causes old roots to rot, hinders new root growth, makes branches weak and leaves smaller. In severe cases, the entire plant turns yellow and withers, with branches wilting and eventually dying	The pathogen overwinters as mycelium on infected plant residues and spreads through soil and contact transmission. It initially infects young and tender roots, then gradually expanding to lateral roots, main roots, and the root collar. Continuous cropping fields, waterlogged and compacted soil are more prone to the disease
Phytophthora Blight (*Phytophthora cactorum*)	The disease can damage stems, leaves, buds, and other parts of peonies. In the early stages, water-soaked lesions appear on the stems, expanding into long spots. Young stems near the ground are affected, causing the whole branch to turn black and die. Lower leaves are more severely affected, with water-soaked irregular lesions, initially dark green, then turning blackish-brown. Diseased leaves wither and droop	The pathogen primarily overwinters as oospores in soil or infected plant debris, spreading by air currents. Under high soil moisture conditions, it produces numerous sporangia, forming central diseased plants. Plots with insufficient nutrient supply and low-lying terrain are more severely affected, and planting susceptible varieties is the main reason for the epidemic of the disease

(continued)

Table 28.1 (continued)

Disease (Pathogen)	Symptoms	Occurrence pattern
Anthracnose (*Colletotrichum siamense*, *C.* gloeosporioides)	Initial symptoms appear as small brown spots on the leaves, gradually enlarging into large, round, or irregular patches, often bounded by main veins and large lateral veins. Lesions are typically brown. In the later stage, the center of the lesion cracks, sometimes forming perforations. Black acervuli, arranged in a wheel shape, develop on the lesions, and reddish-brown viscous masses overflow when humidity is high	The pathogen overwinters as mycelium and acervuli on infected plant debris. In the next spring when temperature is suitable and after rain, a large number of conidia spread through air currents and rain, and directly invade leaves. High temperature and rainy weather favor disease progression, with peak incidence from August to September
Southern Blight (*Sclerotium rolfsii*)	Primarily damages the root collar of peonies. Early symptoms appear as water-soaked to dark brown lesions. The cortical tissue gradually sinks, and white silk-like mycelium grows. When humid, sclerotia grow. The sclerotia are like rapeseed, initially white, then turn yellow, and finally become brown	The pathogen overwinters as sclerotia, which become the initial infection source in the next year. Waterlogged fields after rain and continuous cropping fields are more severely affected
Root-Knot Nematode (*Meloidogyne hapla as the main species*)	Primarily affects fibrous roots of peonies, causing the formation of mung bean-sized, nearly round root knots. Affected plants exhibit stunted growth, yellowed leaves, and wilting. In severe cases, the entire plant wilts, leading to crop failure	The nematodes overwinter in the soil or as larvae, adults and galls attached to diseased roots, which are the initial infection source in the next year. Continuous cropping fields and sandy loam soil are more severely affected. One egg sac can lay 150 eggs, and 4 generations can be completed in a year
Leaf spot (*Piliidium concavum*)	This disease affects leaves, branches, and buds. Initial spots on leaves are round or nearly round with gray centers and dark brown edges, which can cause leaves to wither and fall early. Stem lesions are grayish-brown, with soft tissue and slightly raised. Both leaves and stems can produce orange-red stick-like small spots, namely acervuli	The pathogen overwinters as mycelium, pycnidia, or stromata on infected branches. The mature conidia produced in the overwintering pycnidia in the next year are the initial infection source. Continuous cropping fields and older peonies are more severely affected

(continued)

Table 28.1 (continued)

Disease (Pathogen)	Symptoms	Occurrence pattern
Canker (*Botryosphaeria dothidea*)	It primarily affects the stems. The lesions are oval, fusiform or irregular in shape, appearing purple-brown with a grayish-white center and a sunken, ulcer-like appearance. As the disease progresses, numerous needle-like small black spots form beneath the cortex of the affected area, which are the pycnidia or stromata of the pathogen. In severe cases, the epidermis of the lesion cracks, and branches may even die. For plants with mild infection, the scale buds germinate slowly, the flower buds are smaller, flowering is delayed, and the flowers are smaller	The pathogen overwinters as mycelium, pycnidia, or stromata on infected branches. The mature conidia produced in the next year are the initial infection source. Continuous cropping fields and older peonies are more severely affected
Viral Discasc (PYV, PMV, PRSV, TRV)	Yellowing virus disease and mosaic virus disease are common in China. Their field symptoms include leaves turning greenish-yellow, developing mottling, becoming deformed and curlcd. Additionally, varying sizes of rings or wheel-patterned spots may form, which are sometimes irregular. Infected plants grow poorly and are stunted	Viruses can be transmitted by aphids and through sap friction. Poor management conditions and drough can exacerbate the severity of the disease
Yellow Spot Disease (*Phyllosticta commonisii*)	The disease initially presents as small, round or nearly round spots that are yellowish-brown or yellow-white and slightly sunken. As the disease progresses, the spots expand and merge, becoming yellow to brown, with no perforation occurring. Under humid conditions, small black dots emerge on the lesions	The pathogen overwinters as pycnidia and mycelium in diseased residues. Conidia are produced in the next spring and spread through wind and rain. Dry weather slows disease progression, while high humidity accelerates it. The disease generally peaks from August to September

(continued)

Table 28.1 (continued)

Disease (Pathogen)	Symptoms	Occurrence pattern
Circular Spot Disease (*Pestalotiopsis paeoniae*, *Pseudocercospora variicola*)	The lesions are brown or grayish-brown with dark centers, appearing circular or nearly circular, large in size, and with distinct concentric rings. Gray- black mold and small black spots arranged in a wheel-like pattern are present on the lesions, which often crack and perforate	The pathogen overwinters as mycelium or acervuli on infected plant parts or diseased tissues remaining on the soil surface. Under favorable conditions the following year, conidia are produced, which are dispersed by wind and rain. These conidia germinate in water droplets and invade the host through wounds or vulnerable areas. Once lesions form, new conidia are produced, enabling multiple rounds of reinfection that progressively aggravate the disease. The disease tends to intensify when leaves are aged or damaged, with heavy outbreaks occurring during the rainy summer and autumn seasons

Chapter 29
The Diseases of Schisandra

Schisandra Michx., a member of the Magnoliaceae family, is a valuable medicinal plant species native to China. The *Pharmacopoeia of the People's Republic of China* contains two species of *Schisandra chinensis* (*S. chinensis* (Tuecz.) Bail.) and (*S. sphenanthera* Rehd. et Wils.). The fruit is harvested in autumn when ripe, then dried, with stems and impurities removed, and the dried fruit is subsequently used in medicinal applications. According to the Tang dynasty's *Newly Revised Materia Medica*, *Schisandra* embodies "five flavors – sweet and sour in the skin and flesh, pungent and bitter in the seed, and salty overall," hence its name, which translates as *Schisandra*.

Schisandra is a perennial woody vine cultivated for its medicinal properties. *S. sphenanthera*, often referred to as "Southern *Schisandra*," is primarily found in provinces such as Jiangsu, Anhui, Zhejiang, Jiangxi, Fujian, Hubei, Hunan, Guangdong, Guangxi, Sichuan, and Yunnan. However, *S. chinensis*, commonly known as "Northern *Schisandra*" or "Liaoning *Schisandra*," is mainly distributed across Heilongjiang, Jilin, Liaoning, Inner Mongolia, Hebei, Shanxi, Ningxia, Gansu, and Shandong, with concentrated cultivation in northeastern China, where its quality is considered the highest. *Schisandra* fruit is rich in active components, including lignans, terpenes, polysaccharides, glycosides, and organic acids. The fruit is renowned for its effects of promoting attention, benefiting kidney energy, calming nerves, boosting immunity, and countering aging.

The primary diseases affecting *S. chinensis* include damping-off, seedling blight, leaf blight, basal stem rot, powdery mildew, ear rot, frost damage, sunburn, and herbicide damage, with leaf blight and basal stem rot posing particularly significant threats.

© China Agriculture Press Co., Ltd. 2026

T. Wang et al., *Atlas of Chinese Medicinal Plant Diseases*,
https://doi.org/10.1007/978-981-95-6326-5_29

29.1 *Schisandra chinensis* Damping-Off Disease

Damping-off is a common fungal disease affecting *S. chinensis* nursery seedlings. It is widespread across nursery fields in the northeastern regions, with clear infection centers and rapid spread, often resulting in large areas of seedling death within a short period. Typical infection rates range from 5% to 10%, but in severe cases, they can exceed 30%, significantly impacting *S. chinensis* nursery production.

Symptoms

Damping-off primarily affects the stem base or underground root portions of seedlings, initially forming oval or irregular dark brown lesions. Early symptoms in infected plants include daytime wilting and nighttime recovery. The affected area gradually becomes sunken and constricted, darkening to a black-brown hue. When lesions encircle the stem, seedlings wither and die, remaining upright or collapsing. Mildly infected plants show only brown, sunken lesions without withering. Distinct infection centers are evident in the field and spread rapidly, resulting in patches of dead seedlings (Fig. 29.1). In high humidity conditions, faint brown and web-like fungal growth may be visible on affected areas.

Pathogen

The pathogen responsible for *S. chinensis* damping-off is *Rhizoctonia solani* Kühn, a basidiomycete fungus in the genus Rhizoctonia. On PDA medium, mycelium grows radially, expanding rapidly; initial colonies are light gray, gradually darkening to yellow-brown or brown with time. The hyphae branch at right angles, with slightly constricted bases and septa, measuring 8–12 μm in diameter. In later stages, the mycelium thickens, with prominent septa and a darker hue, interweaving to form sclerotia. The sclerotia are brown to dark brown, irregular in shape, with diameters of 1–3 mm. The surface cells of the sclerotia are short, loose, and rough (Fig. 29.2).

Occurrence Pattern

The pathogen overwinters as mycelium and sclerotia in soil or on infected plant residues, serving as a primary infection source the following year. *Rhizoctonia solani* is a typical soil-dwelling fungus with strong saprophytic properties, capable of surviving in soil for 2–3 years. The pathogen spreads via rain, irrigation, and farming operations, entering the seedling stem base or root wounds, or directly penetrating the host epidermis. In northeastern China, disease onset typically begins in spring (April–May) when soil temperatures exceed 15 °C. The disease peaks in June, with spread slowing or halting in July and August. In autumn (September), as temperatures decrease, unprotected fields may experience a resurgence of infection. Heavy soils, high planting density, low-lying terrain, waterlogging, or over-watering predispose fields to Schisandra damping-off.

Control Measures

1. Seed Disinfection: Before sowing, treat seeds with 0.1–0.3% (w/w) of 50% carbendazim WP or 50% procymidone WP.

Fig. 29.1 Symptoms of *Schisandra chinensis* Damping-Off Disease. (**a**, **b**) Comparison of healthy (left) and infected (right) plants; (**c**, **d**) Field symptoms

2. Soil Treatment with Fungicides: In heavily infected areas, apply 50% carbendazim WP, 75% chlorothalonil WP, 50% thiram WP, 50% iprodione WP, or 65% zineb WP at 10–15 g/m^2, incorporating it into the top 5 cm of soil for disinfection.
3. Scientific Management of Seedbeds: Choose fertile, loose, well-aerated soil, preferably sandy loam, for seedbeds. Create raised beds to prevent waterlogging and ensure proper drainage during rainy seasons. After germination, loosen soil frequently to increase soil temperature and maintain aeration. Use clean straw or pine needles as mulch, applied lightly. Ensure that farm manure is fully decomposed to avoid root rot or stem scorch. Pre-water the seedbed thoroughly before sowing, then water sparingly and frequently after germination, reducing water frequency later to avoid excessive soil moisture.
4. Immediate Removal of Infected Plants: Regularly inspect nursery fields and promptly remove or isolate infected plants. Treat infected spots with 250–500-fold diluted 50% carbendazim WP or 200-fold diluted 40% hymexazol WP, or apply lime to prevent further spread.
5. Early-Stage Chemical Control: At the onset of infection, irrigate the bed surface with 50% carbendazim WP, 200–300-fold diluted 10% amino acid-copper AS, or 800–1,000-fold diluted 30% metalaxyl-hymexazol solution, ensuring penetration of 3–5 cm into the soil. Increase irrigation in infected areas.

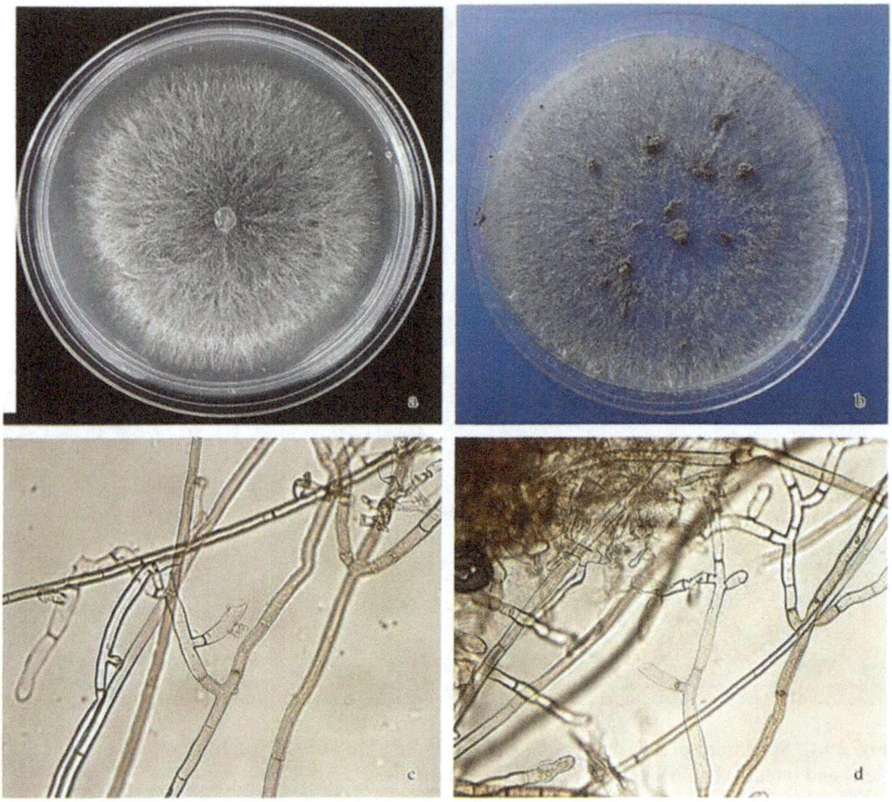

Fig. 29.2 Morphology of *Schisandra chinensis* damping-off pathogen. (**a**) 3-day culture characteristics; (**b**) 7-day culture characteristics; (**c**, **d**) Mycelium

29.2 *Schisandra chinensis* Seedling Damping-Off Disease

Damping-off disease is one of the common diseases affecting Schisandra nurseries. It is prevalent across nursery fields, frequently causing widespread seedling death with distinct disease centers. The infection rate is around 5–10%, often occurring alongside seedling blight, which exacerbates the overall disease impact. Therefore, it is crucial to implement preventive measures in nursery fields.

Symptoms

This disease primarily targets the basal stem of seedlings, initially forming light brown, water-soaked lesions. As the lesions spread, they encircle the stem base, causing shrinkage and brown rot. The stems and leaves above the affected areas initially remain green, but soon exhibit wilting symptoms, resulting in cluster death with clear disease centers. Under high humidity conditions, white, cottony mycelium may be observed on the infected areas and soil surface (Fig. 29.3).

Fig. 29.3 Field symptoms and typical symptoms of *Schisandra chinensis* Damping-Off Disease

Pathogen

The pathogen responsible for *S. chinensis* damping-off is *Pythium debaryanum* R. Hesse, from the Pythiaceae family within the Oomycetes order. The pathogen forms white, cotton-like, branched mycelium that lacks septa and measures 2–6 μm in diameter. Zoosporangia are terminal or intercalary, spherical to nearly spherical, or irregularly lobed, with a diameter of 15–25 μm; mature sporangia do not easily detach. During germination, a discharge tube forms from the zoosporangium, expanding into a vesicle that contains 30–38 zoospores. When the vesicle ruptures, kidney-shaped, motile zoospores measuring (4–10) μm × (2–6) μm are released with two lateral flagella. Each oogonium contains a single spherical, light yellow oospore.

Occurrence Pattern

The pathogen overwinters as oospores or mycelium in soil or plant debris, where it can survive for extended periods. It primarily spreads through rainwater, sprinkler irrigation, and contaminated organic fertilizers or tools. The pathogen reproduces most rapidly at soil temperature of 15–16 °C, with 10 °C being optimal for disease development. Therefore, the disease tends to occur in early spring when nursery bed temperatures are low and humidity is high. Conditions of inadequate sunlight, excessive sowing density, and weak seedling growth increase disease severity. Areas with water accumulation after irrigation and low-lying regions are particularly prone to infection, often becoming disease centers.

Control Measures

1. Site Selection: Select nursery sites with higher elevations, level ground, and good drainage.

2. Field Management: Promptly remove diseased seedlings and apply fertilizers as appropriate to encourage robust seedling growth. Ensure timely drainage of nursery beds to reduce soil moisture. Implement proper plant spacing and maintain ventilation and sunlight penetration to lower canopy humidity, which are essential for minimizing disease occurrence.
3. Chemical Control: After removing diseased seedlings, disinfect affected areas with quicklime or apply fungicides such as a 500-fold dilution of 70% mancozeb WP, a 500-fold dilution of 58% metalaxyl-mancozeb WP, a 600-fold dilution of 34% kasugamycin-dimethomorph SC, or an 800–1,000-fold dilution of 30% metalaxyl-dimethomorph solution.

29.3 *Schisandra chinensis* Leaf Blight

Leaf blight is a significant leaf disease of *S. chinensis*, widely prevalent and severe across major production areas such as Liaoning, Jilin, and Heilongjiang. The disease affects nearly 100% of fields, with leaf infection rates between 20% and 30%, and in severe cases, exceeding 50%. This often results in premature leaf drop, dieback of new shoots, weakened tree vigor, reduced fruit quality, and lower yields. Severe outbreaks can also induce sunscald in *S. chinensis*.

Symptoms
S. chinensis leaf blight primarily damages leaves. Early symptoms appear as lesions that start at the leaf tips or edges, although they may also occur in the central part of the leaf. Typically, symptoms begin on lower leaves and gradually spread upward. As the disease progresses, lesions extend along the leaf margins and midribs, causing large areas of leaf tissue to wither. If lesions occur in the center of the leaf, they are generally circular or oval, with a brown, reddish-brown, or dark brown color, often displaying faint or prominent concentric rings. Under high humidity, a dark brown mold layer may appear on the lesions, while in dry conditions, affected areas desiccate, and leaf tips or edges may crack, curl, and wrinkle, with diseased areas prone to detaching. In severe cases, lesions can spread over half of the leaf or even kill the entire leaf, leading to shriveled fruit and premature fruit drop (Fig. 29.4).

Pathogen
The pathogen causing *S. chinensis* leaf blight is *Alternaria tenuissima* (Nees: Fr.) Wiltshire, a fungus in the genus *Alternaria* (Deuteromycota). On PDA medium, *A. tenuissima* colonies appear gray-green with abundant aerial mycelium, growing at a fast rate. Conidiophores are mostly solitary, though occasionally clustered in small groups, standing upright or slightly curved, pale brown to dark brown, with slight basal swelling and septa, measuring (25.0–70.0) μm × (3.5–6.0) μm. Conidia are brown, mostly clavate, with fewer ovoid or near-oval forms, containing three to seven transverse septa and one to six vertical (or oblique) septa, with constrictions at the septa, measuring (22.5–47.5) μm × (10.0–17.5) μm. The beak is columnar, pale brown, septate, measuring (40–35.0) μm × (3.0–5.0) μm (Fig. 29.5).

Fig. 29.4 Field symptoms of *Schisandra chinensis* leaf blight and typical leaf symptoms

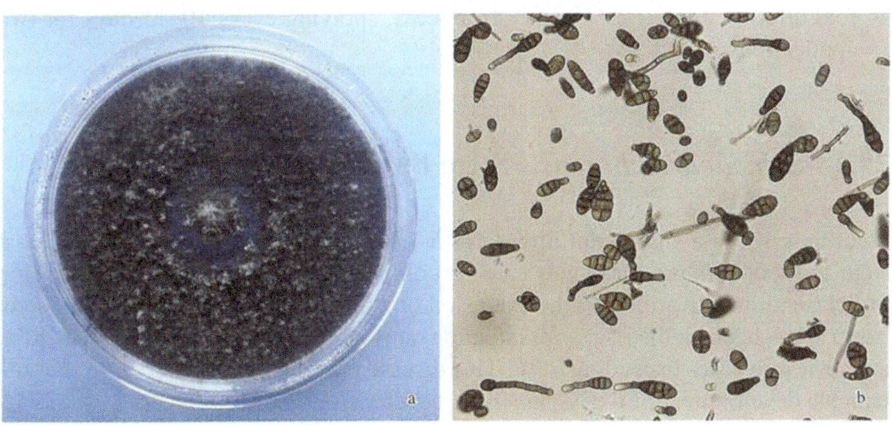

Fig. 29.5 Morphology of *Schisandra chinensis* leaf blight pathogen. (**a**) Colonial morphology; (**b**) Conidiophore and Conidia structure

Occurrence Pattern

S. chinensis leaf blight typically begins in late May, with high temperature and humidity serving as main factors for disease development. During rainy weather from July to September, the disease spreads rapidly, marking the peak period. Plants with weak growth or excessive fruit loads are more severely affected. Within the same orchard, areas with low-lying terrain, poor drainage, and sprinkler irrigation show higher disease incidence. Older orchards with advanced tree age, excessive nitrogen fertilization, or poor ventilation are also more susceptible. Severe outbreaks lead to extensive leaf necrosis, significantly weakening the tree and triggering a large incidence of sunscald on subsequent fruits.

Control Measures

1. Scientific Site Selection: Since Schisandra has a long planting period, select well-drained sites with deep, loose soil high in organic matter, preferably on sunny, gently sloping or flat ground. Avoid low-lying, saline-alkaline, wind-exposed, or gravelly areas.
2. Enhanced Cultivation Management: Choose an appropriate trellis system based on local conditions, with single-trellis systems recommended for easier management. Prune carefully to ensure a balanced distribution of branches, prevent canopy crowding, and improve ventilation and sunlight penetration. Balance fertilization, increasing the proportion of phosphorus and potassium to enhance plant resistance.
3. Field Sanitation: Remove infected leaves as soon as they appear to reduce disease spread. In late autumn or early winter, clear out diseased leaves, branches, weeds, and fallen fruits, and dispose of them outside the orchard to reduce the initial source of infection for the following year.
4. Chemical Control: Spray a 1:1:100 Bordeaux mixture for prevention in late May. At disease onset, apply a 500–600-fold dilution 50% mancozeb WP every 7–10 days, for 2–3 consecutive sprays. Alternatively, use a 200-fold diluted 3% pyrimethanil AS, a 1,000–1,500-fold diluted 10% kasugamycin WP, or a 1,000–1,500-fold dilution diluted 25% azoxystrobin SC, spraying every 10–15 days for two applications.

29.4 *Schisandra chinensis* Stem Base Rot

Stem base rot is a significant disease in *S. chinensis* production, leading to basal stem and root bark rot, ultimately causing complete plant death. With the expansion of Schisandra cultivation and the extension of planting years, the incidence of stem base rot has been on the rise, with infection rates typically ranging from 2% to 40%, and in severe cases exceeding 70%. This disease presents a major threat to the sustainable development of the *S. chinensis* industry.

Symptoms

S. chinensis stem base rot primarily affects the basal stem and root. It occurs across Schisandra plants of all ages but is most common in 2- to 3-year-old plants. Early symptoms include wilting and drooping of leaves that resemble water stress but do not recover, followed by gradual leaf desiccation, ultimately resulting in complete wilting of the aerial parts. In the early stages, peeling the bark of the basal stem reveals slight yellow-brown discoloration in the cortex. In advanced stages, the cortex decays, darkens, and becomes easily detached. Vertical sectioning of the affected area shows black-brown vascular bundles. Under favorable conditions, lesions spread upward and downward, leading to root bark decay and detachment (Fig. 29.6). In humid conditions, pink or white fungal layers may appear on the lesions, which, under microscopic examination, reveal an abundance of *Fusarium conidia*. Disease incidence tends to increase annually in older fields following the fruiting period, sometimes leading to extensive field loss.

Pathogen

The pathogen causing *S. chinensis* stem base rot belongs to the *Fusarium* genus in the Ascomycota phylum. According to Xue Caiyun (2017), four *Fusarium* species are responsible for stem base rot in northeastern China: *F. oxysporum* Schltdl. ex Snyder et Hansen, *F. solani* (Mart.) Appel et Wollenw. ex Snyder et Hansen, *F. equiseti* (Corda) Sacc., and *F. semitectum* Berk. et Ravenel. These pathogens can be isolated from infected plants, although their frequency varies significantly by region. Among them, *F. oxysporum* shows the highest frequency and strongest pathogenicity. On PDA medium, these fungi exhibit robust red, woolly growth, with the underside of the colony appearing felt-like and dark purple. Pathogens generally produce large and small conidia. Small conidia are oval with zero to one septum and borne on elongated conidiophores, while large conidia are slightly curved, blunt-tipped, and typically have three to five septa, measuring (25–30) μm × (3.8–4.9) μm. Under stressful conditions, near-spherical, darkened chlamydospores may also form (Fig. 29.7).

Fig. 29.6 Field symptoms and typical damage symptoms of *Schisandra chinensis* stem base rot

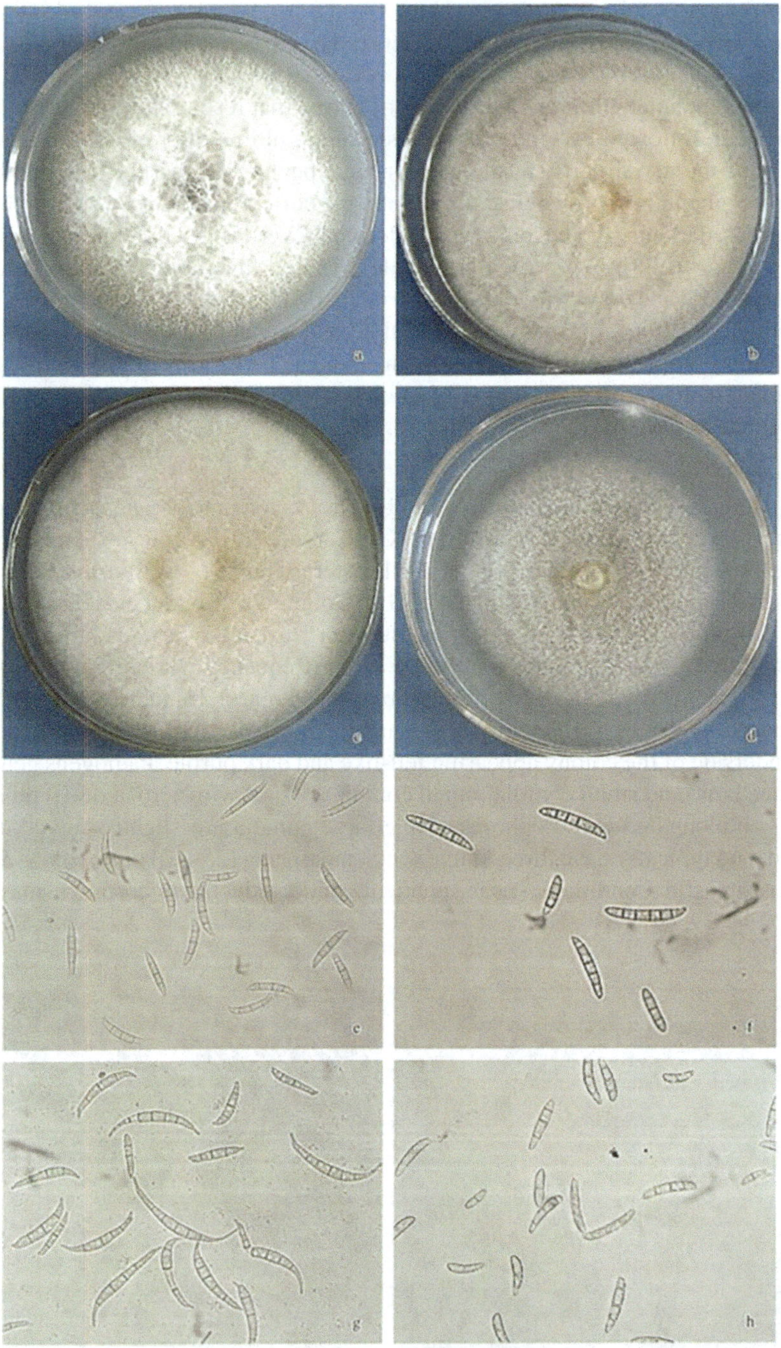

Fig. 29.7 Morphology of *Fusarium* spp. causing *Schisandra chinensis* stem base rot. (**a**) *Fusarium oxysporum* colony morphology; (**b**) *F. semitectum* colony morphology; (**c**) *F. equiseti* colony morphology; (**d**) *F. solani* colony morphology; (**e**) *F. oxysporum* Conidia; (**f**) *F. semitectum* Conidia; (**g**) *F. equiseti* Conidia; (**h**) *F. solani* Conidia

Occurrence Pattern

The pathogen, a typical soil-dwelling organism, primarily spreads through soil but can also be transmitted by infected seedlings. Disease outbreaks typically occur from early May to late August, with initial symptoms in early May and peak infection in early June. The disease thrives in high-temperature, high-humidity, and rainy conditions, worsening when sunny weather follows rain. Factors such as soil pests, nematodes, and transplant injuries, as well as poor root development, further predispose plants to infection. Low winter temperatures that cause frost damage increase disease severity in the following season. *S. chinensis* plants in low-lying, waterlogged areas are particularly susceptible. In practice, infected seedlings often carry the pathogen into new fields, where it spreads through transplant wounds and during the extended recovery period post-transplant. In newly established fields, 2-year-old Schisandra plants show the highest disease incidence. As plants mature, thickening of the phloem and increased robustness reduce pathogen susceptibility. However, in older orchards—particularly those over 10 years old—stem base rot incidence increases significantly following the peak fruiting period.

Control Measures

1. Scientific Site Selection: Since *S. chinensis* has a long cultivation period, it is best to select sites with deep, loose, humus-rich soil on sunny, gently sloping, well-drained terrain with moderate sand content. Avoid planting in low-lying, saline-alkali, windy, or gravelly soils.
2. Field Management: Maintain field cleanliness, promptly remove and destroy infected plants, and disinfect affected areas with a 600-fold dilution of 50% carbendazim WP. Apply balanced nitrogen with additional phosphorus and potassium to improve plant resistance. Drain water promptly after rain to prevent waterlogging, and avoid planting *S. chinensis* in areas with a history of severe root rot.
3. Seedling Disinfection: Select healthy, disease-free seedlings and soak them in a 600-fold dilution of 50% carbendazim WP or 80% mancozeb WP for 4 h before planting.
4. Chemical Control: Preventive measures are essential for controlling Schisandra stem base rot. Before disease onset, apply a 600-fold dilution of 50% carbendazim WP, ensuring the solution flows down the stem into the soil. At the initial stage of infection, treat roots with a 600-fold dilution of 50% carbendazim WP, 200–300-fold dilution of 10% mixed amino acid copper AS, or a 1,500–2,000-fold dilution of 70% thiophanate-methyl SP. Repeat root treatments every 7–10 days, for two to three applications.

29.5 *Schisandra chinensis* Powdery Mildew

Powdery mildew is a common disease in *S. chinensis* production. It occurs in both nursery and production fields across Liaoning, Jilin, and Heilongjiang provinces. The disease is relatively mild during the seedling stage but becomes more severe as the fruit matures. In serious cases, the infection rate of diseased fruits can reach 10–25%, significantly affecting both yield and quality.

Symptoms

S. chinensis powdery mildew primarily affects leaves, young fruits, and new shoots, but it can also damage petioles, fruit stems, and tender stems. Young leaves and fruits are especially susceptible. The disease often leads to leaf desiccation, shoot dieback, and fruit drop. Initially, infected leaves develop small pinpoint spots on the underside, gradually covered with a white powdery substance consisting of fungal mycelium, conidia, and conidiophores. In severe cases, the entire leaf becomes coated, changing from green to yellow, with edges curling upward, wilting, or even dropping off. The disease usually begins near the spike axis on young fruits and gradually spreads to cover the entire fruit cluster, stunting growth and leading to small, shriveled fruits or fruit drop. After ripening, the impact of disease decreases. When new shoots are heavily infected, they may fail to unfold or may even die (Fig. 29.8). In later growth stages, small black dots may appear on infected areas, indicating the formation of chasmothecia by the pathogen.

Fig. 29.8 Symptoms of *Schisandra chinensis* powdery mildew. (**a**) Fruit symptoms; (**b**) Leaf symptoms; (**c**) Young leaf symptoms; (**d**) Young stem symptoms

Pathogen

The causative agent of *S. chinensis* powdery mildew is *Erysiphe schisandrae* (Sawada) U. Braun & S. Takam. (synonym: *Microsphaera schisandrae* Sawada), from the Erysiphales order in the Ascomycota phylum. The white powdery substance on affected areas consists of mycelium, conidia, and conidiophores. Mycelium grows on both sides of the leaf and on petioles. Conidia are solitary, colorless, and either oval, ovoid, or nearly cylindrical, measuring 24.2–38.5 μm × 11.6–18.8 μm. Chasmothecia are scattered to clustered, compressed-spherical, dark brown, and 92–133 μm in diameter. They typically have 7–18 appendages, mostly 10–14, each measuring 93–186 μm in length, which is 0.8–1.5 times the diameter of the chasmothecia. The appendage bases are 8.0–14.4 μm thick, either straight or slightly curved, and occasionally knee-shaped. The basal portion of the chasmothecia's outer wall is rough, becoming smoother toward the top, with no septa or occasionally one below the midpoint. The apex divides into four to seven dichotomous branches, typically five to six. Each chasmothecium contains (four to eight) asci, which are round, ovoid, or broadly ovoid, measuring (54.4–75.6) μm × (32.0–48.0) μm. Ascospores are (three to seven) per ascus, colorless, and oval or ovoid, measuring (20.8–27.2) μm × (12.8–14.4) μm (Fig. 29.9).

Occurrence Pattern

High temperatures and dry conditions favor powdery mildew development. In Northeast China, disease onset is typically in late May to early June, with peak incidence from late June to late July. Conditions such as dense branches, excessive nitrogen fertilization, and poor ventilation contribute to disease progression. The pathogen overwinters as mycelium, ascospores, and conidia in infected plant debris. In the following spring, as the average temperature reaches 15–20 °C between mid-May and early June, overwintering conidia begin to germinate, aided by rainfall and dew, initiating infection in the field. Mid-July marks the peak period for conidia dispersal, leading to a rapid increase in infected leaves and stems, with a high rate

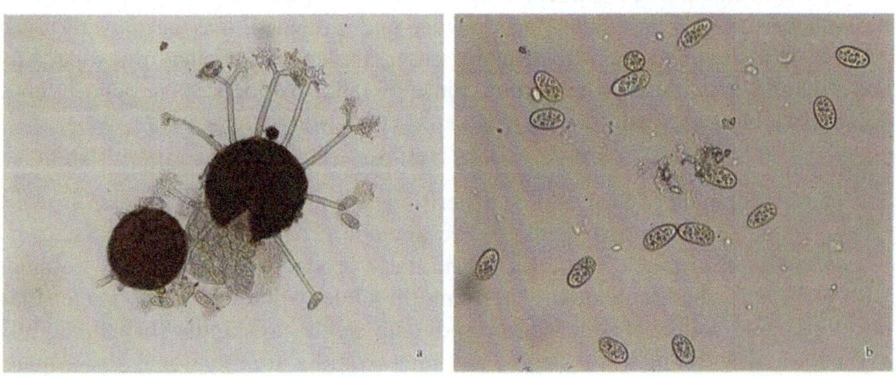

Fig. 29.9 Morphology of *Schisandra chinensis* powdery mildew pathogen. (**a**) Appendages, Chasmothecia, Asci, and Ascospores; (**b**) Conidia

of fruit infection. By mid-October, as temperatures drop, *S. chinensis* leaves wither and fall, leaving infected debris in the field, where the pathogen enters dormancy.

Control Measures

1. Enhanced Cultivation Management: Maintain balanced branch distribution and prune to improve airflow and light penetration. Increase the proportion of phosphorus and potassium fertilizers to strengthen plant resistance and overall vigor.
2. Removal of Infection Sources: Before budbreak, remove diseased branches and leaves. In the early infection stage, promptly prune infected clusters and gather fallen diseased fruits to reduce sources of infection.
3. Chemical Control: In late May, apply a 0.3–0.5 Baumé lime sulfur mixture, *Bacillus subtilis* WP at 1,350–1,800 g/ha (100 billion CFU/g), 1000–2000-fold diluted 1% osthole solution, 800–1000-fold diluted 25% triadimefon WP, or 800–1000-fold diluted 70% thiophanate-methyl WP for prevention. At the early stage of disease onset, use a 400–500-fold diluted 40% sulfur SC, 1500–2000-fold diluted 15% triadimefon EC, 1500-fold diluted 25% azoxystrobin SC, or 3000–4000-fold diluted 50% pyraclostrobin DS. Spray every 7–10 days for two consecutive treatments.

29.6 *Schisandra chinensis* Cluster Rot Disease

Cluster rot is a newly emerging disease affecting *S. chinensis* production, frequently occurring in older cultivation areas. The disease generally impacts 3–10% of clusters, causing fruit rot and drop, significantly affecting *S. chinensis* yield.

Symptoms

S. chinensis cluster rot primarily affects the fruit, with symptoms occurring throughout the fruiting period and becoming most severe during ripening. In the early stages, tiny brown or yellow-brown spots appear at the tips or central sections of the fruit clusters, expanding to infect the entire fruit. This leads to shriveled, rotten, easily detached fruit, which later becomes dry and hardened, leaving only the seed encased in a shriveled peel. In advanced cases, the entire fruit may rot, especially under humid conditions, when white, pink, or reddish fungal spores and hyphae become visible on the affected areas (Fig. 29.10).

Pathogen

The causal pathogen of Schisandra cluster rot is a fungus from the genus *Fusarium* (Ascomycota). Isolated pathogenic species include *F. tricinctum* (Corda) Sacc. and *F. equiseti* (Corda) Sacc. *F. tricinctum* initially produces white, woolly colonies with well-defined edges and an underlying pink hue, while the colony center may develop a pale yellow shade. Macroconidia are sickle- to spindle-shaped, slightly curved, often with one to two septa, and microconidia are round and non-septate.

Fig. 29.10 Field symptoms of *Schisandra chinensis* cluster rot. (**a**) Early symptoms; (**b**) Late symptoms

F. equiseti produces abundant, woolly or cotton-like mycelium, initially white to pink and later light tan, with a pale yellow to tan colony base. Few microconidia are present, which are ovoid, while macroconidia are sickle-shaped and curved, with the central cells distinctly swollen. The apical cell is tapered into a cone shape, with three to seven septa, most commonly three to five, measuring (39.62–45.72) μm × (3.44–4.78) μm (Fig. 29.11).

Occurrence Pattern

In northeastern China, *S. chinensis* cluster rot disease begins from late May to mid-June and can recur throughout the fruit ripening period. Field surveys indicate that weaker plants or those with an excessive fruit load show higher infection rates. High incidence of early-season leaf blight or sunscald aggravates cluster rot incidence. Disease spread is accelerated in densely planted fields and during periods of prolonged rainfall.

Control Measures

1. Field Management: Ensure proper plant spacing to promote ventilation and light penetration. Apply a balanced mix of nitrogen, phosphorus, and potassium fertilizers, with additional potassium to enhance disease resistance. In autumn, remove diseased leaves and fruits from the field to reduce the pathogen reservoir.
2. Leaf Disease Prevention: Effective control of early-season leaf blight and sunscald can help reduce cluster rot incidence later.
3. Chemical Control: At the onset of the disease, apply a 600-fold dilution of 50% carbendazim WP, a 200-fold dilution of 3% nucleoside antibiotic AS, or a 1,000–1,500-fold dilution of 10% polyoxin WP.

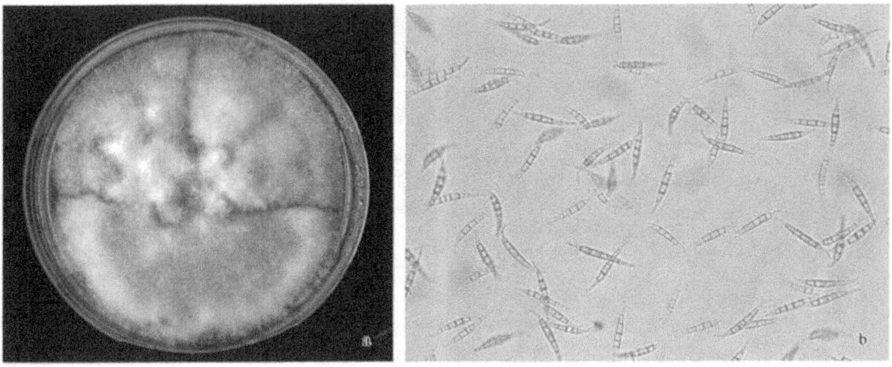

Fig. 29.11 *F.equiseti* morphology. (**a**) Colony morphology; (**b**) Conidia

29.7 *Schisandra chinensis* Frost Injury

Frost injury is a prevalent noninfectious disease in *S. chinensis* production areas across Northeast China, occurring to varying degrees every year from late April to mid-May, especially in colder regions such as Jilin and Heilongjiang. Differences in orchard location, topography, cultivation management, and climatic conditions lead to considerable variability in frost damage across regions and years. If frost occurs during the flowering period, it can result in extensive frost damage to leaves and flower clusters, which severely affects *S. chinensis* yields for that year, significantly reduces tree vigor, and often leads to more severe infestations of pests and diseases later in the season.

Symptoms
Frost injury affects the entire Schisandra plant, with the most severe impact on leaves, buds, and flower clusters. In mild cases, shoot tips experience frostbite, and leaves develop brown or yellow-brown spots, with the foliage color fading. Severe frost can cause large areas of leaves to suffer water-soaked necrosis, with edges curling. Peduncles and fruit stalks may shrivel and die, and flower clusters may wilt, ultimately leading to the death of the entire plant. Later in the disease, young shoots experience severe dehydration and wilting, tissue desiccation and necrosis, and the drying and shedding of leaves, flower clusters, and buds, which weakens the tree significantly (Fig. 29.12).

Cause of Disease
The main cause of frost injury in *S. chinensis* is low spring temperatures. After *S. chinensis* begins budding in spring, sudden temperature drops at night can cause water vapor to condense as frost, resulting in frostbite to tender plant parts. Frost injury is also closely related to topography, as cold air, being heavier, accumulates more readily in low-lying areas than in flat regions, with such areas experiencing larger and longer-lasting drops in temperature. *S. chinensis* orchards located in valleys or depressions where cold air can pool are particularly prone to late frost damage.

Fig. 29.12 *Schisandra chinensis* frost injury symptoms

Occurrence Pattern

In Northeast China, *S. chinensis* frost injury typically occurs from late April to mid-May, with isolated years in Jilin and Heilongjiang experiencing frost as late as early June. Cold tolerance varies among *S. chinensis* varieties, with early-maturing varieties being more susceptible to frost damage and thus suffering greater yield losses. Tree structure and vigor also influence frost tolerance, with weaker trees being more severely affected. Mature, lignified branches, with lower water content, higher cell sap concentration, and greater starch reserves, exhibit stronger cold resistance. Moreover, management practices affect frost susceptibility: orchards with higher soil moisture and those using sprinkler irrigation experience less damage, whereas nonirrigated areas are more severely affected. Since the peak frost period coincides with the flowering stage, preventive measures in spring are essential to mitigate potential yield losses.

Control Measures

1. Scientific Site Selection: Choose sites on sunny slopes or flat areas for orchard establishment, avoiding frost-prone valleys and depressions to reduce late frost damage.
2. Ground Covering: Use straw or corn stalks to cover *S. chinensis* roots, delaying soil warming in spring to postpone leaf-out and flowering and thus avoid frost damage.
3. Smoke Insulation: After *S. chinensis* budbreak, monitor local weather forecasts. On nights when frost is likely, ignite piles of damp branches, leaves, wood chips, or artemisia, covering them lightly with soil to prolong combustion. Place smoke piles around the orchard and along operational paths, with additional smoke piles on the windward side to ensure smoke envelops the orchard quickly.
4. Sprinkler Irrigation for Insulation: Based on weather forecasts, irrigate the ground heavily or use canopy sprinkling to maintain warmth.
5. Foliar and Fertilizer Sprays: Apply nitrogen fertilizer during the growing season to promote robust branch growth and overall tree vigor. Later in the season, apply appropriate amounts of phosphorus and potassium fertilizers to promote early cessation of branch growth, tissue maturation, and extended nutrient accumulation, thereby enhancing cold resistance.

29.8 *Schisandra chinensis* Sunscald Disease

Sunscald disease is a prevalent noninfectious issue in *S. chinensis* production, leading to significant losses every year. Typically, sunscald disease affects 10–30% of crops, resulting in scorched and sometimes rotting fruits, which seriously impacts both the yield and quality of *S. chinensis*.

Symptoms
Sunscald mainly affects *S. chinensis* fruits, leading to blistered, sunken, hardened leathery spots on affected fruit surfaces. Early symptoms appear as white, yellow, or pink discolorations, which later darken to yellow-brown or brown (Fig. 29.13). In severe cases, necrotic sunken spots develop in the fruit tissue, with portions of the pulp decaying. These injured areas are particularly vulnerable to secondary infections, accelerating fruit decay and increasing field losses.

Causes of Disease
Schisandra sunscald primarily results from heat stress and UV radiation. Heat stress occurs when high surface temperatures on the fruit lead to scorching without direct sunlight, whereas UV damage is caused by ultraviolet exposure that compromises cellular integrity. Several factors influence sunscald, including temperature, light, relative humidity, wind speed, variety, fruit development stage, and tree health, with temperature and light intensity being the key factors.

Temperature: Air temperature significantly influences sunscald risk for Schisandra fruit. During hot, sunny summer days, fruit surface temperatures can reach 40–50 °C, far exceeding the ambient air temperature. Studies indicate that sunscald occurs at temperatures around 30–32 °C, with higher temperatures shortening the onset time and increasing severity.

Fig. 29.13 *Schisandra chinensis* sunscald symptoms in field

Light: Light intensity and UV exposure also contribute to sunscald. Under natural conditions, sunlight absorbed by the fruit raises its surface temperature. This thermal effect, combined with ambient heat, brings the fruit surface to a critical threshold, triggering sunscald.

Occurrence Pattern

Sunscald in *S. chinensis* typically occurs from June through September, peaking in July and August during high-temperature, high-sunlight periods. Observationally, the peak sunscald period aligns with the hottest time of year. Clear, sunny days following rainy spells exacerbate sunscald risk, whereas cooler sunny days result in lower incidence. Higher relative humidity can reduce sunscald rates, while wind influences fruit temperature by regulating transpiration. Light breezes can lower surface temperature, thus decreasing sunscald incidence. Sunscald sensitivity varies by variety, and fruit at different development stages also show different resistance levels, with susceptibility decreasing as fruit ripens. Within the same orchard, more vigorous trees are less susceptible, while weaker trees experience higher sunscald incidence. Fields with severe early-season leaf blight also tend to have higher sunscald rates.

Control Measures

1. Enhanced Cultivation Management: Adjust the leaf-to-fruit ratio by retaining more foliage during pruning to shield fruit from direct sunlight. Optimize branch-to-fruit and leaf-to-fruit ratios, and conduct timely thinning of flowers and fruit. Avoid excessive nitrogen fertilization, and prioritize organic fertilizers to improve soil moisture and nutrient retention, promote deeper root growth, and increase drought resilience.
2. Sprinkler Irrigation for Cooling: Prior to expected high temperatures, sprinkler irrigation can effectively lower fruit surface temperature, reducing sunscald incidence.
3. Bagging for Sun Protection: Bagging fruit can help reduce light exposure and fruit surface temperature, effectively lowering sunscald risk.
4. Spray foliar fertilizer and plant growth regulators:Once mild phytotoxicity is found in *Schisandra chinensis*, targeted foliar fertilizers and plant growth regulators should be sprayed in a timely manner. Plant growth regulators have a good stimulating effect on the growth and development of crops. At the same time, they can also use micro-fertilizers such as zinc, iron, molybdenum, and foliar fertilizers to promote crop growth and effectively reduce pesticide damage. In general, after drug damage occurs, quick acting fertilizers such as 1–2% urea and 0.3% potassium dihydrogen phosphate can be sprayed to promote the growth of *Schisandra chinensis* and improve its resistance to drugs. The commonly used plant growth regulators mainly include gibberellin, brassinolide, etc. In severe cases, 10-40 mg/kg gibberellin or 1 mg/kg brassinolide can be sprayed two to three times in a row, and timely fertilization and watering can effectively accelerate the recovery and growth of *Schisandra chinensis*.

Chapter 30
The Diseases of Honeysuckle

Honeysuckle (*Lonicera japonica*Thunb.) is a perennial, semievergreen climbing plant from the *Lonicera* L. family, whose stems, branches, and flowers are widely used in herbal medicine. It is cultivated across most regions of China, with Shandong, Hebei, and Henan serving as primary production areas. Honeysuckle is valued for its properties in clearing heat, detoxifying, and dispelling wind-heat, while honey suckle vines have similar uses in detoxifying, dispersing wind, and unblocking meridians. The primary diseases affecting honeysuckle include root rot, brown spot, powdery mildew, white mold, and anthracnose. Among them, root rot, powdery mildew, and brown spot causing the most significant damage.

30.1 Honeysuckle Root Rot Disease

Root rot is one of the most serious diseases affecting honeysuckle cultivation. This common root disease is characterized by complex pathogenic factors and high levels of severity, making it difficult to control. Field surveys in honeysuckle-growing regions of Shandong, Hebei, and Henan show varying incidence rates: trees younger than 5 years generally have an infection rate of around 10%; those aged 5–10 years have rates between 10% and 25%; and trees over 10 years can experience rates as high as 35%, with severe cases leading to total crop loss. This disease is widespread, with documented cases in the provinces of Shandong, Henan, Hebei, Jiangsu, Zhejiang, Guizhou, and Guangdong.

Symptoms
The disease primarily targets the plant's conductive tissues. In the early stages, yellow-brown lesions appear in the lower sections of the roots, causing wilting of branches and leaves. As the disease progresses, branches and leaves wither and fall off, the roots turn brown and rot, and eventually separate from the pith, leading to the entire plant's death (Fig. 30.1).

© China Agriculture Press Co., Ltd. 2026

T. Wang et al., *Atlas of Chinese Medicinal Plant Diseases*,
https://doi.org/10.1007/978-981-95-6326-5_30

Fig. 30.1 Root rot symptoms in Honeysuckle. (**a**) Infected plant in the field, (**b**) Widespread field infection (plants uprooted for disposal), (**c**) Withered branches, (**d**) Withered leaves, (**e**–**g**) Root symptoms

Pathogen

The causative agent of honeysuckle root rot is fungi from the *Fusarium* genus, involving single or multiple infections by different strains. Identified pathogens include *F. oxysporum* Schltdl. ex Snyder et Hansen, *F. solani* [Mart. Appel et Wollenw. ex Snyder et Hansen], and *F. incarnatum* [Rob. ex Desm.) Sacc.]. Among these, *F. incarnatum* has been identified as the dominant strain in Shandong. Colonies of *F. incarnatum* appear circular, with abundant woolly aerial mycelia initially white, later turning yellow, while the reverse side of the colony appears buff-colored. Conidiophores develop on the aerial mycelia, with spores varying significantly in shape and size. Large conidia are sickle-shaped, tapering at both ends with distinct basal cells and three to five septa, measuring 23.28–36.93 μm × 3.36–5.86 μm. Medium-sized conidia are spindle-shaped with three to five septa, measuring 8.16–12.97 μm × 2.15–4.37 μm. Small conidia are oval and aseptate, measuring 4.05–8.76 μm × 1.58–3.93 μm (Fig. 30.2).

Occurrence Pattern

The severity of honeysuckle root rot varies by tree age, region, and plot conditions. The pathogen overwinters in soil or plant debris, typically spreading through soil and irrigation from April to June, entering through the roots and moving upward. The disease peaks from June to August, with high temperatures, humidity, poor soil drainage, and older tree age increasing susceptibility.

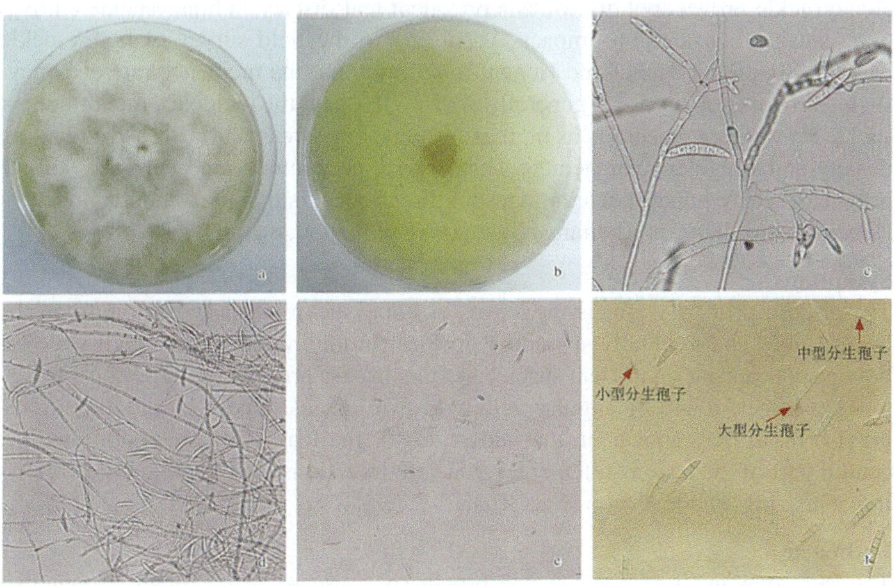

Fig. 30.2 Morphology of *Fusarium incarnatum*. (**a**, **b**) Colony morphology, (**c**) Conidiophore, (**d**) Mycelium, (**e**) Conidia, (**f**) Various conidia types

Control Measures

1. *Scientific Site Selection:* Choose sunny hillsides or slopes with well-drained sandy soil for cultivation.
2. *Seedling Management:* Use sterile nursery methods to cultivate strong seedlings before transplanting, enforce strict quarantine measures to prevent infected seedlings, and handle seedlings carefully to minimize injury during transport and planting.
3. *Field Management:* Ensure drainage ditches remain unobstructed, promptly drain excess water during rainy seasons, reduce soil humidity, and prune weak or overly vigorous branches to improve airflow and light penetration. Infected plants should be uprooted and destroyed immediately, and quicklime should be applied to disinfect the surrounding soil.
4. *Biological Control:* During March–April and June–September, apply a 400-fold dilution of *Bacillus polymyxa* (5 billion CFU/g) with a 200-fold dilution of amino acid fertilizer to the roots once per season, applying 2–4 kg per plant. In early stages of infection, apply a 100 mL/kg solution of 2% pyrimidine nucleoside antibiotic every 15 days for three to four applications.
5. *Chemical Control:* Apply a 500-fold dilution of 50% carbendazim WP to affected plants for effective control.

30.2 Honeysuckle Brown Spot Disease

Honeysuckle brown spot disease is a prevalent leaf disease in honeysuckle cultivation, causing considerable impact. According to the field survey of honeysuckle bases in Shandong, Hebei and Henan, it is found that the incidence rate of honeysuckle brown spot disease is generally 10% ~ 20%, and the incidence rate of honeysuckle brown spot disease is as high as 30% in some areas with excessive planting density and irregular management This disease is common, and currently known provinces where honeysuckle brown spot disease occurs include Shandong, Henan, Hebei, Gansu, Jiangsu, Zhejiang, Guangdong, Guizhou, etc.

Symptoms

The disease primarily affects the leaves of honeysuckle, usually beginning on the lower leaves and gradually spreading upward. In the initial stages, small yellow--brown spots appear on the leaf surface. As the disease progresses, these spots merge to form larger, round or angular yellow-brown lesions. Under humid conditions, grayish mold may appear on the undersides of the leaves; in dry conditions, the central part of the lesions often cracks. Severe infections lead to premature yellowing, withering, and leaf drop (Fig. 30.3).

Pathogen

The causative agent of honeysuckle brown spot disease is the fungus *Phomopsis capsici* (Magnaghi) Sacc., an asexual form within the ascomycete group. In the early stages, fungal hyphae appear white or colorless, with sparse, weak, cottony

Fig. 30.3 Honeysuckle Brown Spot Disease symptoms. (**a**, **b**) Field symptoms (**c**) Early symptoms (**d**) Merging lesions (**e**) Cracking lesions (**f**) Leaf withering

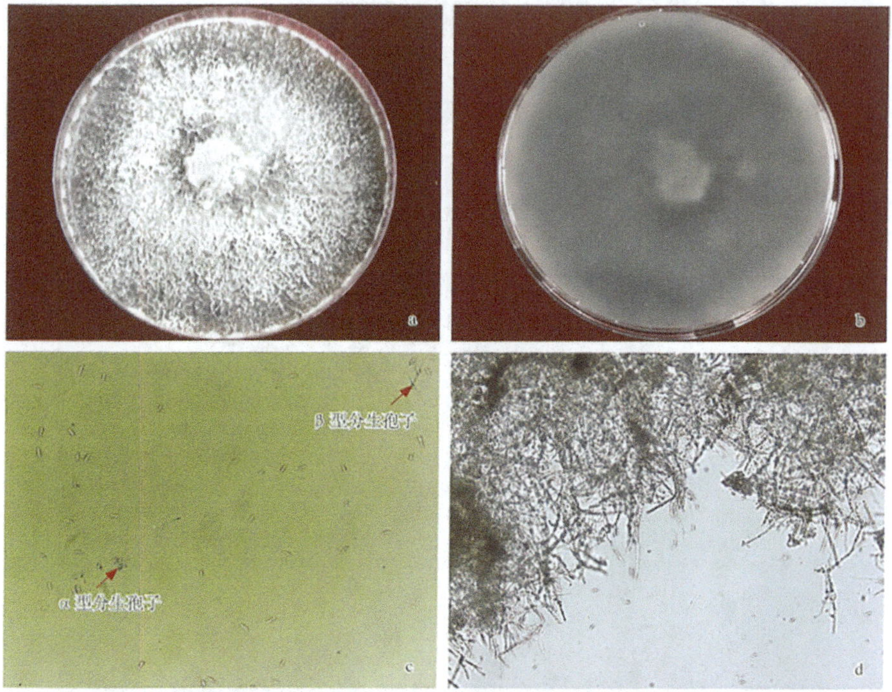

Fig. 30.4 Honeysuckle brown spot pathogen morphology. (**a, b**) Culture characteristics (**c**) Alpha and Beta Conidia (**d**) Hyphae

aerial growth that clings closely to the PDA medium. After about 15 days of culture, dense, wool-like hyphae develop, with spore-producing structures appearing later. Microscopic examination reveals two types of conidia. The alpha conidia are single--celled, colorless, nonseptate, typically ovoid or elliptical, with prominent oil droplets at both ends, measuring 5.43–9.31 μm × 1.05–2.15 μm. Beta conidia are also single-celled, colorless, and nonseptate, but they are more curved, often hook-shaped, without visible oil droplets, measuring 16.56–25.43 μm × 0.75–1.14 μm (Fig. 30.4).

Occurrence Pattern

The severity of honeysuckle brown spot disease varies across regions and cultivation sites. The pathogen overwinters in infected leaves, producing conidia the following June, which spread via wind and rain, infecting leaves through stomatal openings. The disease peaks from July to September, often leading to significant leaf drop in early autumn. High temperatures, high humidity, and dense foliage promote disease spread.

Control Measures

1. *Improved Field Management:* Combine winter pruning with the removal of infected branches and leaves, destroying these to reduce sources of infection.

During the rainy season, ensure proper drainage to lower soil humidity. Optimize planting density, and prune weak or excessively long branches to maintain adequate airflow and light penetration. Use organic bio-fertilizers. For mature honeysuckle plants, apply 4–5 kg of organic fertilizer per plant one to two times per year, either in early spring, after the first flower harvest, or in autumn, to boost plant immunity and resistance.

2. *Biological Control:* In early spring, apply a 3–5° Baumé lime–sulfur solution to honeysuckle fields as a preventive measure. During disease outbreak, use *B. methylotrophicus* (30 billion CFU/g) WP at a 500-fold dilution, spraying every 7–10 days for two to three applications.

3. *Chemical Control:* At the first signs of disease, apply Bordeaux mixture (1:15:300), 50% carbendazim WP at a 600-fold dilution, or 50% mancozeb wettable powder at a 500-fold dilution, rotating different fungicides as needed. Apply treatments every 10–15 days, with shorter intervals of 7–10 days during rainy periods, for a total of two to three treatments.

30.3 Honeysuckle Powdery Mildew

Powdery mildew is one of the primary leaf diseases affecting honeysuckle, commonly causing significant damage. According to the field survey of honeysuckle bases in Shandong, Hebei and Henan, it is found that the incidence rate of honeysuckle powdery mildew is generally 15% ~ 20%, and the incidence rate of honeysuckle in some honeysuckle planting areas can exceed 20%.. Currently, powdery mildew is widespread in several provinces, including Shandong, Hebei, Henan, Gansu, Ningxia, Jiangsu, Zhejiang, Hunan, Guizhou, and Guangdong.

Symptoms

The disease primarily infects honeysuckle leaves, though it may also spread to stems and flowers. Early symptoms appear on leaves as small white spots, gradually expanding into a powdery white coating across the leaf surface. In advanced stages, the entire leaf may be covered with a white powder, leading to yellowing, deformation, and even defoliation in severe cases. Stem lesions appear as irregularly shaped brown patches with white mildew, while flowers may become distorted and drop prematurely in severe infections (Fig. 30.5).

Pathogen

The causal agent of honeysuckle powdery mildew is *Erysiphe lonicerae* DC. Fr. [synonym: *Microsphaera lonicerae* (DC..) G. Winter], a species within the *Erysiphe* genus. The chasmothecia appear scattered, spherical, dark brown, and measure 63.34–102.51 μm in diameter, with 5–15 appendages that are 51.35–132.21 μm long, colorless, with or without a septum, and bifurcate three to five times. Each chasmothecium contains three to seven asci, which are ovoid or elliptical, measuring 35.52–56.21 μm × 28.64–46.53 μm, and contain two to five ascospores, which

Fig. 30.5 Powdery mildew symptoms in Honeysuckle. (**a**, **b**) Field symptoms, (**c**) Early-stage infection, (**d**) Late-stage infection (upper leaf surface), (**e**) Late-stage infection (underside of the leaf), (**f**) Yellowing and deformation of leaves

are generally elliptical. Conidiophores are erect, measuring 49.83–91.21 μm × 7.64–10.25 μm, with two to three conidia often occurring in chains, though some are solitary. Conidia are round or cylindrical, measuring 30.67–46.75 μm × 13.19–19.31 μm (Fig. 30.6).

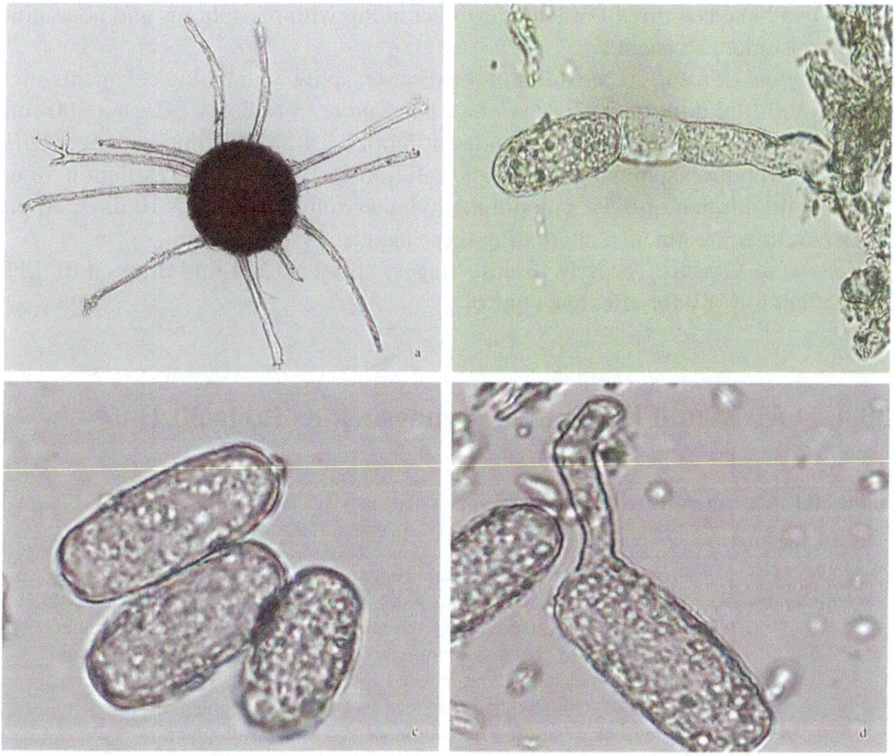

Fig. 30.6 Morphology of the powdery mildew pathogen, (**a**) Chasmothecia, (**b**) Conidiophore, (**c**) Conidia, (**d**) Germination of conidia

Occurrence Pattern

The severity of powdery mildew in honeysuckle varies by region and environmental conditions. The pathogen overwinters as chasmothecia in plant debris, releasing ascospores the following March as buds develop, initiating primary infections. By April, infections spread rapidly to young leaves, with peak disease prevalence occurring from May to September. Infected plants generate conidia that enable secondary infections until late autumn. Conditions of dense canopy cover, alternating dry and wet environments, and poor ventilation further promote the disease.

Control Measures

1. *Cultivating Healthy Seedlings:* Select branches from vigorous, high-quality mother plants, cut into 25–30 cm sections. Trim the lower end to a slant, keeping three nodes intact. Remove the lower leaves, retaining two to four leaves at the top. Soak cuttings in a 50–75-fold dilution of *Bacillus polymyxa* (5 billion CFU/g) for 10 min, allow to air dry, and then proceed with planting.
2. *Enhanced Field Management:* Maintain appropriate planting density and conduct timely pruning to improve ventilation and light penetration, removing any infected branches or buds. Avoid excessive nitrogen fertilization and instead

apply a balanced mix of organic fertilizer along with phosphorus and potassium to boost plant resistance.

3. *Biological Control:* At the onset of the disease, spray with biological agents such as a 300-fold dilution of *Trichoderma harzianum* (5 billion CFU/g), a 300-fold dilution of 6% pyrimidine nucleoside antibiotics, a 400-fold dilution of *B. subtilis* WP (100 billion spores/g), a 300-fold dilution of 5% polyoxin D solution, or an 800-fold dilution of 0.5% emodin methylether solution every 7–10 days. Apply two to three treatments, alternating these agents.
4. *Chemical Control:* At early disease stages, apply a 1200-fold dilution of 15% triadimefon WP for effective control.

30.4 Occasional Diseases of Honeysuckle (Table 30.1)

Table 30.1 Characteristics of occasional diseases of Honeysuckle

Disease (pathogen)	Symptoms	Occurrence pattern
White Mold (*Sclerotium rolfsii*)	Primarily affects the base and root areas of the plant. In the early stages, brown rot spots appear 5–10 cm above the root collar. As the disease progresses, these spots expand, covering the entire root base, with a layer of white hyphae. This causes the honeysuckle bark to gradually decay, leading to the plant's eventual death	Typically begins in early April, with peak incidence from May to September. High temperatures, abundant rainfall, and low light encourage disease development
Anthracnose (*Colletotrichum gloeosporioides*)	Primarily affects leaves, petioles, and stems, occurring simultaneously. In the initial stages, small brown spots appear, which then expand. In the advanced stage, the lesions turn dark brown with clear borders, and under humid conditions, dark spots appear in the lesion centers	Predominantly occurs in spring (March–April), summer (June–August), and autumn (September–October). High humidity, poor ventilation, and low-lying terrain worsen disease severity

Chapter 31
The Diseases of *Ligusticum chuanxiong*

Ligusticum chuanxiong (*L. sinense* cv. Chuanxiong) is a perennial herbaceous plant in the Apiaceae family, with its roots used medicinally. Primarily cultivated in Sichuan province, it is also grown in other regions such as Yunnan, Guizhou, and Guangxi province, favoring mild climates. Known for its aromatic, warming properties and invigorating effects, *L. chuanxiong* promotes blood circulation, alleviating ailments associated with blood stasis, including migraines and rheumatic pain.

Primary diseases affecting *L. chuanxiong* include root rot, brown spot, nematode, powdery mildew, sclerotinia, anthracnose, and rust. Among these, root rot, brown spot, and nematode are prevalent and require targeted control measures.

31.1 Root Rot in *Ligusticum chuanxiong*

Compared to mature plants, seedlings of *L. chuanxiong* are highly susceptible to root rot, especially when grown in high-clay, high-acidity soils with poor drainage or polluted water. Disease incidence is higher in seasons with heavy rainfall, low light, high humidity, and warm temperatures. Root rot is widespread across Sichuan and Yunnan province, with general infection rates of 5–10%, reaching over 30% in severe cases.

Symptoms

Root rot primarily affects the roots and rhizomes of *L. chuanxiong*, generally entering through wounds in the roots or stems. This disease can occur at any growth stage, with the most severe impact at the seedling stage. Pre-emergence symptoms include seed rot, while early-stage infection presents as yellowing of central leaves and localized browning within the lower rhizome. Affected roots display water-soaked lesions, reduced root growth, and slower plant development, with leaf tips and edges gradually drying out. In advanced stages, the plant stops growing,

© China Agriculture Press Co., Ltd. 2026
T. Wang et al., *Atlas of Chinese Medicinal Plant Diseases*,
https://doi.org/10.1007/978-981-95-6326-5_31

yellows, and wilts, with rhizomes rotting entirely, leading to plant death. Vascular tissues in the stem turn brown (Fig. 31.1).

Pathogen

The causative pathogen of root rot in *L. chuanxiong* is *Fusarium oxysporum* Schltdl. ex Snyder et Hansen, an ascomycete fungus in the *Fusarium* genus. On PDA medium, colonies are circular with woolly aerial mycelia, initially rose-red, gradually turning purplish-red, purple, or pale lavender. After 4 days, the colony diameter reaches 3 cm, filling the Petri dish by day 10. Numerous small conidia are elongated, measuring (9–10) μm × (4–5) μm, and form singly or in pairs on flask-shaped conidiophores. Larger conidia are often three-septate, occasionally with five septa, and measure (26–36) μm × (4–6) μm (Fig. 31.2).

Occurrence Pattern

Root rot can occur throughout the growth period of *L. chuanxiong*. In primary production areas of Sichuan province, such as Pengzhou, Dujiangyan, and Meishan cities, early signs of root rot appear sporadically in February. Infection rates increase during the rhizome expansion phase in April, peaking in May, particularly during hot, rainy weather conducive to disease spread.

Control Measures

1. Agricultural Control: Dry rhizome seeds in a ventilated, shaded area to reduce pathogen transmission; during propagation in high altitudes and lowland cultivation, remove any diseased "nursery rhizomes" and rotten "seed rhizomes." Remove infected plants from fields and destroy them immediately to prevent spread; implement crop rotation with dry crops and ensure adequate field drainage. Adjust fertilization to promote balanced plant growth.

Fig. 31.1 Symptoms of root rot in *Ligusticum Chuanxiong*. (**a**) Field symptoms, (**b**) Root symptoms

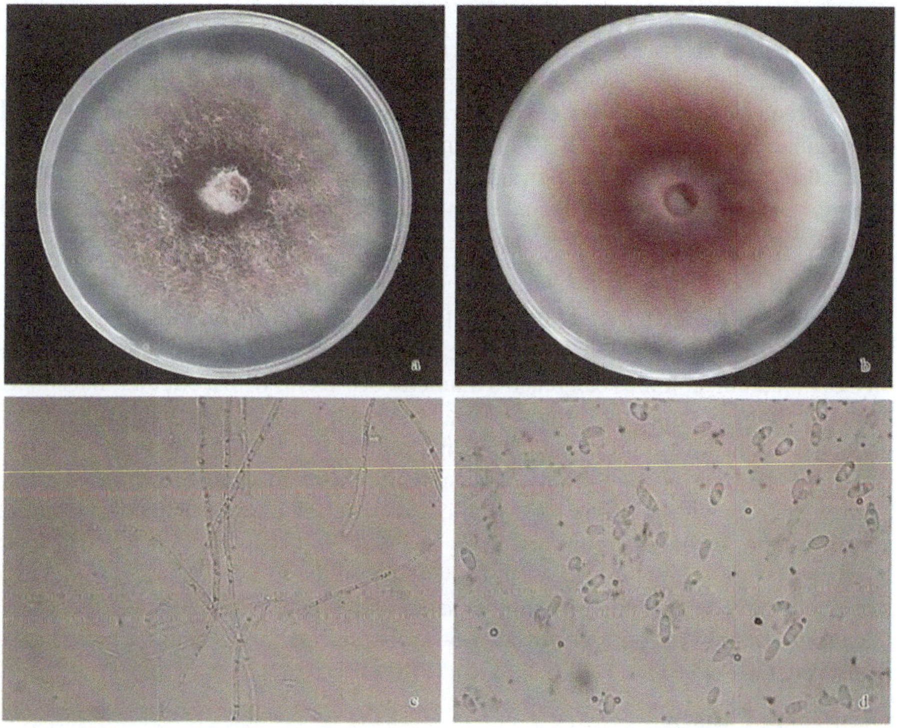

Fig. 31.2 Morphology of root rot pathogen in *Ligusticum Chuanxiong*. (**a**, **b**) Colony characteristics, (**c**) Hyphal morphology, (**d**) Conidia

2. Biological Control: Studies show that *Trichoderma*, *Bacillus*, and *Streptomyces* strains have antagonistic effects on *F. oxysporum*, effectively controlling root rot while mitigating environmental impacts associated with chemical pesticide use.
3. Chemical Control: Apply a 500-fold dilution of 75% mancozeb WG or 50% carbendazim WP for foliar spraying every 10 days, with a total of two applications. Soil drenching with an 800-fold dilution of 50% carbendazim WP is also effective.

31.2 Blight Spot Disease of *Ligusticum chuanxiong*

Blight spot disease is a common disease affecting the cultivation of *L. chuanxiong*, with an incidence rate typically ranging from 5% to 20%. In severe cases, it can exceed 30%. The disease causes early leaf wilting, and when it occurs consecutively over multiple years, it weakens the plant's growth, leading to stunted plants, fewer and smaller flowers, and even complete plant death, significantly impacting yield.

Symptoms

Blight spot disease primarily affects older leaves, occasionally appearing on stems and petioles, while new leaves are rarely affected. Initially, the leaf spots appear as small, circular brown dots. These spots gradually expand, becoming spindle-shaped or irregular, with lengths of 1–10 mm and widths of 1–5 mm. The spots are surrounded by reddish-brown or purplish-brown halos. Over time, the centers of the spots thin out, becoming brittle and prone to cracking or perforation. In advanced stages, multiple spots may merge, causing the leaves to wither and fall off (Fig. 31.3).

Pathogen

The pathogen responsible for blight spot disease is *Alternaria alternata* (Fr.) Keissl., an asexual form of the ascomycete in the *Alternaria* genus. After 6 days of culture on PDA medium, colonies can reach a diameter of 7.0 cm, with a circular, flat shape. The mycelium is radiating, gray to dark gray, with a white edge, and the colony underside is black with prominent pigment production. On naturally infected spots, conidiophores are mostly solitary, occasionally clustered, pale brown, straight or slightly curved, with multiple septa and occasional branches near the top. Conidia vary in shape, including round, oval, or inverted club-shaped, with a broad base tapering upward. They are yellowish-brown, with three to six transverse septa, and some cells have longitudinal or oblique septa with slight constrictions at the septa. Conidia measure (16.5–27.9) μm × (7.6–11.4) μm, and most conidia have beaks measuring (4.8–22.6) μm × (3.6–5.6) μm (Fig. 31.4).

Occurrence Pattern

This disease typically begins to appear in early May, becoming widespread in open fields from late May to early June, just before harvest. It becomes more severe in July and August as the leaves of *L. chuanxiong* mature. The pathogen is prevalent in soil, air, and various organic materials used as growing media, and its spores spread through the air. Insufficient sterilization, high soil moisture, and elevated

Fig. 31.3 Symptoms of Blight Spot Disease on *Ligusticum chuanxiong* leaves

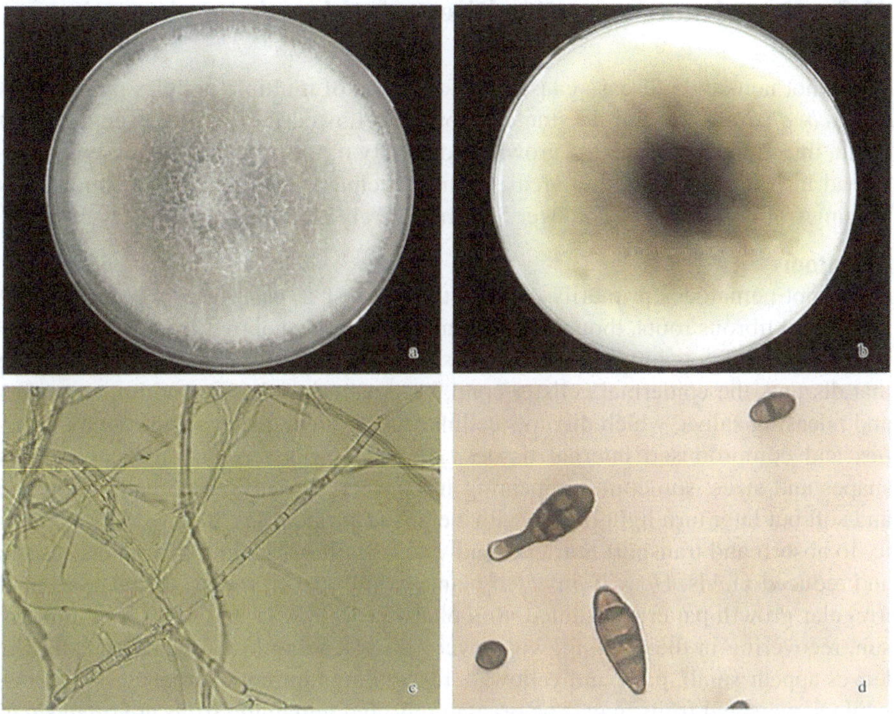

Fig. 31.4 Morphology of Blight Spot Disease pathogen of *Ligusticum chuanxiong*: (**a**, **b**) Colony morphology; (**c**) Mycelium; (**d**) Conidia

temperatures favor disease development. Poor field management, low soil fertility on slopes, and relatively dry conditions also increase disease susceptibility.

Control Measures

1. Agricultural Control: Early removal and destruction of diseased plant residues and weeds are essential. During the growing season, infected plants should be removed and destroyed immediately. Crop rotation with nonhost crops and avoiding continuous cropping or rotation with other susceptible hosts can help control the disease. Proper field management to maintain balanced soil moisture is crucial; avoid both excessive moisture and drought, and maintain even soil moisture levels.
2. Biological Control: Research has shown that a mixed fermentation solution of *Trichoderma koningii* and *Rhizopus nigricans* can effectively control *Alternaria* species and enhance the activity of defensive enzymes in plant leaves, providing an induced resistance effect.
3. Chemical Control: At the early onset of the disease, spray plants with 50% carbendazim WP at a dilution of 1,000-fold. Reapply after 7–10 days to effectively suppress pathogen development and prevent further disease spread. Bordeaux mixture (1:1:100) or 45% zineb AS diluted 800-fold can also be sprayed every 10 days for two to three consecutive treatments.

31.3 Root-Knot Nematode Disease in *Ligusticum chuanxiong*

Root-knot nematode disease causes the formation of multiple nodules on the roots of *L. chuanxiong*, leading to stunted growth, yellowed leaves, and eventual plant death, thus impacting both the growth and quality of the crop. This disease is widespread in primary production areas, such as Sichuan, with general infection rates ranging from 5% to 30%, and over 50% in severely affected regions.

Symptoms
Root-knot nematodes primarily damage the roots of *L. chuanxiong*, especially the lateral and fibrous roots, though symptoms also manifest above ground. The nematodes harm the roots in two primary ways: first, through direct mechanical damage that disrupts the epidermal cells; second, by puncturing the host with their stylets and releasing saliva, which disrupts cellular metabolism, resulting in root deformities and compromised internal tissues. Affected roots form nodules of various shapes and sizes, sometimes appearing in clusters. Initially, these galls are white and soft but later turn light brown with a cracked surface. The damaged roots' ability to absorb and transport nutrients and water declines, leading to weak seedlings and reduced yields. Heavily infected plants exhibit poor growth, varied sizes, and irregular growth patterns. Stunted, thin plants grow slowly and wilt in the midday sun, recovering in the evening, with severe cases leading to the plant's death. The leaves appear small, pale, and yellowed, resembling nutrient deficiency symptoms, while flowers and fruits may drop prematurely, and remaining fruits are often small and misshapen (Fig. 31.5).

Pathogen
The causative agent of root-knot nematode disease in *L. chuanxiong* is the Southern Root-Knot Nematode (*Meloidogyne incognita*). Its life cycle consists of three stages: egg, juvenile, and adult. When temperatures exceed 10 °C, eggs hatch, with juveniles primarily active at a soil depth of 5–30 cm. In greenhouses, *M. incognita* can produce around ten generations annually, with each female laying 300–800 eggs. At optimal temperatures of 25–30 °C, a complete generation can occur within 25 days. Ideal soil relative humidity ranges from 40% to 70%, and suitable soil pH is 4.0–8.0. Nematodes become inactive at soil temperatures above 40 °C or below 10 °C, with a lethal temperature of 55 °C. The second-stage juvenile is the infective stage, entering the root tip and moving between cells to complete infection, stimulating accelerated cell division and forming root galls or knots in affected areas (Fig. 31.6).

Occurrence Pattern
The Southern Root-Knot Nematode overwinters in the soil as eggs or other stages and can survive up to 3 years without a host plant. Its activity range in the soil is limited, moving less than 1 meter annually. The primary sources of initial infection are contaminated soil, infected seedlings, and irrigation water. The nematodes spread over longer distances through water, wind, soil transport, and contaminated farm equipment. Soil moisture is a key factor affecting egg hatching and

Fig. 31.5 Symptoms of Root-Knot Nematode Disease in *Ligusticum Chuanxiong*

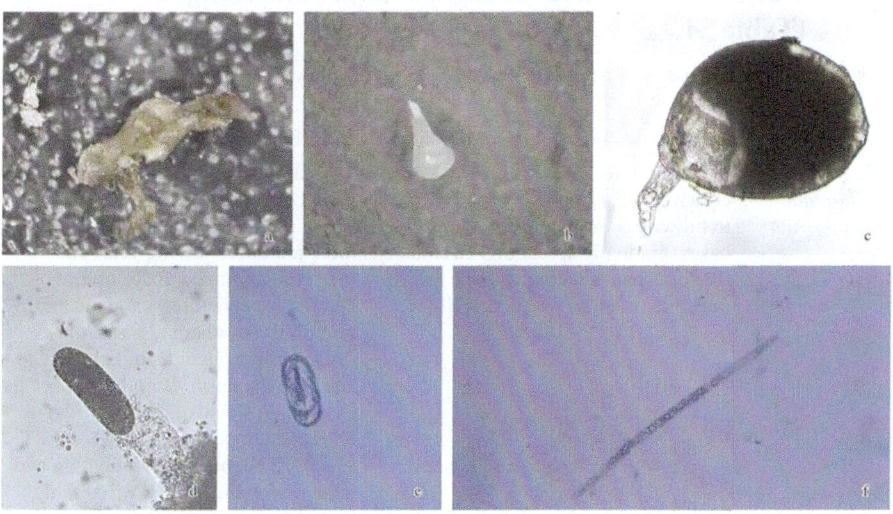

Fig. 31.6 Morphology of the Southern Root-Knot Nematode. (**a**–**c**) Adult female nematodes, (**d**, **e**) Nematode eggs, (**f**) Juvenile nematodes

reproduction; wet seasons promote hatching and infection, while dry or overly wet soils suppress nematode activity. Preferred soil conditions include pH 4.0–8.0, elevated terrain, loose-textured soil, and low salinity, with sandy soils more susceptible than clay soils, and higher incidence in monoculture fields.

Control Measures

1. Agricultural Control: Maintain clean fields by removing infected plant debris from open fields and greenhouses, reducing initial sources of infection. Use appropriate irrigation and fertilization practices to cultivate robust seedlings that resist nematode infection. Implement crop rotation with nonhost plants to effectively reduce soil-residual larvae and eggs and lower disease incidence.
2. Physical Control: Flood the soil to reduce oxygen levels, causing nematode suffocation. In greenhouses, use high temperatures by sealing the space on hot days to kill nematode eggs and larvae in the soil.
3. Biological Control: The endophytic fungus *Paecilomyces lilacinus* is widely used to control *Meloidogyne incognita*, effectively reducing nematode numbers and egg sacs.
4. Chemical Control: Before sowing or planting, apply 0.3–0.5 g/m^2 of 98% dazomet GR diluted to 3,500–7,500-fold and distribute it evenly over the soil surface for soil disinfection. Cover the treated soil with plastic film for 48–72 hours when possible to enhance fumigation efficacy. During the growth period, when disease symptoms appear, apply 1.8% avermectin emulsion diluted 3,500–7,000-fold around the base of the plants.

31.4 Occasional Diseases of *Ligusticum chuanxiong* (Table 31.1)

Table 31.1 Characteristics of occasional diseases of *Ligusticum chuanxiong*

Disease (pathogen)	Symptoms	Occurrence pattern
Stem Rot (*Sclerotinia sclerotiorum*)	Primarily affects leaves and root crowns. The root crowns rot, and black to brown lesions appear on the stem bases, gradually spreading. Eventually, the rot expands significantly, causing the entire plant to wither and die	The pathogen overwinters as sclerotia in soil or crop residues attached to seeds, serving as the primary source of infection the following year. The sclerotia germinate under moist conditions, spreading via wind and rain. In the southern regions, the disease is prevalent from February to April; in the north, from November to December. Relative humidity above 85% and temperatures between 15°C and 20°C encourage germination and hyphal growth, while low light, high humidity, and prolonged rains in spring and autumn promote outbreaks and spread

(continued)

Table 31.4 (continued)

Disease (pathogen)	Symptoms	Occurrence pattern
Powdery Mildew (*Erysiphe polygoni*)	Primarily damages leaves, beginning on the lower leaves and gradually spreading upward. White powdery mildew appears, eventually covering the leaf surfaces. As the disease progresses, small black dots appear on the infected areas, and severely infected plants die	This disease typically occurs annually from June to July, with high humidity and warm conditions promoting its spread
Gray Mold (*Botrytis cinerea*)	The disease occurs during both the seedling and mature plant stages. In seedlings, infection typically begins from the upper parts of the plant or from areas that have previously sustained damage, such as mechanical injuries or frost. The infected areas exhibit a grayish-brown rot with a dense layer of gray mold on the surface. In mature plants, the disease usually starts at the leaf tips or edges, progressing in a V-shaped pattern toward the inner leaf. Initially, light brown, water-soaked necrotic spots appear. Under high humidity, these spots quickly expand into irregular shapes, forming large lesions with alternating light and dark ring patterns, with gray mold covering the surface	The pathogen overwinters and oversummers in the soil or infected plant debris as sclerotia. Optimal growth occurs between 20°C and 30°C, and the pathogen is tolerant to low temperatures, producing a large quantity of spores at 7-20°C. Conditions such as extended overcast rain, cold fronts with high winds, inadequate ventilation, excessive plant density, elongated seedlings, and root or leaf damage during transplanting can significantly increase disease incidence

Chapter 32
The Diseases of *Angelica sinensis*

Angelica sinensis, also known as Min Gui, Qin Gui, Xi Dang Gui, and Chuan Gui, is a perennial herbaceous plant in the Apiaceae family. It is primarily cultivated in the provinces of Gansu, Sichuan, and Yunnan, with some production in Shaanxi, Guizhou, and Hubei. The largest output and highest quality come from Min County in Gansu Province. Harvested in late autumn, the root of *A. sinensis* is used medicinally. It has a sweet and spicy flavor, with a warm nature that affects the heart, liver, and spleen meridians. Its properties include nourishing and invigorating blood, warming and relieving pain, and moistening the intestines to relieve constipation. Angelica is widely used in clinical settings to treat symptoms like irregular menstruation, dysmenorrhea, blood deficiency, blood stasis amenorrhea, blood deficiency headaches, constipation from blood deficiency, anemia, and rheumatic pain. Known in Europe as the "Ginseng for Women," it is also used to treat abscesses, injuries from falls, and chronic coughs.

The primary diseases affecting *A. sinensis* include Root Rot Nematode Disease (Ma Kou Disease), root rot, water rot, leaf spot, powdery mildew, anthracnose, and dwarf virus disease. Among these, Root Rot Nematode Disease is the most significant.

32.1 Angelica Root Rot Nematode Disease

Angelica Root Rot Nematode Disease (commonly referred to as "Ma Kou Disease") is the most prevalent root disease affecting *A. sinensis*. With complex pathogenic factors and severe impact, it poses significant challenges for control. This disease has an annual incidence rate of 60–70%, substantially reducing both yield and quality. It is widespread and affects Angelica-growing regions nationwide.

© China Agriculture Press Co., Ltd. 2026

T. Wang et al., *Atlas of Chinese Medicinal Plant Diseases*,
https://doi.org/10.1007/978-981-95-6326-5_32

Symptoms

This disease primarily targets the root of *A. sinensis,* with no significant symptoms on the plant's aerial parts. Early symptoms are typically found on the petiole bases below the soil surface, manifesting as reddish-brown spots or streaks with clear borders that distinguish them from healthy tissue. In severe cases, the petioles break, leading to gradual leaf yellowing, wilting, and detachment from the bottom upward, though the plant remains alive. When roots are infected, initial symptoms on the outer skin may be subtle; however, a longitudinal cut reveals localized brown, bran-like necrosis. As the disease progresses and roots thicken, brown vertical cracks of 1–2 mm depth appear on the root epidermis. Affected plants often develop excessive, malformed root hairs. In advanced stages, the cortex of the root head takes on a brown, bran-like decay. This decay typically does not exceed the cambium depth, but in severe cases, the stem base may turn brown, with necrosis extending to the vascular bundles. Mildly infected plants display no above-ground symptoms, whereas severely infected plants may show stunted growth, smaller, wrinkled leaves, and other deformities (Fig. 32.1). This disease often co-occurs with root rot.

Pathogen

The pathogen causing Angelica Root Rot Nematode Disease is *Ditylenchus destructor* Thorne, a migratory endoparasitic nematode commonly known as the Potato Tuber Rot Nematode or Sweet Potato Tuber Rot Nematode. Adult nematodes, both male and female, are elongated and cylindrical, with females typically larger than males, measuring 996.67–1650.00 μm in length. Female nematodes have blunt anterior ends, a smooth lip region, and a long, conical tail that tapers to a rounded tip. The nematode's cuticle shows fine annulations, with six lateral lines and a stylet approximately 12–14 μm long for feeding. The esophagus is club-shaped, with a median bulb extending along the anterior intestine. Female reproductive structures include a slightly protruding vulva and a post-vulval sac extending roughly two-thirds of the vulval region. Females lay 7–21 eggs per cycle, each egg being approximately 60.33 μm × 26.39 μm. Male nematodes possess copulatory spicules about 22.37 μm long, which are broad at the base, tapering toward the front, with two finger-like projections in the center. The copulatory bursa covers about two-thirds of the tail (Fig. 32.2). This nematode infects a wide range of host plants, with over 90 known species, including tubers, bulbs, and roots of potato, sweet potato, ginseng, dahlia, and iris. It is classified as a quarantine pest in several countries and regions.

Occurrence Pattern

This nematode overwinters as adults or mature juveniles in soil, on residual Angelica plants, or in plant debris, serving as primary infection sources for the following year. Throughout the Angelica growth period (April to September), the nematode can invade young fleshy roots, with peak invasion and reproduction occurring from May to July, coinciding with the disease's peak field incidence. Infected seedlings, contaminated soil, water, and farming tools can facilitate nematode spread. Heavy infestations by other underground pests can exacerbate the occurrence of Root Rot Nematode Disease.

Fig. 32.1 Symptoms of Angelica Root Rot Nematode Disease (Ma Kou Disease). (**a**) Field symptoms; (**b**) Root symptoms; (**c**) Early symptoms; (**d**) Late symptoms

The incidence of this disease correlates with nematode population density, soil temperature, and the Angelica growth stage. The nematode population is highest within the top 10 cm of soil, with root tissue especially attractive to nematodes, resulting in significant root head damage. Optimal temperatures for nematode activity range from 2 to 35 °C, with peak activity around 26 °C. Extremes of high or low temperatures reduce nematode mobility. Relative humidity below 46% restricts survival. In Min County, Gansu Province, the nematode completes 6–7 generations annually, with each generation requiring 21–45 days depending on soil temperature. Regions such as Min County, Weiyuan County, and Zhang County in Gansu are heavily affected, making this disease a major cause of reduced Angelica yields.

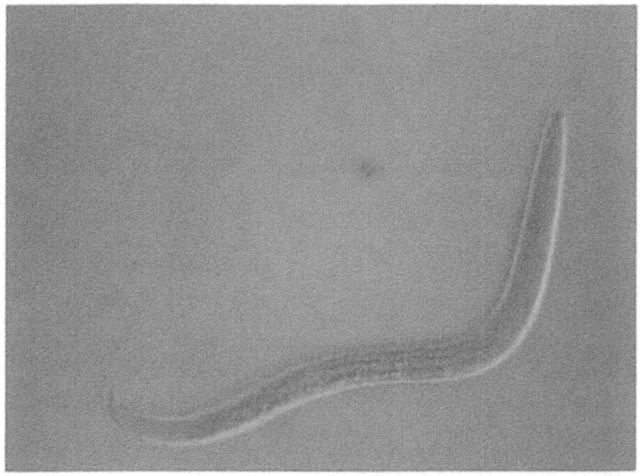

Fig. 32.2 Morphology of Ditylenchus destructor

Control Measures

1. Scientific Land Selection

Choose planting areas that can be plowed and left exposed to the sun for over a month during peak summer to reduce the density of pathogens and parasitic nematode eggs, thus lowering disease incidence the following year.

2. Cultivation and Farming Practices

   (a) Disease-Resistant Varieties: Select resistant varieties from the Min Gui series, such as Min Gui No. 1, No. 2, or No. 3, based on disease types and climate in the Angelica-growing region.
   (b) Ridge Planting: Use ridge planting with plastic film mulching, aligning ridges with slopes to aid in drainage during rainy seasons and reduce field humidity. Use black polyethylene film thicker than 0.01 mm to suppress weeds, and avoid damaging roots and seedlings during weeding.
   (c) Healthy Seedling Cultivation: Choose high-altitude (over 2000 m) uncultivated land for seedling growth to minimize initial infection. Soil treatment in seedling fields involves mixing 5–6 kg of 98% dazomet per acre with 3 kg of fine soil, distributing it over the soil surface, and plowing it to a depth of 20 cm. After 20 days, the soil is loosened for planting. Select smooth, undamaged seedlings with minimal lateral roots for transplanting.
   (d) Crop Rotation and Fertilization: Practice rotation with nonhost crops like wheat or rapeseed, avoiding rotation with potatoes, broad beans, alfalfa, or red clover. Apply well-decomposed chicken manure as an organic fertilizer.
   (e) Field Sanitation: Remove decayed roots and plant debris promptly to reduce infection sources. Weed frequently and remove uprooted weeds to prevent contamination.

3. Soil Disinfection

Before planting, apply 3% phoxim granules mixed into fine soil at 3 kg per square meter, or spray the planting furrows with 2000-fold diluted 1.8% Avermectin EC or 200-fold diluted 50% sulfur suspension. Alternatively, apply 41.25 kg/ha of *Paecilomyces lilacinus* granules (containing 5 billion CFU/g) to the planting holes.

4. Seedling Disinfection

Use hot water treatment for seed disinfection. Three to four days before sowing, soak seeds in room temperature water for 15 min, then transfer to water at 50–55 °C (three times the seed volume) for 10–15 min, stirring continuously. Remove and soak in 20–30 °C water for 12 h, then drain and cover at room temperature, stirring once daily. Sow when a few seeds begin to sprout.

5. Root Dipping with Pesticide Solution

Dip roots in a 2000-fold solution of 1.8% Avermectin EC for 30 min before planting.

32.2 Angelica Root Rot

Root rot is one of the main diseases affecting Angelica production, occurring throughout the entire growth period, from seedling transplanting to harvest. It is a common root disease and is widely prevalent in all Angelica-growing regions nationwide.

Symptoms

Angelica root rot can occur throughout the entire growth season. In the early stages, only a few lateral and fibrous roots are infected. As the disease progresses, it gradually spreads to the main root. Initially, infected plants may not exhibit noticeable symptoms above ground. However, as root decay intensifies, the leaves begin to wilt during the day but may recover at night. After several days, wilting becomes irreversible, even at night. When uprooted, the main root appears rust-colored, decayed, and fibrous, making it easy to pull from the soil. Infected plants are stunted, with leaves showing elliptical brown spots. In severe cases, the leaves turn yellow, droop, and eventually, the entire plant dies (Fig. 32.3).

Pathogen

Angelica root rot is caused by infection from the *Fusarium* species (asexual type of the ascomycete fungi) through either single or multiple strains. Pathogenic species include *Fusarium oxysporum* (Schltdl. ex Snyder et Hansen), *F. avenaceum* [(Fr.) Sacc.], *F. solani* [(Mart.) Appel et Wollenw. ex Snyder et Hansen], *F. redolens* (Wollenw.), *F. equiseti* [(Corda) Sacc.], and *F. sporotrichioides* [(Sacc.) Nirenberg]. These fungi can all infect Angelica roots, though the dominant pathogen may vary by region. In Gansu Province, *F. oxysporum* has been identified as the dominant species.

On PDA medium, *F. oxysporum* grows with a colony diameter of 3.1 mm after 3 days at 25 °C, expanding to 4.8 mm after 6 days, and 5.7 mm after 9 days. The colony has a powdery to peach-pink center, with white, dense mycelium around the edges. On SNA medium, colonies reach 2 mm in diameter after 3 days, 4.1 mm after 6 days, and 5.2 mm after 9 days, with white, fuzzy mycelium on the surface and a

Fig. 32.3 Symptoms of Angelica Root Rot. (**a**) Aboveground symptoms; (**b**) Root symptoms

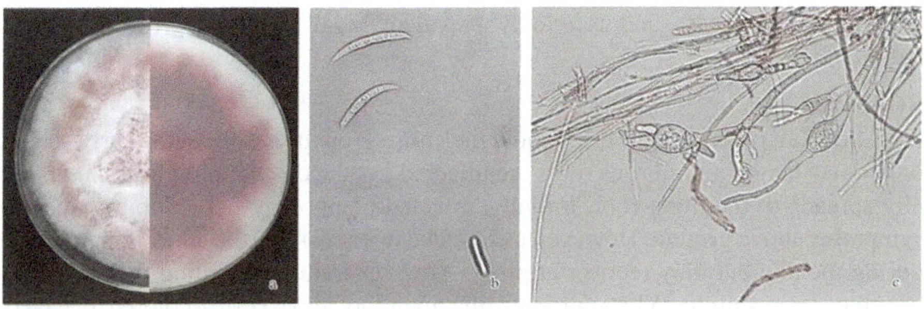

Fig. 32.4 Morphology of *Fusarium oxysporum*. (**a**) Colony characteristics; (**b**) Macroconidia; (**c**) Chlamydospores

light pink center on the reverse. On CLA medium, colonies grow to 2.4 mm after 3 days, 7.0 mm after 6 days, and 8.2 mm after 9 days, with white and pink mycelium interspersed, and feather-like mycelial tufts at the edges.

On PDA medium, *F. oxysporum* produces large, sickle-shaped macroconidia with blunt or tapering tips; microconidia are elliptical, kidney-shaped, spindle-shaped, or hook-shaped, borne on single phialides. Chlamydospores are irregularly shaped. On SNA medium, macroconidia are also sickle-shaped with tapering tips; microconidia are elliptical, kidney-shaped, spindle-shaped, or hook-shaped, with chlamydospores forming in irregular chains. On CLA medium, macroconidia have blunt tips and measure 27.5–47.5 μm × 3.75–7.5 μm with one to five septa; microconidia are kidney-shaped, measuring 3.75–15 μm × 2.5–7.5 μm with zero to three septa; chlamydospores are chain-forming, measuring 2.5–3.75 μm × 2.5–5 μm (Fig. 32.4).

Occurrence Pattern
The pathogen overwinters in the soil and on seedlings, becoming a primary infection source in the following year. Initial infections generally appear in early May, intensify in June, peak in July and August, and continue until harvest. Factors such as root injury from soil insects, excessive irrigation, waterlogging after rain, and poor root development all exacerbate the disease. This disease often co-occurs with Angelica root rot nematode infestations.

Control Measures

1. Scientific Site Selection: Choose suitable fields and deep-till the soil. Angelica is typically grown in cool, moist areas with late summer and autumn harvests. After harvesting the previous crop, promptly select the plot for the next year's planting. Choose well-drained, sandy loam soil, and deep-till to eliminate root rot pathogens and reduce weed pressure.
2. Cultivation Practices: Implement crop rotation with grass or cruciferous crops. Remove and destroy infected plants, disinfect the planting site with quicklime, and clear the field of diseased residues after harvest to reduce primary infection sources.
3. Root Dipping in Fungicides: Soak seedlings in a 1:1:150 Bordeaux mixture for 10–15 min or in a 50% carbendazim wettable powder solution (1,000-fold dilution) for 30 min, then air-dry before transplanting.
4. Soil Treatment: Before sowing in nursery or planting fields, apply 200–300 g of 20% copper acetate wettable powder mixed with 30 kg of fine soil per mu, incorporating it into the soil. Alternatively, apply 3 kg of 3% phoxim granules mixed with fine soil per 667.7 m^2 to control both Angelica root rot and root rot nematode infestations.
5. Seedling Disinfection: Prior to transplanting, soak seedlings in a 1,000-fold dilution of 70% thiophanate-methyl WP or a 500-fold dilution of 25% carbendazim WP for 10 min to eliminate pathogens on the seedling surface. Additionally, soil disinfection with 1% ferrous sulfate during pre-transplant plowing or seedling disinfection with Bordeaux mixture or 40% sodium ethyl sulfate WP (800-fold dilution) for 10 min before transplanting is recommended.

32.3 Angelica Water Rot Disease

Water Rot Disease has become a prevalent root disease in Angelica cultivation in recent years, resulting in extensive plant mortality. In severely affected areas, disease incidence can reach up to 60%, significantly impacting Angelica yield and quality. Currently, the primary occurrence of Angelica Water Rot Disease is noted in Gansu Province.

Symptoms

The pathogen initially infects the junction between the root and stem of Angelica. In early stages, water-soaked lesions appear at the petiole base, while the above-ground parts of the plant appear normal. As the disease progresses, soft rot develops at the petiole base, and pressing on the crown of the Angelica plant reveals liquid with bubbles. The entire plant eventually wilts, with above-ground parts appearing dried, while the underground root structure rots (Fig. 32.5).

Pathogen

The causative agent of Angelica Water Rot Disease is *Pseudomonas fluorescens* (Trev.) Migula, a fluorescent *Pseudomonas* bacterium within the prokaryotic domain. On NA medium, this bacterium forms opaque colonies that appear slightly yellow after 4–5 days, with white, snowflake-like indentations in the center. On KB medium, the colonies are transparent, producing a diffusible yellow-green fluorescent pigment. Cells are rod-shaped, measuring approximately (0.7–0.8) μm × (2.3–2.8) μm, and are Gram-negative. Optimal growth temperature for *P. fluorescens* is 25–30 °C. The bacterium is motile, sensitive to salt, and can dissolve in 3% KOH solution, emitting a yellow-green fluorescence under UV light. It is strictly aerobic, can hydrolyze starch, is negative for nitrate reduction, positive for catalase, and capable of metabolizing glucose, maltose, inositol, D-mannose, and glycerol, but not L-sorbitol (Fig. 32.6).

Fig. 32.5 Symptoms of Angelica Water Rot Disease. (**a**, **b**) Field symptoms; (**c**) Root symptoms

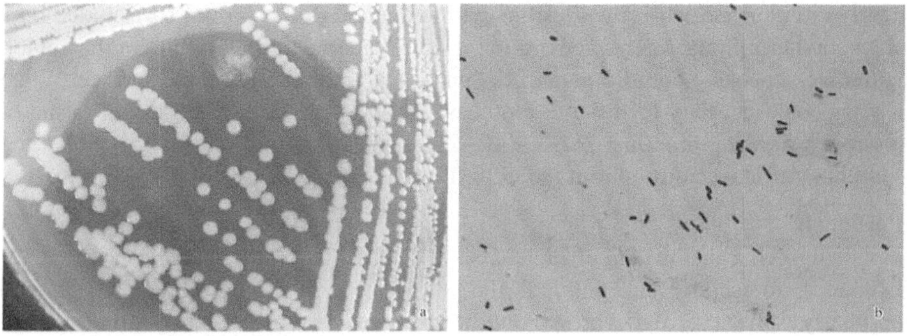

Fig. 32.6 Morphology of Angelica Water Rot pathogen. (**a**) Colony characteristics; (**b**) Bacterial cell morphology

Occurrence Pattern

The pathogen overwinters in soil within infected plant debris or in contaminated Angelica seedlings, acting as the primary infection source the following year. When contaminated seedlings are planted, early infection can develop in young plants. After infection onset, the bacterium spreads via rainwater, insects, and farming activities. Humid conditions favor pathogen growth and intensify disease severity.

Control Measures

1. Cultivation Measures: Rotate heavily infected fields with nonhost crops for at least 2 years. Remove diseased plants and residues after harvest. Prior to planting, treat soil with a 400-fold dilution of 30% copper gluconate WP or a 600-fold dilution of 3% kasugamycin WP.
2. Chemical Control: At the onset of the disease, apply a 400-fold dilution of 77% copper hydroxide WP, a 1,200-fold dilution of 50% chlorobromoisocyanuric acid SP, a 400-fold dilution of 30% copper gluconate WP, a 600-fold dilution of 3% kasugamycin WP, or a 600-fold dilution of 0.3% tetracycline AS as a root drench.

32.4 Angelica Leaf Spot Disease

Leaf spot disease is one of the common leaf diseases affecting Angelica, widely distributed across Angelica-growing regions in China. It occurs with particular severity in regions like Taibai County, Baoji City, Long County, Feng County, and Pingli County in Shaanxi Province, as well as Min County, Weiyuan County, and Zhang County in Gansu Province.

Symptoms

The disease primarily affects the leaves and petioles of Angelica. Initially, small brown spots appear on the leaf surface, which gradually expand into polygonal or nearly circular reddish-brown lesions with a diameter of 1–3 mm, surrounded by a chlorotic halo. In the later stages, some lesions become grayish-white in the center, with black specks—conidiomata of the pathogen—appearing on the surface. When lesions coalesce, they form large blotches, and in some cases, the central tissue of lesions falls out, creating perforations. Severe infections cause browning and scorching of leaves throughout the field (Fig. 32.7).

Pathogen

The pathogen causing Angelica leaf spot disease is *Septoria anhrisci* (Pass. & Brunaud), an ascomycete fungus in the genus *Septoria*. Conidiomata are flattened or nearly spherical, dark brown, with a diameter of 67.2–103.0 μm (average 84.5 μm) and a height of 62.7–89.6 μm (average 78.1 μm). Conidia are needle-shaped or filamentous, straight or curved, colorless, with indistinct septa, measuring 22.3–61.2 μm × 1.3–2.9 μm (average 44.2 μm × 1.7 μm). On PDA medium, *S. anhrisci* forms raised, dense, dark brown, velvety colonies, reaching a diameter of 6.3 cm after 20 days. Optimal conditions for mycelial growth, conidial germination, and sporulation include temperatures of 5–30 °C (optimal 15–25 °C), 5–30 °C (optimal 20 °C), and 5–25 °C (optimal 15 °C), respectively. Continuous light exposure promotes growth, germination, and sporulation. Conidia germinate best in water at relative humidity above 75%, and growth is fastest at pH 5.5, with optimal sporulation occurring at pH 6.0. Angelica leaf extracts and glucose solutions enhance spore germination, while sucrose and soil extracts inhibit it. Glucose and D-galactose serve as effective carbon sources, while glutamic acid supports the fastest growth on nitrogen-supplemented media. Glycine, proline, and peptone enhance sporulation (Fig. 32.8).

Occurrence Pattern

The pathogen overwinters as mycelium and conidiomata in infected plant debris in the soil, with conidia causing primary infections the following year. Conidia produced during the growth period spread by wind and rain, resulting in secondary infections. Warm, humid, and low-light conditions favor disease progression.

Fig. 32.7 Symptoms of Angelica Leaf Spot Disease. (**a**, **b**) Field symptoms; (**c**, **d**) Leaf symptoms

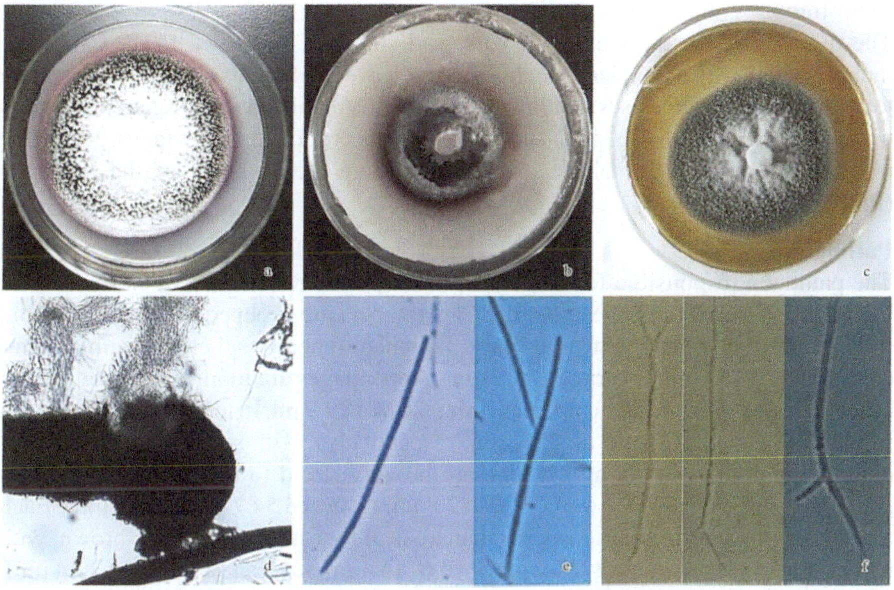

Fig. 32.8 Morphology of Septoria anhrisci. (**a**) Growth on PDA medium (**b**) Growth on OA medium (**c**) Growth on MEA Medium (**d**) Conidiomata (**e**) Conidial structure (showing septa) (**f**) Germinated Conidia

Symptoms typically appear in late May, intensify from July to August, and persist until harvest. High initial pathogen load and elevated humidity levels exacerbate the disease.

Control Measures

1. Cultural Practices: In early winter, thoroughly remove infected plant residues from the field to reduce primary infection sources, and implement crop rotation.
2. Chemical Control: At the early onset of symptoms, apply a 200-fold dilution of 70% propineb WP, a 600-fold dilution of 70% thiophanate-methyl WP, or a 600-fold dilution of 10% difenoconazole WG. These treatments have demonstrated control efficacies exceeding 71% and also support increased yield. Spray every 7–10 days for 2–3 consecutive applications, rotating fungicides to delay resistance development.

32.5 Angelica Powdery Mildew

Powdery mildew is a recurring foliar disease in Angelica cultivation, causing premature leaf senescence and presenting as a significant issue in Angelica production. This disease is widespread across Angelica-growing regions in China, with infection rates reaching 40–85% in Gansu Province's Weiyuan, Min, and Zhang counties.

Symptoms

The disease affects the leaves, flowers, and stems of Angelica. Early symptoms include small white powdery patches on the leaves, which expand to cover entire leaf surfaces in severe cases, leading to yellowing. Advanced infections cause leaves to become narrow, misshapen, and eventually necrotic. In later stages, the white powdery layer develops black specks, which are cleistothecia of the pathogen (Fig. 32.9).

Pathogen

The pathogen responsible for Angelica powdery mildew is *Erysiphe heraclei* DC., belonging to the Erysiphaceae family. Cleistothecia are grouped or scattered within the mycelial layer, dark brown to black, flattened to nearly spherical, with a diameter of 76.0–147.8 μm (average 103.2 μm). Appendages are filamentous, sometimes branched once or twice at the tip, measuring 26.9–129.9 μm in length and 4.7–5.9 μm in width (average 50.8 μm and 5.3 μm, respectively), with septa. Each cleistothecium contains four to six asci, which are broadly ovoid to elliptical, with a short stalk, measuring 51.7–61.2 μm × 35.3–42.3 μm (average 54.7 μm × 39.1 μm). Each ascus contains four to six ascospores that are oval to ovate, pale yellow-brown, with a thick wall, measuring 15.3–21.2 μm × 10.6–14.1 μm (average 18.4 μm × 12.4 μm). Conidia are barrel-shaped or drum-shaped, single-celled, colorless, measuring 25.9–38.8 μm × 12.9–16.5 μm (average 32.9 μm × 15.1 μm) (Fig. 32.10).

Occurrence Pattern

The pathogen overwinters as cleistothecia and mycelium in plant debris. In the following year, overwintered cleistothecia release ascospores, initiating primary infection, while overwintered mycelium produces conidia, continuing the infection cycle. Airborne conidia facilitate secondary infections throughout the season. Optimal conditions for conidial germination include temperatures of 18–30 °C and relative humidity above 75%, with an incubation period of 2–5 days. Loose management practices and weak plant growth favor disease development, typically

Fig. 32.9 Symptoms of Angelica Powdery Mildew. (**a**) Leaf symptoms; (**b**) Stem symptoms

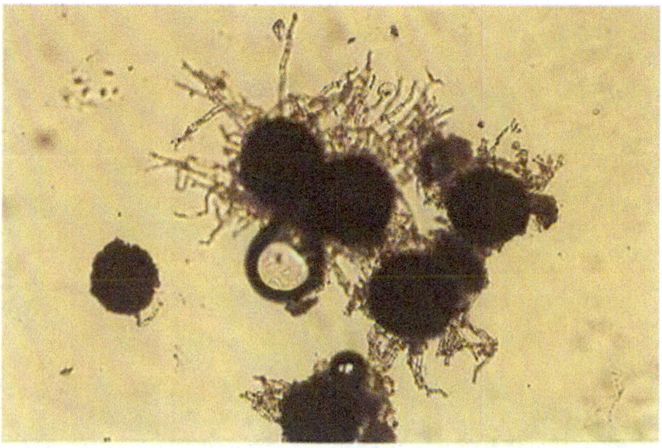

Fig. 32.10 Morphology of *Erysiphe heraclei*

emerging in early August, peaking from late August to early September, and producing cleistothecia by mid-September.

Control Measures

1. Cultural Practices: After harvest, thoroughly remove plant debris to reduce initial inoculum. Implement crop rotation to avoid continuous cultivation.
2. Chemical Control: At the onset of infection, apply a 1,000–2,000-fold dilution of 25% prochloraz EC, a 1,200-fold dilution of 40% procymidone WP, a 1,200-fold dilution of 50% iprodione WP, or a 1,000–1,500-fold dilution of 50% flusilazole-manganese WP.

32.6 Angelica Anthracnose Disease

Anthracnose disease is a primary issue in Angelica production, mainly affecting the stems. This disease is widespread and significantly impacts Angelica yield and quality, particularly in the primary production areas such as Western Hubei and Dingxi City, Gansu Province. In severe outbreak years, infection rates can reach 44–85% in major production regions, including Weiyuan, Zhang, and Min counties in the Gansu Province.

Symptoms
The disease primarily attacks the stems of Angelica plants. Initially, light brown spots appear on the outer stems, which gradually expand into dark brown, elongated lesions. Affected plants show yellowing and leaf dieback. In advanced stages, both stems and leaves dry out progressively, from the outer layers inward, and black specks appear on the stems—these are the conidial spore masses of the pathogen. Eventually, stems decay and turn gray to off-white, resulting in complete plant death (Fig. 32.11).

Fig. 32.11 Angelica Anthracnose Disease symptoms. (**a**) Field symptoms; (**b**) Stem symptoms

Pathogen

The causative agent of Angelica anthracnose is *Colletotrichum dematium* (Pers.:Fr.) Grove, a fungus in the *Colletotrichum* genus of the Ascomycota phylum. Conidial spore masses are dark brown, spherical or disc-shaped, with a diameter of 50–400 μm. These structures have brown setae around them, which stand upright, varying in length from 45 to 200 μm, and tapering at the tips. The pathogen has two types of spores: crescent-shaped spores that are pointed at both ends, colorless, transparent, with a single oil globule, measuring (18.0–24.5) μm × (3.5–5.0) μm; and oval or ellipsoidal spores that are colorless, transparent, and measure (9.7–16.5) μm × (2.5–4.0) μm (Fig. 32.12).

Optimal growth temperature for *C. dematium* hyphae and spore germination is 25 °C, with an ideal sporulation temperature of 20 °C. Spores germinate effectively at relative humidity above 95%, with the best results in water. The optimal pH levels for hyphal growth and sporulation are 11.0 and 10.0, respectively. The pathogen grows quickly on media with carbon sources like glucose, sucrose, lactose, maltose, mannitol, and D-arabinose, while mannose, D-galactose, and chloralose are less favorable. Nutrient sources such as soybean protein peptone and L-leucine support hyphal growth, and sucrose solutions promote spore germination.

Occurrence Pattern

The pathogen overwinters in soil and infected plant debris, serving as the primary infection source for the following year. When conditions are suitable, the pathogen infects stems through wounds, roots, or natural openings in above-ground parts. This disease typically begins to appear in mid-to-late June, initially in isolated plants with inconspicuous symptoms. By July, plants approximately 20 cm high exhibit classic symptoms, with some plants showing severe symptoms even before reaching 30 cm in height. By late August to early September, disease incidence peaks. Disease severity correlates with relative humidity and temperature, with higher humidity and warmer temperatures promoting disease development.

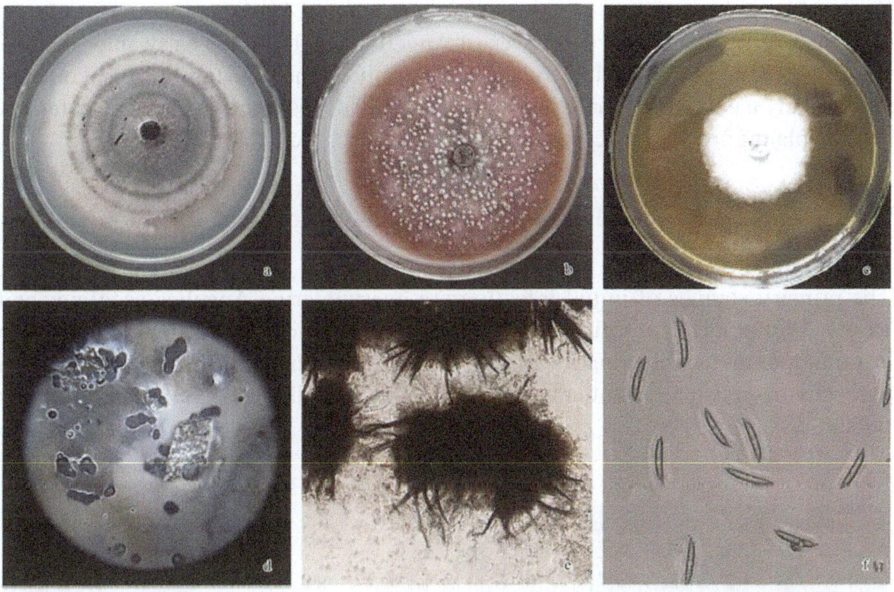

Fig. 32.12 Morphology of Angelica Anthracnose pathogen. (**a**) Growth on PDA medium; (**b**) Growth on OA medium; (**c**) Growth on MEA medium; (**d**, **e**) Conidial spore mass; (**f**) Conidia

Control Measures

1. Cultivation Practices: After harvest, remove diseased plants and plant debris promptly, deep-plow the soil to reduce infection sources, and employ crop rotation to reduce pathogen buildup, especially in fields previously planted with Angelica. Rotation with cereal or cruciferous crops helps decrease pathogen accumulation in the soil.
2. Chemical Control: At the first sign of disease, apply 43% tebuconazole SC at a 4,000-fold dilution, 30% pyraclostrobin WP at a 1,200-fold dilution, 70% thiophanate-methyl WP at an 800-fold dilution, 50% carbendazim WP at a 600-fold dilution, 10% prochloraz WP at a 1,000-fold dilution, or 40% difenoconazole EC at an 8,000-fold dilution.

32.7 Angelica Dwarf Virus Disease

Angelica dwarf virus disease has emerged as a new disease in recent years, showing distinctive symptoms and causing significant damage. It primarily affects the Angelica production regions in Min County, Gansu, and Yunnan, with no reports of occurrence in other Angelica-growing areas.

Symptoms

This disease mainly harms the leaves of Angelica plants. In the early stages, a few leaves on the upper plant display slight deformities and develop blister-like wrinkles. In advanced stages, new leaves fail to unfold, resulting in severe clustering, stunted plant growth, and leaves that exhibit a dark green color (Fig. 32.13).

Pathogen

The causative agent of Angelica dwarf virus disease is Konjac mosaic virus (KoMV).

Occurrence Pattern

The primary sources of initial infection are overwintering Angelica seedlings and various host plants. The virus spreads through mechanical transmission via friction or farm operations, infecting the plant. After entering the host, it reproduces within parenchyma cells and eventually spreads to vascular tissues, resulting in systemic infection.

Fig. 32.13 Angelica Dwarf Virus Disease symptoms. (**a**) Field symptoms; (**b**, **c**) Leaf symptoms

Control Measures

1. Cultivation of Virus-Free Seedlings: Use meristem tip culture to produce virus-free seedlings, or propagate using seeds to obtain virus-free plants. Establish disease-free seed production bases in high-altitude areas. Before thinning or transplanting, wash hands thoroughly with soap, and promptly remove any infected plants.
2. Pest and Disease Control: At the onset of aphid infestation, apply insecticides such as 10% imidacloprid WP diluted to 1,500-fold, 40% cypermethrin EC diluted to 6,000-fold, or 25% thiamethoxam EC diluted to 5,000-fold.
3. Chemical Control: In the early infection stages, apply a 1.5% tridecanol-copper sulfate-sodium dodecyl sulfate EC diluted 1,000-fold, or spray with a 2% Ningnanmycin AS at an effective dose of 90–120 g/ha per hectare to prevent and mitigate disease progression.

32.8 Occasional Diseases of Angelica (Table 32.1)

Table 32.1 Characteristics of occasional diseases of Angelica

(Pathogen)	Symptoms	Occurrence pattern
Sclerotinia Rot (*Sclerotinia* spp.)	Primarily affects stems. Initially, water-soaked lesions appear at the stem base, which then becomes hollow. Infected areas may display white, cotton-like mycelium and black sclerotia; leaves wither, and ultimately the entire plant dies	The pathogen overwinters in plant residues and soil. Under suitable conditions in the second year, it initiates infection. This disease thrives in cool, moist environments, particularly in unheated greenhouses, where it is more prevalent
Angelica mosaic virus disease (Tomato Mosaic Virus: ToMV)	Mainly affects leaves, causing symptoms such as mosaic patterns, leaf curling, and mottling, with three main symptom types	The virus primarily infects Angelica as well as other host plants. It spreads through contact and insect vectors. Agricultural tools can also transfer the virus, which then invades plant cells, disrupting cellular structure
Gray Mold (*Botrytis cinerea*)	Affects stems and leaves. Infected tissues become soft and watery, and affected areas develop a layer of gray mold. Extensive lesions on stems cause wilting and plant death. Leaves exhibit U-shaped or V-shaped lesions	The pathogen overwinters in plant residues and soil. Under favorable conditions, it produces spores on plant debris, spreading through wind and rain to initiate infection, followed by secondary infections. The disease peaks from June to early July, especially under conditions of high humidity, low temperatures, and prolonged rainfall. Poor ventilation and low light levels also promote the development of the disease
Bacterial leaf spot (pathogen pending identification)	Primarily affects leaves, causing small, irregularly shaped, oily spots	The pathogen overwinters in soil and plant residues. Humid and rainy conditions exacerbate the disease. Symptoms appear in the field from June, peaking from July to September, disease incidence is 45%. The severity level is 1–2 weeks

Chapter 33
The Diseases of *Saposhnikovia divaricata*

Saposhnikovia divaricata (Turcz.) Schischk., a perennial herbaceous plant belonging to the Apiaceae family, is harvested in spring and autumn for its roots, after removing rootlets and soil. The plant is primarily distributed across China in provinces like Heilongjiang, Jilin, Liaoning, Inner Mongolia, Hebei, Ningxia, Gansu, Shaanxi, Shanxi, and Shandong. It contains active ingredients like essential oils, chromones, coumarins, polysaccharides, and organic acids, providing therapeutic effects such as dispelling wind, relieving pain, and stopping convulsions.

Major diseases affecting *S. divaricata* include powdery mildew, leaf blight, gray mold, root rot, and parasitic Cuscuta species.

33.1 Powdery Mildew of *Saposhnikovia divaricata*

Powdery mildew is a major disease in *S. divaricata* cultivation, widespread across main production areas with an average incidence rate of 20–30%, reaching over 50% in severe cases. White powdery patches often cover the leaves, significantly hindering photosynthesis, disrupting normal metabolism, and causing premature leaf death. This greatly affects seed maturation and root yield.

Symptoms

The disease primarily affects leaves and tender stems. Initially, white, near-circular powdery spots appear on the leaves and tender stems. These spots gradually expand to cover the entire leaf and tender stem with a white powdery substance, which are the conidia of the pathogen. In later stages, numerous small black dots, representing the sexual stage chasmothecia of the pathogen, appear on diseased leaves and stems. Severe infections lead to leaf shrinkage, distortion of young shoots, flower bud suppression, defoliation, and stem wilting, ultimately reducing yield (Fig. 33.1).

© China Agriculture Press Co., Ltd. 2026

T. Wang et al., *Atlas of Chinese Medicinal Plant Diseases*,
https://doi.org/10.1007/978-981-95-6326-5_33

Fig. 33.1 *Saposhnikovia divaricata* Powdery Mildew Symptoms

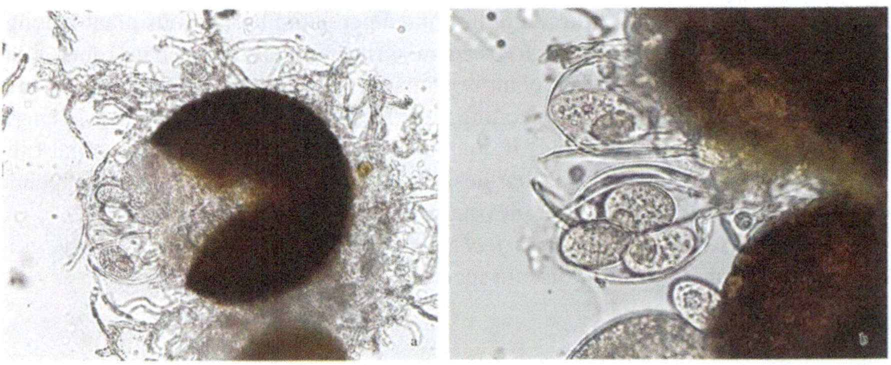

Fig. 33.2 *Saposhnikovia divaricata* powdery mildew pathogen Morphology. (a) Chasmothecia and appendages. (b) Asci and Ascospores

Pathogen

The pathogen responsible for *S. divaricata* powdery mildew is *Erysiphe heraclei* DC., a member of the *Erysiphaceae* family within the Ascomycota phylum. Mycelium develops on both sides of the leaves and may disappear or persist. Conidia are mostly cylindrical, occasionally barrel-shaped, measuring (20.3–40.6) μm × (12.7–17.8) μm. Chasmothecia are scattered to loosely clustered, dark brown, flattened, and spherical, measuring 75–120 μm in diameter. Appendages are filamentous, dichotomous or irregularly branched, measuring (30–125) μm × (3.8–8.9) μm, with a smooth to slightly rough surface and zero to three septa. Asci are near-ovoid to spherical, with or without short stalks, measuring (45.7–83.8) μm × (38.1–49.5) μm. Ascospores are ovoid to elliptical, measuring (19.1–27.9) μm × (12.7–16.3) μm (Fig. 33.2). This pathogen also infects other plants, such as *Cnidium monnieri*, *Coriandrum sativum*, *Daucus carota*, *Oenanthe javanica*, *Sium suave,* and other species.

Occurrence Pattern

The pathogen overwinters as chasmothecia in diseased plant debris. In spring, under favorable temperature and humidity conditions, the chasmothecia release ascospores, which directly penetrate the host's epidermis, causing initial infection. Conidia produced on infected plants spread through wind and rain, leading to repeated infections. Disease occurrence is closely related to temperature and humidity, with optimal development at 20–24 °C and high relative humidity. Poor field management practices, such as excessive watering, over-application of nitrogen fertilizer, excessive plant growth, shade, and poor field ventilation, can exacerbate the disease.

Control Measures

1. Remove Plant Debris: Before winter, remove plant debris and centralizedly destroy it to reduce sources of infection. Compost with diseased plant material must be fully decomposed before application.
2. Enhance Field Management: Rotate with gramineous crops; improve cultivation practices with proper spacing, ventilation, and light penetration in the field. Increase phosphorus and potassium fertilizer applications, and avoid planting in low-lying areas.
3. Chemical Control: At the onset of disease, apply 0.3-0.5° Baumé sulfur lime suspension, or spray 90-120 g per mu of 10 billion CFU/g Bacillus subtilis WP, or use 1% osthol soluble solution (SL) at 1,000-2,000-fold dilution, 15% triadimefon WP at 1,000-fold dilution, 50% carbendazim WP at 600-fold dilution, or 25% tebuconazole EW at 1,000-1,500-fold dilution. Spray every 7–10 days, for a total of two to three applications, depending on disease severity.

33.2 *Saposhnikovia divaricate* Spot Blight

Spot blight is one of the primary leaf diseases affecting *S. divaricate* production. This disease is widespread, typically with an incidence rate of 10–20%, and can lead to premature leaf death and defoliation. In severe cases, the disease can cause extensive leaf dieback, significantly impacting both yield and seed harvest.

Symptoms

The disease primarily affects *S. divaricata* leaves, but can also impact petioles, stems, and inflorescences. Initially, small brown spots appear along the leaf edges. These lesions are circular or nearly circular, with a diameter of 25 mm, and later spread across the entire leaf, merging into large, irregularly shaped brown lesions with dark brown edges and lighter centers. Tiny black dots appear within the lesion centers—these are the pycnidia of the pathogen. In severe cases, the lesions coalesce, leading to complete leaf necrosis. The disease is most prevalent in autumn. When petioles are infected, they develop long, brown necrotic streaks that extend along both the petiole and the stem, causing leaves and stems to wither and die. Stems initially exhibit fusiform lesions that are gray-brown, which gradually enlarge into elongated black streaks densely populated with black pycnidia. The disease can also affect the lower part of the flowering stem, gradually spreading upward, causing the stems to wilt and die, which severely impacts seed production (Fig. 33.3).

Pathogen

The causal agent of *S. divaricate* spot blight is *Septoria saposhnikoviae* G.Z. Lu & J.K. Bai, a member of the genus Septoria in the Ascomycota phylum. Pycnidia form on both sides of the leaves, either scattered or clustered beneath the host epidermis. Initially embedded, they later break through the surface with slightly protruding ostioles. Pycnidia are spherical to subspherical, light brown, 65–169 μm in diameter, and 60–110 μm in height, with membranous walls that are light brown, while the inner layer is colorless and contains conidiogenous cells. Conidia are needle-shaped with a rounded base, colorless, transparent, straight or slightly curved, and taper to a pointed apex, generally containing one to four septa, most commonly three. Conidia measure (20–45) μm × (1.5–2.5) μm (Fig. 33.4).

Occurrence Pattern

The pathogen overwinters as pycnidia in infected plant debris. Under favorable conditions in the following year, the pycnidia absorb moisture and release conidia, causing primary infections. The conidia produced on lesions are dispersed by wind and rain, causing secondary infections. High temperatures, high humidity, and prolonged rainy periods promote disease development. In northeastern regions, the disease typically begins in late June, with peak incidence from July to August.

Control Measures

1. Field Sanitation: Before winter, clean the fields thoroughly, removing and destroying infected plant debris to reduce the overwintering inoculum.
2. Improved Cultivation Practices: Maintain optimal planting density to avoid overcrowding. Ensure prompt drainage after rainfall to reduce field humidity, and weed as necessary.

Fig. 33.3 Symptoms of *Saposhnikovia divaricata* Spot Blight. (**a**) Leaf symptoms (**b**) Stem symptoms

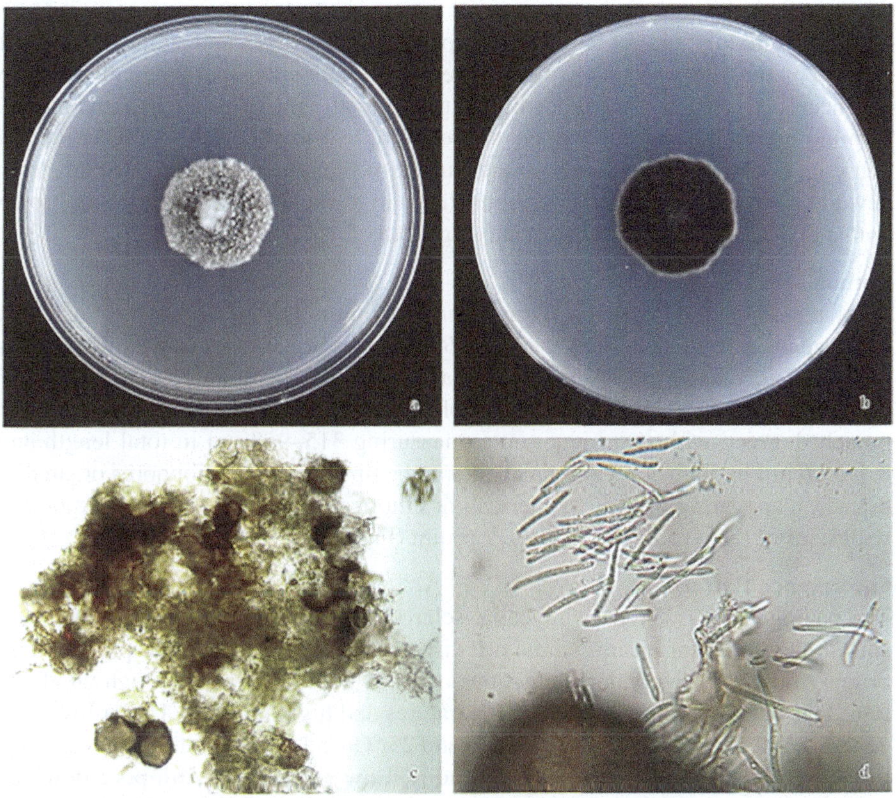

Fig. 33.4 Morphology of *Saposhnikovia divaricata* Leaf Blight pathogen. (**a**, **b**) Colony characteristics. (**c**) Pycnidia. (**d**) Conidia

3. Chemical Control: At the onset of disease symptoms, promptly remove infected leaves and apply fungicides. Recommended treatments include spraying with 50% carbendazim WP at a 500-fold dilution, 30% benzothiazole-propiconazole EC at a 1,500–2,000-fold dilution, 77% copper hydroxide WP at a 500-fold dilution, or 70% mancozeb WP at a 500-fold dilution.

33.3 Gray Mold of *Saposhnikovia divaricata*

Gray mold has recently emerged as a significant disease in *S. divaricata* cultivation, with a general incidence rate of 5% to 10%, and it is particularly severe in mountainous areas. This disease primarily affects the leaves and stems and spreads rapidly, potentially causing extensive leaves and stems death in a short period.

Symptoms

Gray mold mainly attacks aboveground parts such as leaves and stems. Initially, it appears on the leaves as inconspicuous yellow circular spots that gradually develop into necrotic areas, eventually leading to tissue death. The spots enlarge and merge, forming brown lesions that can cover the entire leaf surface. In humid weather, a dense layer of gray mold forms on the surface of the affected areas. Infected stems develop fusiform, water-soaked lesions that start as dark green, later turning brown with a gray mold layer on the surface (Fig. 33.5).

Pathogen

The pathogen responsible for *S. divaricata* gray mold is *Botrytis cinerea* Pers.:Fr., of the genus Botrytis. On PDA medium, *B. cinerea* colonies start as pale white, later becoming gray, and may produce sclerotia. Hyphae are transparent, with a relatively uniform diameter of 5–6 μm. Conidiophores are grouped, either unbranched or branched, erect, with transverse septa, measuring 315–958 μm in total length and 8.4–12.6 μm in diameter. Conidia cluster at the tips of the conidiophores or smaller branches, appearing obovate, spherical, or elliptical, smooth, and nearly colorless, with sizes of (8.4–15.8) μm × (6.3–12.6) μm (Fig. 33.6).

Occurrence Pattern

The pathogen overwinters as mycelia, sclerotia, and conidia in crop residues and soil. In the following year, conidia are dispersed on leaves by raindrops, initiating primary infection. Secondary infections occur as conidia spread through wind and rain. This pathogen thrives in low temperatures and high humidity. Under low-light conditions, temperatures between 15 °C and 25 °C, and rainfall with relative humidity above 90%, the disease develops readily. Low-lying areas with poor drainage and dense plantings are more prone to severe outbreaks.

Control Measures

1. Remove of Field Debris: Promptly clear fallen leaves and plant debris from the field and destroy them to reduce sources of overwintering pathogens.

Fig. 33.5 Symptoms of *Saposhnikovia divaricata* Gray Mold. (**a**) Leaf symptoms (**b**) Stem symptoms

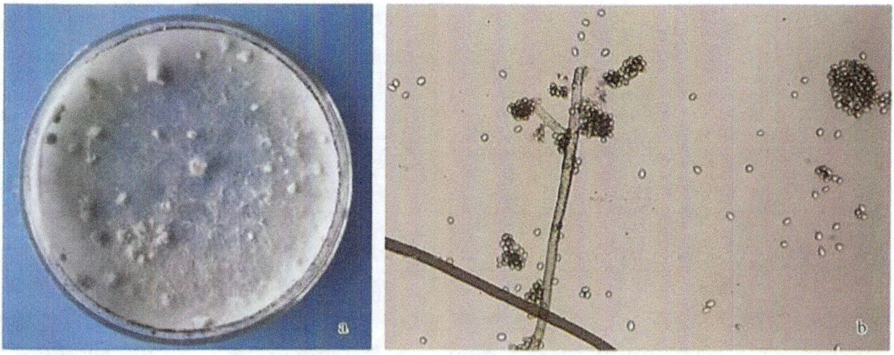

Fig. 33.6 Morphology of the *Saposhnikovia divaricata* Gray Mold pathogen. (**a**) Colonial morphology (**b**) Conidiophores and Conidia

2. Improve Field Management: Select fertile, well-drained sandy loam for cultivation, and avoid dense planting. Apply a balanced mix of nitrogen, phosphorus, and potassium fertilizers to improve plant disease resistance.
3. Chemical Control: At the onset of the disease, apply 40–60 g per mu of *Bacillus subtilis* WP containing 100 billion CFU/g, 86-120 mL per mu of 0.3% eugenol SL, 25-35 mL per mu of 20% fludioxonil SC, or 125-150 g per mu of Trichoderma WG. Alternatively, spray with fungicides such as 43% iprodione WP diluted to 800-fold, 70% thiophanate-methyl WP diluted to 600-fold, 50% iprodione WP diluted to 1,000-fold, or 40% pyrimethanil SC diluted to 800-1,000-fold.

33.4 *Saposhnikovia divaricata* Root Rot Disease

Root rot is a major disease impacting *S. divaricata* production, with average infection rates ranging from 2% to 10% and in severe cases, exceeding 15%. The disease results in the decay of both aerial and root structures, with advanced cases causing the complete breakdown of root systems, which significantly reduces *S. divaricata* yield.

Symptoms

This disease primarily affects the root system and can occur throughout the entire growth period. In the early stages, wilting of the aerial parts appears during hot weather, with recovery observed in the morning and evening. As the disease progresses, basal leaves gradually yellow, overall growth weakens, and the entire plant eventually wilts and dies. Affected plants are easily uprooted, showing reduced fibrous roots, partial decay of primary and secondary roots, and, in severe cases, complete rot below the stem base (Fig. 33.7). In humid conditions, white or pinkish fungal threads and conidia may be observed on the infected areas.

Fig. 33.7 Symptoms of *Saposhnikovia divaricata* Root Rot Disease. (**a**) Field symptoms (**b**) Typical symptoms

Pathogen

The pathogen responsible for *S. divaricata* root rot is a complex of fungi from the *Fusarium* genus (*Fusarium* spp.). Root rot can result from infections by several *Fusarium* species, including *Fusarium solani* (Mart.) Appel et Wollenw. ex Snyder et Hansen, and *F. equiseti* (Corda) Sacc., both of which can cause substantial root decay. *F. solani* initially forms white colonies on PDA medium, which later turn yellowish-white, with sparse aerial mycelium and dense, orderly hyphae. Macroconidia are sickle-shaped or elongated, measuring (21.23–32.69) μm x (3.73–6.19) μm; microconidia are ovoid or elongated, measuring (5.67–12.46) μm x (2.51–5.36) μm. *F. equiseti* colonies initially appear pink and later become pale yellow, with irregular white edges and dense mycelium. Macroconidia are sickle-shaped, thin at both ends, with three to five septa, and measure (42.74–52.13) μm x (3.28–4.47) μm (Fig. 33.8).

Occurrence Pattern

The pathogen primarily overwinters as mycelium or chlamydospores in soil, infected roots, or plant debris, which serve as initial infection sources for the following year. After winter, the pathogen invades roots through injuries to primary or secondary roots or directly penetrates the tissues. The disease typically initiates in late May, with peak incidence occurring in July and August, followed by a decline in September. High temperatures, humidity, and continuous rain are conducive to disease progression. *Fusarium* spp., as soilborne fungi, can survive in soil for 3–5 years. Infection centers in fields are often distinct, and the pathogen spreads through rain, irrigation, and farming practices, with subterranean pests and nematodes intensifying infection. Fields with dense, compacted soil and poor drainage are more prone to severe infection.

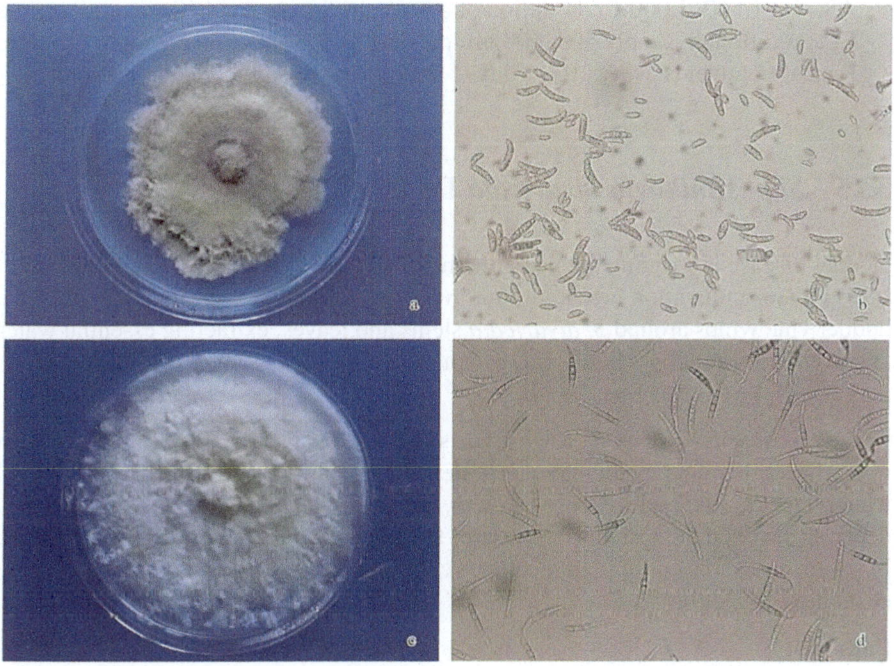

Fig. 33.8 Morphology of *Saposhnikovia divaricata* Root Rot pathogens. (a) *Fusarium solani* colony morphology, (b) *F. solani* conidia, (c) *F. equiseti* colony morphology, (d) *F. equiseti* conidia

Control Measures

1. Site Selection: Select well-drained sandy loam fields, avoid low-lying, dense, or poorly drained soils.
2. Crop Rotation: Avoid consecutive plantings in the same location. Rotate with nonhost crops, such as corn or rice, with a rotation interval of at least 3 years to reduce pathogen levels and lower infection risk.
3. Removal of Diseased Plants: After harvest in autumn and winter, thoroughly clean fields, bury or burn diseased plant residues deeply, and disinfect soil around infection sites with lime. Alternatively, drench with 50% carbendazim wettable powder (300x) or 50% thiophanate-methyl wettable powder (500x) to limit disease spread.
4. Enhanced Cultivation Practices: Use appropriate fertilization practices to improve plant resistance, including increased organic matter and phosphorus-potassium applications. During prolonged rainfall or high soil moisture, practice timely soil loosening to enhance aeration.
5. Chemical Control: At the early disease stage, apply root treatments with 500–1,000-fold dilution of *Bacillus subtilis* microcapsule SC containing 1 billion CFU/g, a 500-fold dilution of a 50% carbendazim WP, a 500-fold dilution of a 50% thiophanate-methyl WP, a 100–150-fold dilution of a 3% pyrimethanil

AS, or a 1,500–3,000-fold dilution of a 30% etridiazole emulsifiable concentrate EC. Additionally, apply 500–600 mL per acre of 5% garlic oil ME after dilution. Repeat treatments every 10–15 days for a total of two to three applications.

33.5 *Saposhnikovia divaricata* Dodder Disease

Dodder disease is a significant parasitic threat in *S. divaricata* production, prevalent across major cultivation areas with an average infection rate exceeding 30%. The dodder vine wraps around *S. divaricata* stems and leaves, extracting essential water and nutrients through its haustoria, which severely impacts yield.

Symptoms
Dodder primarily infests *S. divaricata* stems. It forms threadlike, yellow-green vines that coil around medicinal plants, developing haustoria to absorb nutrients and water. Due to its rapid, dense growth, dodder interferes with photosynthesis, deprives the plant of nutrients, and causes the leaves to turn yellow, dry, and weaken. In severe cases, the entire plant becomes entwined in dodder, turns yellow, and may even die (Fig. 33.9).

Pathogen
The main species affecting *S. divaricata* is *Cuscuta japonica*, a member of the Convolvulaceae family. This parasitic herb lacks true roots and leaves, or has scale-like leaves with no chlorophyll, making it a full parasite. Dodder stems are filamentous, yellow-white, or slightly purplish-red. Its flowers are inconspicuous, white, yellow, or pink, forming small spherical clusters. The fruit is a dehiscent, spherical capsule containing two to four seeds, which are small and lack cotyledons and radicles (Fig. 33.9).

Fig. 33.9 Dodder Infestation Symptoms on *Saposhnikovia divaricata*

Occurrence Pattern

Mature dodder seeds overwinter in the soil or within seed mixtures. The seeds germinate in May–June when the host grows, forming a thin threadlike sprout anchored to soil particles, while the other end elongates in search of a host. Upon contact, dodder attaches to the stem, forming haustoria that penetrate the host and connect to its vascular system, drawing water and nutrients. Once the parasitic connection is established, the dodder detaches from the ground. Seed maturity is staggered, leading to a continual spread of new parasitic sites in the field. Dodder's regenerative ability is robust; even a single growth point can sustain parasitism. Germination is enhanced by spring rains, and during rainy summers, its spread is accelerated, causing severe damage.

Control Measures

1. Seed Screening: Thoroughly screen *S. divaricata* seeds before planting to remove any dodder seeds. Use well-composted manure or other fertilized organic matter to prevent introducing dodder seeds into the field.
2. Crop Rotation: In heavily infested areas, rotate *S. divaricata* with nonhost crops, such as grasses, and use deep tilling to bury dodder seeds.
3. Field Management: Remove dodder plants upon detection, ideally uprooting them along with affected plants before flowering. Complete removal is essential, as dodder fragments can continue to spread.
4. Chemical Control: Applying biological agents can have some effects. Treatments are more effective when applied after breaking dodder vines, particularly during or after rainfall.

Chapter 34
The Diseases of *Paeonia lactiflora* Pall.

Peony (*Paeonia lactiflora* Pall.) is a perennial herbaceous plant from the Ranunculaceae family, valued both as a medicinal herb and ornamental plant. Its root, referred to as "Baishao" in traditional Chinese medicine, is commonly used for therapeutic purposes. Peonies are primarily distributed across China, Korea, Japan, Mongolia, and Russia (Siberia). Within China, they are found in provinces such as Jiangsu, Shaanxi, southern Gansu, Sichuan, Guizhou, Anhui, Shandong, and Zhejiang, as well as the northeastern and northern regions, where they are also cultivated in urban parks. In northeastern China, they grow on mountain slopes and forest areas at altitudes of 480–700 m, while in other provinces, they thrive on grassy slopes at altitudes of 1000–2300 m. The peony root, once boiled and peeled, appears pale white or light reddish-brown, smooth or with longitudinal wrinkles and small root scars, sometimes retaining traces of brown outer skin. It is traditionally used to nourish the blood, regulate menstruation, astringe sweat, soothe the liver, relieve pain, and stabilize liver yang, addressing conditions such as blood deficiency with pallor, menstrual irregularities, spontaneous and night sweating, hypochondriac pain, abdominal pain, limb spasms, and dizziness.

The main diseases affecting peonies include black spot, anthracnose, southern blight, viral diseases, powdery mildew, and blight. Among these, black spot, anthracnose, southern blight, and viral diseases occur more frequently and require focused prevention and control.

34.1 Peony Leaf Spot Disease

Peony leaf spot disease is one of the most common and significant diseases affecting cultivated peonies, prevalent across major cultivation regions such as Jiangsu, Sichuan, Shandong, and Anhui provinces. This disease severely impacts both cut flower yield and the production of "Baishao" roots. Generally, the infection rate

© China Agriculture Press Co., Ltd. 2026

T. Wang et al., *Atlas of Chinese Medicinal Plant Diseases*,
https://doi.org/10.1007/978-981-95-6326-5_34

ranges from 10% to 50%, with some regions experiencing severe outbreaks where the rate can exceed 80%.

Symptoms

In severe cases, entire leaves become scorched. During autumn, lesions turn dark brown, becoming brittle and prone to cracking. In the later stages, the undersides of lesions may develop a dark green mold-like layer. The disease also affects young stems, producing small, oval, purple-brown spots that may appear raised. These lesions spread slowly, but in severe cases, they can merge to cover large areas. The disease's persistent occurrence over multiple years weakens plant vigor, leading to stunted growth, fewer and smaller flowers, and potentially resulting in the death of the entire plant (Fig. 34.1).

Typically, infection begins on the lower leaves and gradually spreads to the entire plant. Early symptoms on infected leaves appear as small, needle-sized, circular spots.

Fig. 34.1 Symptoms of Peony Leaf Spot Disease. (**a**) Field symptoms; (**b**, **c**) Leaf symptoms

Pathogen

Peony leaf spot disease is caused by *Alternaria alternata* (Fr.) Keissl, an asexual type from the *Alternaria* genus under Ascomycota.

On PDA culture medium, colonies appear gray, with well-developed mycelium and aerial hyphae, initially colorless but gradually darkening to various shades of brown, with septate, highly branched structures. Short spore chains form on the upper sections of conidiophores, each chain typically bearing several conidia. The size and shape of the conidia vary greatly, measuring approximately (20.0–39.0) μm × (8.0–14.5) μm. Under adverse conditions, thick-walled chlamydospores may form, helping the pathogen survive in unfavorable environments (Fig. 34.2).

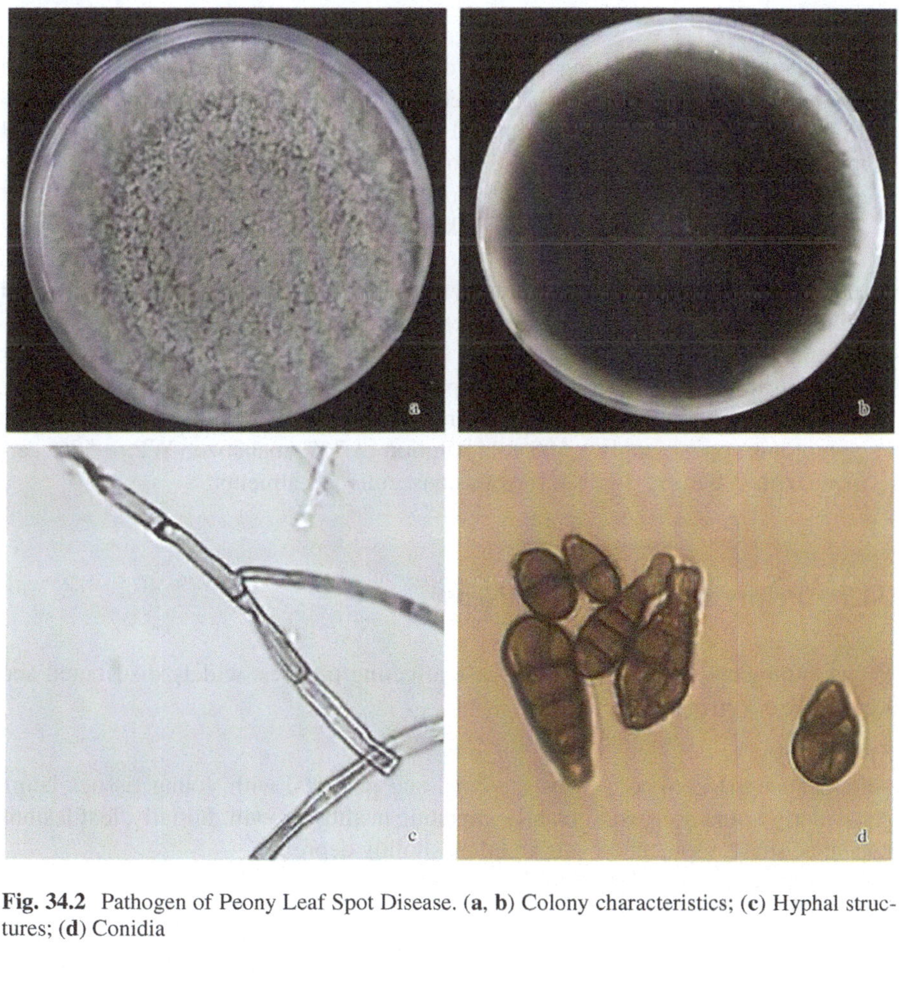

Fig. 34.2 Pathogen of Peony Leaf Spot Disease. (**a**, **b**) Colony characteristics; (**c**) Hyphal structures; (**d**) Conidia

Occurrence Pattern

The timing and severity of peony leaf spot disease vary by region and cultivation base. Mycelium and conidia overwinter on plant debris and infected seed pods, and can persist saprophytically on leftover fleshy roots in nursery beds. Conidia spread through wind, rain, and air currents, and can directly infect healthy plants, with wounds providing an ideal entry point. At 20–24 °C, the incubation period is 5–6 days, while at 8 °C, it extends to 14 days with a lower infection rate. In the field, leaf and stem lesions appear in March, and fruiting bodies form before the rainy season in May to June or in humid conditions during late autumn, a process that can take over 2 months. Secondary infections cause lesions that expand slowly, with the characteristic green mold layer on lesions becoming visible by mid-August. Hot, dry summer weather inhibits the formation of fungal fruiting bodies, spore germination, and mycelial growth. The severity of the disease primarily depends on the quantity of primary infections that survive winter. High-density planting, high humidity, and poor ventilation favor disease development.

Control Measures

1. Resistant Varieties: Planting high-yield, disease-resistant varieties such as "Hero Flower" and "Sparkling Star" can effectively reduce the incidence of leaf spot.
2. Agricultural Practices: In autumn and early spring, thoroughly remove infected plant debris and fallen leaves, and trim residual stem bases. Remove any fleshy roots left after division to prevent reinfection, and add a layer of compost (approximately 15 cm thick) to block potential infections. Rotate crops every 3–4 years when propagating by root division. Manage planting density to improve ventilation, drain excess water promptly after rain to reduce field humidity, and apply balanced fertilizers to regulate plant growth.
3. Chemical Control: Before bud break in early spring, apply a 3–5% lime sulfur solution to the soil surface. Begin spraying in early March, and repeat every 2 weeks for a total of two applications. During the growing season, if disease symptoms appear, apply a 500-fold solution of 75% mancozeb WP or 50% carbendazim WP every 10 days for two consecutive treatments.

34.2 Peony Anthracnose Disease

Peony anthracnose is a common disease affecting peonies, widely distributed and frequently occurring.

Symptoms

Anthracnose affects peony leaves, stems, and petioles, with young tissues being particularly vulnerable, significantly impeding healthy growth. Initially, leaf lesions appear as small, elongated ovals and then slightly depress.

Over time, these lesions expand into large, irregular, dark brown spots. In humid conditions, the surface of the lesions develops sticky, pinkish spore clusters—a mixture of conidia and a gelatinous matrix produced by the pathogen. Severe cases lead to drooping leaves, which become covered in these spore clusters (Fig. 34.3). Similar lesions can form on stems, and severe infections may cause the stem to break.

Pathogen

The pathogen responsible for peony anthracnose is *Colletotrichum gloeosporioides* (Penz.) Penz. et Sacc., an asexual fungus of the *Colletotrichum* genus. On PDA medium, colonies initially appear white, turning grayish-white to gray, and finally dark gray. The aerial mycelium is dense and woolly. After some time, the center of the colony forms orange spore clusters, creating concentric rings. Prolonged cultivation leads to the formation of dark brown rings of ascocarps around the orange spore clusters. The acervuli of *C. gloeosporioides* develop beneath the epidermal

Fig. 34.3 Symptoms of Peony Anthracnose Disease

layer, colorless to brown in hue, with tightly packed, unbranched conidiophores. Setose structures arise in the center or at the edges of the acervuli. Conidia are nearly round or cylindrical, unicellular, and colorless (Fig. 34.4).

Occurrence Pattern

The prevalence and severity of peony anthracnose vary across regions and seasons. The pathogen overwinters as mycelium on infected leaves or stems, producing conidia the following year that spread by wind and rain, infecting plants through wounds. Years with heavy rainfall in August and September experience higher infection rates. In densely potted plants or those watered improperly (especially at night), moisture remains on the leaves, promoting the germination of conidia and increasing disease susceptibility.

Control Measures

1. Select Disease-Resistant Varieties

Cultivate high-yield, disease-resistant varieties to control and mitigate the disease effectively. Varieties such as Zihongkui and Meiju show strong resistance to anthracnose. Selecting, propagating, and cultivating pure seeds will help prevent seed degradation.

2. Agricultural Control
   (a) Use Healthy Seeds: As seedborne pathogens are the primary source of initial infection, selecting and treating seeds is essential. Establish disease-free seed fields in affected areas and treat seeds through sun drying and water selection to discard immature seeds, enhancing germination rates.
   (b) Improve Cultivation Management: Apply a balanced base fertilizer, early seedling fertilizer, and controlled amounts of nitrogen, phosphorus, and potassium. Avoid overuse or delayed application of nitrogen to support robust plant growth and increase disease resistance. Additional agricultural practices such as drainage and crop rotation also reduce disease incidence.

3. Chemical Control.

Seed treatment with a mixture of prochloraz, carbendazim, and difenoconazole is effective against peony anthracnose and other seedling diseases. Alternatively,

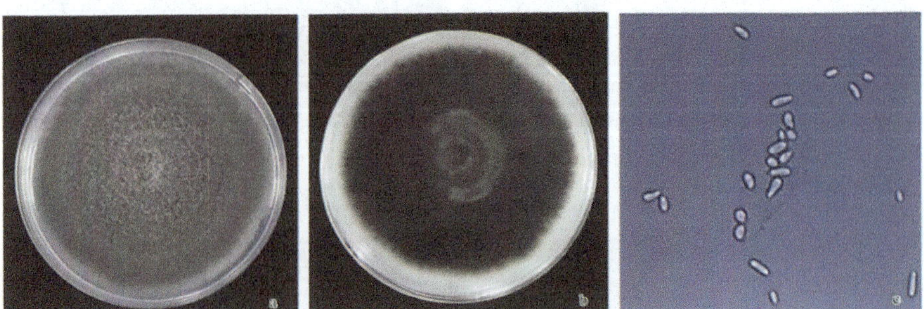

Fig. 34.4 Morphology of Peony Anthracnose Pathogen. (**a**, **b**) Colony characteristics; (**c**) Conidia

seeds can be soaked in a 1:1:150 Bordeaux mixture (copper sulfate: lime: water = 1:1:150) or in a 1,000-fold dilution of a 50% thiophanate-methyl WP for 10 minutes, then air-dried before sowing. At the onset of infection, apply fungicides, such as 70% thiophanate-methyl WP at 1000-1200-fold dilution or 50% carbendazim WP at 800-1000-fold dilution spraying every 10 days for three applications.

34.3 Southern Blight of Peony

Southern blight is a significant disease affecting numerous medicinal plants, with widespread occurrence and severe impacts. Besides peony, it also affects plants such as Atractylodes, Salvia, Gynostemma, Astragalus, Coptis, Aconitum, Jasminum, Aster, Dianthus, Chrysanthemum, Lilium, Iris, Pseudostellaria, Scrophularia, Platycodon, and Rehmannia. In peony-producing regions like Deyang city of Sichuan province, southern blight is prevalent, with general incidence rates of 5–10%, though in severe cases, rates can exceed 30%.

Symptoms

Southern blight in peony typically affects the root collar near the soil line in young seedlings. Initially, the infected area on the cortex turns brown and gradually expands. White, silky fungal mycelium appears on the lesion, often radiating and spreading to nearby soil surfaces. As the disease progresses, clusters of sclerotia, resembling canola seeds, form on the mycelial mat at the plant base or soil surface. Initially, sclerotia are milky white, then turn pale yellow, and finally dark brown. Affected plants develop rot in the cortex of the stem base and roots, which blocks water and nutrient transport. This results in leaf yellowing and wilting, leading to plant death. The dead root collar leaves behind a tangled mass of fibrous tissue that can be easily pulled from the soil (Fig. 34.5).

Pathogen

The causal agent of southern blight in peony is *Sclerotium rolfsii* Sacc., an asexual species of the *Sclerotium* genus in Basidiomycota. Colonies of *S. rolfsii* on PDA medium appear circular, with white, silky, septate, and highly branched mycelium, radiating outward with a diameter of 3.5–6.5 μm.

After 7–10 days, sclerotia form within the colony. Sclerotia appear on the surface, are spherical, initially white, then gradually darken to brown, with a smooth, shiny exterior, and measure 0.5–0.8 mm in diameter (Fig. 34.6).

Occurrence Pattern

After June, when the average soil temperature at a depth of 15 cm exceeds 25 °C, conditions become favorable for the onset and progression of southern blight. The optimal temperature range is typically 30–35 °C, with high risk under conditions of 30 °C soil temperature, abundant rainfall, and high humidity, particularly in July and August. The pathogen generally overwinters in soil, weeds, or plant residues in

Fig. 34.5 Symptoms of Peony Southern Blight. (**a**) Leaf symptoms; (**b**) Root collar symptoms; (**c–e**) Root symptoms; (**f**) Far right: Healthy plant; progressively more severe infection from right to left

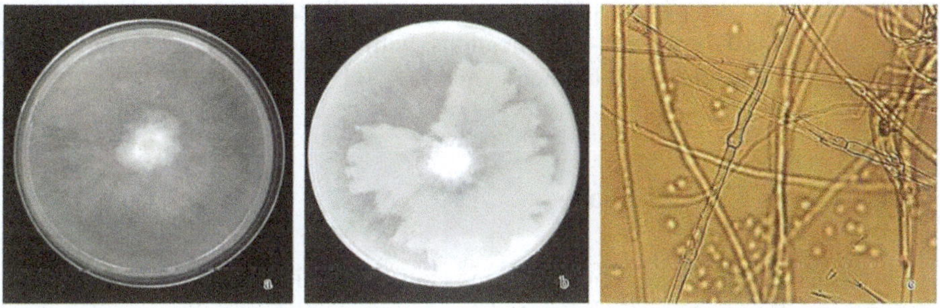

Fig. 34.6 Pathogen of Peony Southern Blight. (**a**, **b**) Colony characteristics; (**c**) Mycelial structures

the form of mature sclerotia and mycelium. Soil-borne pathogens serve as the primary source of initial infection, with subsequent spread via infected seedlings, manure, and irrigation water. High temperatures and humidity promote disease development, with areas of low elevation, poor drainage, excessive soil moisture, and insufficient sunlight being particularly susceptible.

Control Measures

1. Agricultural Control
   (a) Remove and destroy or deeply bury infected plant debris and weeds at the earliest sign of infection; promptly remove infected plants during the growing season.
   (b) Implement crop rotation. Avoid monoculture or rotation with other susceptible hosts; ideally, rotate with Poaceae crops for 3–5 years or alternate with wet-dry crops.
   (c) Strengthen field management. Plant in well-drained, elevated areas, and avoid resurfacing previously tilled soil. Ensure proper drainage to allow fields to dry after rain, and till to a depth of 15 cm before sowing or planting. Use fully decomposed organic fertilizers.
2. Biological Control

Biological control with antagonistic bacteria has proven effective against southern blight in peony. First, prepare *Trichoderma* cultures, which are then mixed with sterilized bran to create a *Trichoderma* preparation. During application, mix the preparation evenly with fine soil and apply to the field, maintaining soil moisture to support *Trichoderma* growth and reproduction, thereby suppressing the southern blight pathogen.

3. Chemical Control
   (a) To prevent disease, apply lime powder to disinfect the soil.
   (b) At the early stage of southern blight, chemical control should be applied promptly as a soil drench around the plant base. Effective fungicides include a 500-fold dilution of 50% carbendazim wettable powder, a 1,500–3,000-fold dilution of 30% etridiazole emulsifiable concentrate, or a 100–150-fold

dilution of 3% pyrimethanil aqueous solution. Treatments should be repeated every 7–10 days for 2–3 applications, with fungicides rotated to reduce the risk of resistance development.

34.4 Peony Viral Disease

Peony viral disease is a common issue in peony cultivation, widespread in production areas such as Jiangsu, Sichuan, and Anhui provinces. Generally sporadic, some fields experience severe outbreaks that significantly affect peony growth.

Symptoms

Infected peony plants show symptoms on leaves, including ring or line-shaped spots, discoloration zones, and various patterns, which can develop into small necrotic spots. Sometimes concentric rings appear, displaying alternating dark and light green bands, along with symptoms of stunting or leaf chlorosis (Fig. 34.7). Different viruses produce distinct symptoms: tobacco rattle virus (TRV) typically

Fig. 34.7 Symptoms of Peony Virus Disease on leaves

causes irregular rings or spots of various sizes; alfalfa mosaic virus (AMV) often results in concentric ring patterns with dark and light green bands, along with small necrotic spots without stunting; tomato spotted wilt virus (TSWV) frequently leads to pronounced stunting, with lower branches weakened or distorted and leaves curled and yellowed.

Pathogen

The viruses causing peony virus disease include tobacco rattle virus (TRV), alfalfa mosaic virus (AMV), and tomato spotted wilt virus (TSWV) (Fig. 34.8).

Occurrence Pattern

TRV spreads through nematodes, dodder, and infected peony seeds. AMV is primarily transmitted by aphids and other insects with piercing-sucking mouthparts, while TSWV can be spread through mechanical contact, seeds, and thrips as persistent carriers. Virus transmission also occurs via vegetative propagation when the virus enters from cut surfaces in bud materials. Storage and peony processing may facilitate further spread. Aphids and thrips serve as primary vectors, with peak infection periods in May and June.

Control Measures

1. Agricultural Control
   (a) Eliminate sources of infection: Remove infected plant residues to reduce the likelihood of infection during the seedling stage. Regularly uproot weeds and wild hosts near seedbeds and remove diseased plants promptly to reduce viral sources.
   (b) Avoid using infected plants for propagation.
   (c) Fertilization management: Avoid using improperly composted manure mixed with infected plant residues.

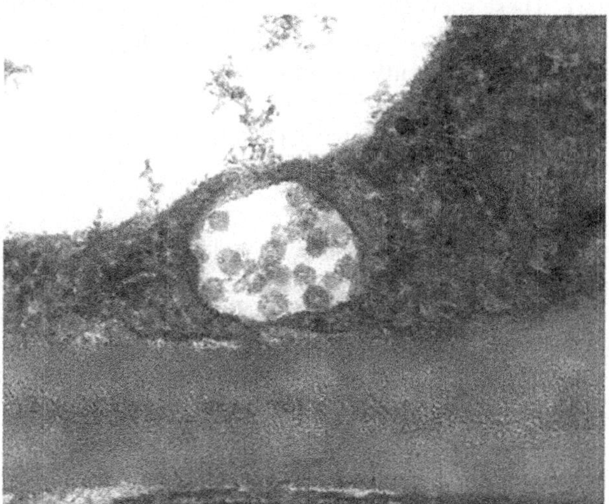

Fig. 34.8 Particles of Tomato Spotted Wilt Virus

(d) Use virus-free, tissue-cultured seedlings to help restore seed quality and enhance resistance, reducing field-accumulated viral load.

2. Physical Control

Using physical methods, such as placing reflective silver mulch in fields, deters aphids and other piercing-sucking pests, reducing virus transmission.

3. Chemical Control

(a) Control aphids and other pests by spraying insecticides such as 1.8% abamectin EC (3,000-5,000-fold dilution), 10% imidacloprid WP (2,000-fold dilution), or 50% pymetrozine WP (1,500-2,000-fold dilution).
(b) Promote early growth with applications of potassium dihydrogen phosphate at a 4,000-fold dilution or 30–50 mL *Bacillus* suspension per 75 L water per acre at the seedling stage.
(c) At symptom onset, apply potassium dihydrogen phosphate or 20% moroxydine hydrochloride SC (500-fold dilution) every 7 days to encourage leaf greening and expansion and to alleviate virus-induced damage. Applications should be discontinued at least 5 days before harvest.

34.5 Peony Powdery Mildew

Powdery mildew is a prevalent disease affecting peony cultivation in northern China, with an incidence rate typically exceeding 50%. It primarily infects peony leaves, causing them to wither and diminishing both the ornamental value and medicinal quality of peony plants.

Symptoms

At the onset, white, nearly circular powdery mildew spots appear on the leaf surface. These spots expand, merging into large, irregular patches with powdery white to grayish-white fungal spores, representing the conidia and conidiophores of the pathogen. Eventually, the entire leaf surface becomes covered in powder, and the leaves dry out. In the later stages, small black dots appear within the powdery layer, representing the pathogen's cleistothecia. Older and mature leaves are more susceptible (Fig. 34.9).

Pathogen

The causative agent of peony powdery mildew is *Erysiphe paeoniae*, a member of the Erysiphaceae family. During sexual reproduction, the fungus forms cleistothecia that produce asci and ascospores following the fusion of antheridia and ascogonia. The cleistothecia are spherical or subspherical, initially yellowish or yellow-brown, turning dark brown to black upon maturation. The cell wall consists of an inner and outer layer, with some cells in the outer layer developing into appendages. The asci are spherical, double-walled, and lack paraphyses, forming singly or in bundles within the fruiting body. The ascospores are ovoid, occurring in numbers of two to eight per ascus, single-celled, and are colorless or slightly yellowish. The conidia,

Fig. 34.9 Symptoms of Peony Powdery Mildew. (**a**) Field symptoms; (**b**, **c**) Leaf symptoms

which are oval or cylindrical and transparent, average 32.3 μm × 17.3 μm. Conidia are borne singly at the tips of unbranched, upright conidiophores (Fig. 34.10). Appressoria form on the fungal hyphae as shallow, lobed structures, with germ tubes emerging from either end of the conidium and forming nipple-shaped appressoria at the tip.

Occurrence Pattern

Powdery mildew on peonies typically emerges in early June, around the end of the blooming period, as temperatures exceed 20 °C. The disease peaks in July and August, especially during periods of abundant rainfall. The pathogen overwinters as mycelium and cleistothecia in plant residues in fields, releasing spores in the following season to initiate primary infection. Conidia are produced on lesions and dispersed by air currents, leading to recurrent infections. Powdery mildew is most favorable under cool, dry, or warm conditions with low relative humidity, although conidia can still germinate in the absence of a water film. Environmental factors such as soil water stress, excess nitrogen, dense foliage, poor airflow, and insufficient light further predispose plants to the disease.

Control Measures

1. Agricultural Control

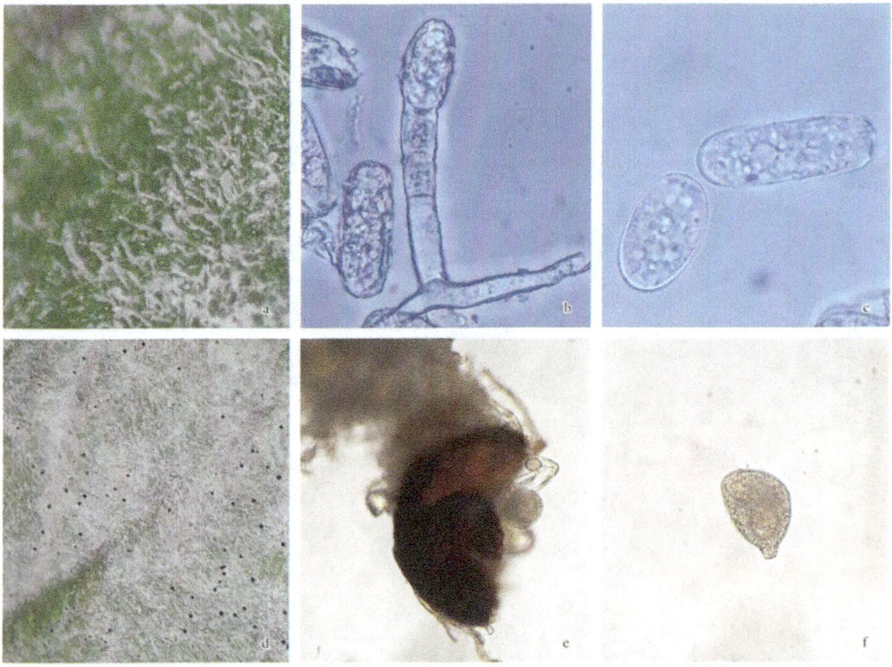

Fig. 34.10 Morphology of the Powdery Mildew Pathogen in Peonies. (**a**) Conidiophore and conidia attached to the peony leaf surface; (**b**) Conidiophore and conidia; (**c**) Conidia; (**d**, **e**) Cleistothecia; (**f**) Ascus

(a) Remove infected leaves early and eliminate plant residues after harvest. In early spring, prune meticulously, removing diseased branches and leaves, and destroy these materials. Conduct thorough garden sanitation in winter to reduce initial sources of infection.
(b) Enhance fertilization practices by limiting nitrogen applications during peak infection periods and increasing phosphorus and potassium to bolster plant resistance.
(c) Thin branches and plants to improve ventilation, drain excess water promptly after rainfall, and manage moisture levels to reduce disease incidence. Both drought management and water drainage are essential.

2. Biological Control

Apply a biological fungicide such as 3% polyoxin WP (150-200-fold dilution) or a 400-fold dilution of a 4% nucleoside pyrimidine antibiotic formulation.

3. Chemical Control

Before flowering, apply 50% carbendazim WP (500-fold dilution) to drench the roots once and spray an 800-fold dilution on leaves twice. After flowering, prune and remove remaining flowers and branches. At the initial disease stage, apply 10% flusilazole SC (1,000-fold dilution) as a foliar spray.

34.6 Phytophthora Blight of Peony

Phytophthora blight is a significant disease in peony production, occurring widely in areas such as Sichuan and Anhui. The disease generally has an incidence rate of 5–20% and can cause substantial damage, affecting the healthy growth of peony plants.

Symptoms

This disease primarily affects the stems, leaves, and buds of peonies. Infected stems initially develop elongated, water-soaked lesions, which later become blackened spots several centimeters in length, with darker centers fading to lighter edges. Infection in young stems near the ground results in blackening and death of entire branches. When the pathogen infects the root collar, it leads to stem rot symptoms. Leaf infection typically begins on the lower leaves, first appearing as dark green, water-soaked spots that turn dark brown as the leaf wilts. Infected buds often exhibit symptoms such as stunted growth, shrinkage, and desiccation (Fig. 34.11).

Pathogen

The causal pathogen of peony Phytophthora blight is *Phytophthora cactorum* (Lebert et Cohn) Schröt., an oomycete in the genus *Phytophthora*. Asexual sporangia are colorless and spherical with a papillary projection at the apex, measuring 51–57 μm × 34–37 μm. The fungus can also produce thick-walled chlamydospores in the central part of hyphae, which have a prolonged survival period. Sexual oospores are spherical, light brown, and measure 27–30 μm in diameter. When temperatures are between 15 and 25 °C and relative humidity is high, sporangia germinate to produce motile zoospores or directly form germ tubes that penetrate the host and initiate infection. Chlamydospores generally require 9–12 months to germinate. The pathogen grows best at 25 °C, with an upper limit of 30 °C and a lower limit of 10 °C (Fig. 34.12).

Occurrence Pattern

The pathogen overwinters in soil as oospores, chlamydospores, or mycelia within plant debris. During the peony growing season, following heavy rainfall, there is often a peak in infection and disease incidence. Continuous rainy days and high precipitation favor disease development, with high temperatures and humidity after rainfall intensifying its severity. The pathogen can develop between 5 and 37 °C, with an ideal range of 25–30 °C; within this temperature range, high humidity is a critical factor for disease occurrence and spread. Conditions such as high greenhouse temperatures, excessive irrigation, high groundwater levels, and excessive humidity increase disease severity.

Control Measures

1. Agricultural Control

   (a) Promptly remove infected plants from the field. After harvest, clear all plant debris to reduce the pathogen source for the following year.

Fig. 34.11 Symptoms of Phytophthora Blight on Peony leaves

(b) Strengthen field management: reduce nitrogen application during infection periods and increase phosphorus and potassium to enhance plant resistance.
(c) Choose well-drained plots or practice ridge cultivation to prevent waterlogging at the stem base. Moderate watering and balanced drainage are essential.

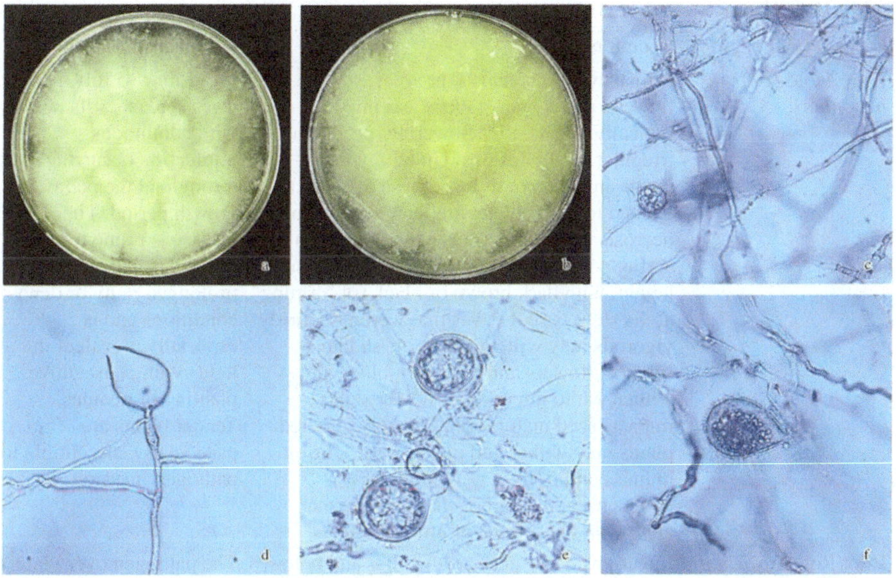

Fig. 34.12 Pathogen Morphology of Peony Phytophthora Blight. (**a**, **b**) Colony characteristics; (**c**) Hyphal structure; (**d**) Sporangia; (**e**) Oospore; (**f**) Sporangium with Zoospores

(d) Maintain suitable planting density, avoid continuous cropping, and implement crop rotation to reduce the likelihood of disease.

2. Biological Control

Trichoderma harzianum and the *Bacillus subtilis* LY-38 strain have demonstrated significant inhibitory and lethal effects on the pathogen, making them effective agents in controlling peony Phytophthora blight.

3. Chemical Control

Promptly remove infected plants and disinfect the site with lime. At the onset of the disease, apply a 400-fold dilution of a 30% metalaxyl-M WP as a soil drench around the plant base and root zone. Applications may be repeated at 7–10-day intervals as necessary, depending on disease severity.

34.7 Common Diseases of Peony (Table 34.1)

Table 34.1 Characteristics of common diseases affecting peony

Disease (pathogen)	Symptoms	Occurrence pattern
Gray Mold (*Botrytis paeoniae*)	Primarily affects leaves, petioles, stems, flowers, and flower stalks. Leaf infection typically occurs after flowering, producing irregular brown lesions. Under favorable environmental conditions, these lesions can expand to cover the entire leaf, leading to necrosis. Stem infections generally begin in spring when stems reach around 30 cm in height, with dark brown or black rot forming at the stem base. Leaves that were previously vigorous may wilt and drop off suddenly. Infected flowers and buds turn black and wither, while infection of flower stalks impacts seed maturation. Severely affected plants may wither and die entirely. During humid conditions, a gray mold layer, representing the pathogen's conidia, may appear on the surface of affected plants	From June to July, this disease reaches its peak, with continuous reinfection occurring throughout the peony growth period. The disease is prone to severe outbreaks under prolonged rainy or dewy conditions and is especially prevalent in fields with consecutive planting of peonies. Tender plants are particularly susceptible to pathogen invasion
Root Rot (*Fusarium solani*)	Infections occur at the root collar and below, affecting the primary, lateral, and fibrous roots, with older roots being particularly vulnerable. Initially, infected root cortexes turn yellow-brown and develop irregular black lesions of varying sizes. Over time, these lesions expand, darkening most of the root, spreading into the xylem, and eventually reaching the pith. In severe cases, fleshy roots are reduced to scattered fragments, leaving only the outer skin, which appears tubular. Advanced infections lead to complete root rot, causing plant wilting and, ultimately, death	The pathogen overwinters in the soil as mycelium and chlamydospores within plant debris, with chlamydospores able to survive in the soil for several years. The primary entry points for the pathogen are wounds caused by soil pests, nematode damage, or mechanical injuries

(continued)

Table 34.1 (continued)

Disease (pathogen)	Symptoms	Occurrence pattern
Leaf Spot (*Pseudocercospora pora* var. *ilicicola*)	Primarily affects leaves. In the early stages, circular or nearly circular lesions appear on the leaf surface, measuring 4–10 mm in diameter. These lesions are numerous, ranging from light brown to grayish-white with brown edges. Older lesions exhibit distinct concentric rings, with a gray-black mold at the center, which is the pathogen's fruiting bodies	The pathogen mainly overwinters in infected plant debris. The disease typically manifests from July to September, with conidia produced the following year to initiate new infections. It spreads via rainwater, contact, and insect transmission, first infecting the lower leaves and gradually progressing upward. Dense planting and poor ventilation increase the severity of the disease, with warm, humid, and rainy conditions further encouraging its spread
Rust (*Cronartium flaccidum*)	Primarily affects leaves. Early symptoms include yellow-green spots on the upper side of the leaves, which are circular, oval, or irregular, with indistinct edges and a diameter of 4–12 mm. Corresponding areas on the leaf underside develop yellow-brown uredinia, and in later stages, brown, columnar hair-like structures about 2 mm in length appear within the gray-brown lesions, indicating the presence of the pathogen's telia	This pathogen is a heteroecious parasite, requiring two different host plants to complete its infection cycle. It overwinters as mycelium on pine trees, producing pycnia and aecia on various pine species. Uredinia and telia form on peonies. Rust spores are generated from April to June, dispersed by wind to infect peonies, followed by the production of urediniospores, which lead to additional infections. Later in the season, telia develop within the uredinia, producing basidiospores that subsequently reinfect pine trees. High temperatures and humidity in July and august, as well as low-lying or waterlogged areas, contribute to more severe outbreaks

(continued)

Chapter 35
The Diseases of Woad

Woad (*Isatis tinctoria* L.), a biennial or annual herbaceous plant in the family Brassicaceae. In traditional Chinese medicine, its root is used as "Ban Lan Gen" (*Radix Isatidis*) and its leaves as "Da Qing Ye" (*Folium Isatidis*). It is cultivated in most areas of China, with major production areas in Hebei, Heilongjiang, Shandong, Henan, Gansu, Anhui, and Jiangxi. The root is cylindrical, slightly twisted, with an epidermis ranging from light grayish-yellow to light brownish-yellow. The cross-section shows a yellowish-white cortex and a yellow xylem. It has a slight odor, initially sweet, followed by a bitter and astringent aftertaste. It has the effects of clearing heat, detoxifying, cooling blood, and relieving sore throat, and is used to treat pestilential toxins, fever with sore throat, scarlet fever with ulcerated throat, and erysipelas.

35.1 Downy Mildew of Woad

Downy mildew is one of the primary diseases affecting woad production, capable of infecting the entire plant. It becomes especially severe during the cool, humid conditions of spring and autumn when the temperature difference between day and night is large. Fields where woad is cultivated for seed experience severe outbreaks in autumn and the following spring, with infection rates around 20%, reaching over 50% in serious cases, which significantly impacts yield and quality. This disease is prevalent across various woad-growing regions, including Beijing, Hebei, Anhui, Gansu, Heilongjiang, Zhejiang, Shanxi, Xinjiang, Henan, Hubei, Guizhou, Shandong, and Liaoning.

Symptoms

This disease primarily affects leaves but can also infect stems, flower stalks, and siliques. In the early stages, lesions appear on the leaf surface as polygonal or irregular patches with indistinct edges, ranging from pale yellow-green to yellow-brown. Affected leaves turn yellow-brown, and a white to light gray downy layer, consisting of zoosporangiophores and zoosporangia, appears on the corresponding area on the underside of the leaves. In severe cases, leaves dry up. When stems, flower stalks, petals, sepals, and siliques are infected, they become discolored and develop

© China Agriculture Press Co., Ltd. 2026

T. Wang et al., *Atlas of Chinese Medicinal Plant Diseases*,
https://doi.org/10.1007/978-981-95-6326-5_35

Fig. 35.1 Symptoms of Downy Mildew on Woad

a white downy layer, and the affected parts become thickened and deformed (Fig. 35.1).

Pathogen

The pathogen causing downy mildew in woad is *Peronospora parasitica* f.sp. *isatidis*, a specialized form of downy mildew belonging to the genus *Peronospora*, family Peronosporaceae, phylum Oomycota. Its sporangiophores are solitary or clustered, colorless, with a total length of 192–332.8 μm, featuring a swollen base. The main stalk measures (123–192) μm × (6.4–16) μm. The top branches have sharp-angled dichotomous branching, mostly three to six times. The terminal branches are curved, tapering to a fine tip, measuring (6.4–25.6) μm × (1.3–3.2) μm. The sporangia are ellipsoidal or nearly spherical, measuring (13.4–28.8) μm × (13.4–25.6) μm. Oospores are yellowish-brown, spherical, with a diameter of 33.6–42 μm, and have a smooth surface, sometimes with wrinkles (Fig. 35.2). This pathogen can infect various Brassicaceae plants, though it has different specialized forms and physiological races.

Occurrence Pattern

The pathogen overwinters and oversummers as oospores within infected plant debris and hypertrophic tissues in the soil. Due to the continuous production cycle of woad or the overlap of early and late plantings, the pathogen can also survive in infected plants as mycelium. Sporangia are dispersed by wind and rain, allowing for multiple cycles of reinfection throughout the growing season. Low temperatures, large diurnal temperature variations, high humidity from rain or heavy dew, and

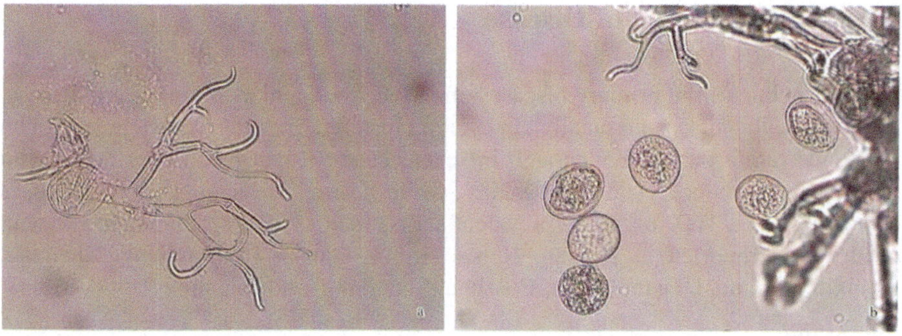

Fig. 35.2 Morphology of the Pathogen of Downy Mildew on Woad. (**a**) Dichotomous branches of sporangiophores, (**b**) Sporangia

misty conditions are conducive to the disease development, which commonly occurs in spring and autumn.

Severe outbreaks occur from April to June, with further spread in September and October. Mild winters, cold springs, frequent rain, and high humidity favor disease occurrence, and the disease becomes particularly serious during bolting and flowering stages.

Control Measures

1. Scientific Site Selection: Choose high and dry fields for planting, avoid consecutive cropping or rotation with Brassicaceae plants that are susceptible to downy mildew, and use raised beds in low, wet areas.
2. Enhanced Cultivation Management: Thoroughly clear and destroy infected plant debris in the fields before winter to reduce overwintering inoculum sources. Use resistant varieties, and ensure appropriate plant spacing to allow for ventilation and adequate sunlight. Adjust planting times as needed. Reduce chemical fertilizer usage, and increase biological organic fertilizers to improve soil microbial environment and fertility, so as to strengthen growth vigor and disease resistance of plants. Water appropriately, ensure good drainage, and reduce field humidity.
3. Chemical Control: At the initial stage of the disease, spray with 69% mancozeb wettable powder (800-fold dilution), 72% cymoxanil mancozeb wettable powder (800-fold dilution), or 72.2% metalaxyl-M hydrochloride aqueous solution (800-fold dilution) at intervals of 10 days, for 2–3 applications. During peak infection periods, spray with 90% fosetyl-aluminum wettable powder (400-fold dilution), and ensure thorough application on the undersides of the leaves.

35.2 Woad Root Rot Disease

Root rot is one of the primary diseases affecting woad and a common root disease. In the field, this disease typically appears in clustered or scattered patches, with an incidence rate around 15%, though in severe cases, it can reach over 30%. Mild infections can reduce yields by 10–20%, while severe cases can cause yield losses exceeding 35%. This disease is prevalent across major woad-producing areas, with occurrences reported in Beijing, Hebei, Anhui, Gansu, Heilongjiang, Zhejiang, Shanxi, Xinjiang, Henan, Hubei, Guizhou, Shandong, and Liaoning.

Symptoms

This disease primarily targets the plant's roots. Initially, lateral or fine roots are infected, turning brown before spreading to the main root. In some cases, infection begins at the main root tip and gradually spreads throughout. The vascular bundles in the root turn dark brown, extending upward to the stem and petiole. As the disease progresses, the root pith undergoes wet rot, becoming dark brown; eventually, sections of the main root transform into a dark brown, hollow shell. Inside, this shell resembles a tangled mass of lignified fibers. Following root infection, branches and leaves above ground begin to wilt, drying gradually from the outer edges inward (Fig. 35.3).

Pathogen

The pathogen causing woad root rot is *Fusarium solani* (Martius) Appel & Wollenweber emend. Snyder & Hansen, belonging to the asexual morph of Ascomycota, in the genus *Fusarium*. The conidiophores and conidia of this fungus are colorless or lightly pigmented. It produces two types of conidia: single-celled, round microconidia, and multi-celled, sickle-shaped macroconidia (Fig. 35.4).

Occurrence Pattern

Infested soil is a major infection source. The disease typically appears in mid-to-late May, peaking from June to July. High field humidity and elevated temperatures are

Fig. 35.3 Symptoms of Woad Root Rot disease

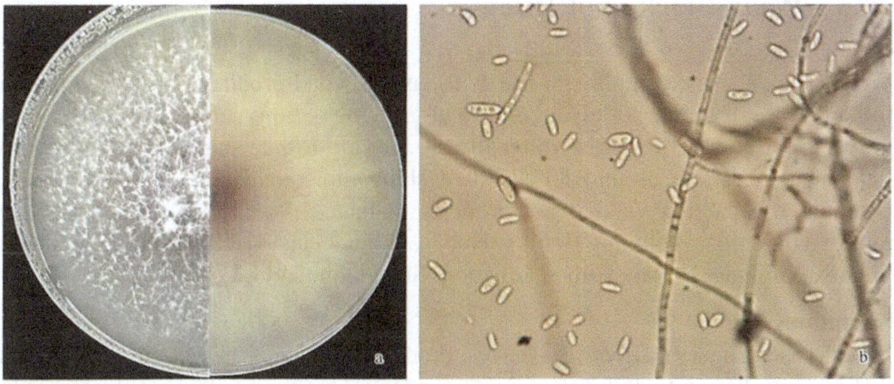

Fig. 35.4 Morphology of the Woad Root Rot pathogen. (**a**) Culture characteristics, (**b**) Conidia

the main factors for the disease occurrence. High humidity, poor drainage, and temperatures between 29 °C and 32 °C favor disease development. Poor cultivation practices and root injuries from underground pests lead to severe cases of root rot. Incidence is generally lower on elevated slopes.

Control Measures

1. Scientific Site Selection.

Choose high-elevation, well-drained sandy loam with a deep soil layer for planting.

2. Cultivation Practices.

Avoid continuous cropping, practice appropriate plant spacing, apply balanced fertilization, reduce chemical fertilizers, increase organic fertilizers, and improve soil microbial conditions to strengthen plant resistance.

3. Chemical Control.

Approximately 15 days before sowing, apply either 70% thiophanate-methyl wettable powder or 50% carbendazim wettable powder at an 800-fold dilution evenly over the soil surface, incorporating it to a depth of about 10 cm. At the initial disease stage, promptly remove and destroy infected plants and debris. Treat the root zone and surrounding plants with 50% thiophanate-methyl wettable powder at a 500-1000-fold dilution to prevent disease spread. During the disease period, apply 75% chlorothalonil wettable powder at a 600-fold dilution or 65% mancozeb wettable powder at a 500-600-fold dilution every 7 days, repeating two to three times.

35.3 Woad Black Spot Disease

Woad black spot is a significant foliar disease in woad production and a common foliar disease. It occurs widely across woad-growing regions and often coexists with gray spot disease. The disease intensifies in the mid-to-late stages of growth, with field incidence rates around 30% and reaching more than 60% in severe cases, potentially impacting yield if not properly managed. Woad black spot has been reported in regions including Beijing, Hebei, Anhui, Gansu, Heilongjiang, Zhejiang, Shanxi, Xinjiang, Henan, Hubei, Guizhou, Shandong, and Liaoning.

Symptoms

The disease primarily affects the leaves, producing round or nearly circular lesions that are gray-brown to dark brown, with concentric rings and often a chlorotic halo around them. These lesions are relatively large, with diameters ranging from 3 to 10 mm. Dark brown mold, consisting of the pathogen's conidiophores and conidia, appears on the lesion surface. When infections are severe, the leaves may turn yellow and wither prematurely (Fig. 35.5). Stems, flower stalks, and seed pods may display similar symptoms when affected.

Fig. 35.5 Symptoms of Woad Black Spot on leaves

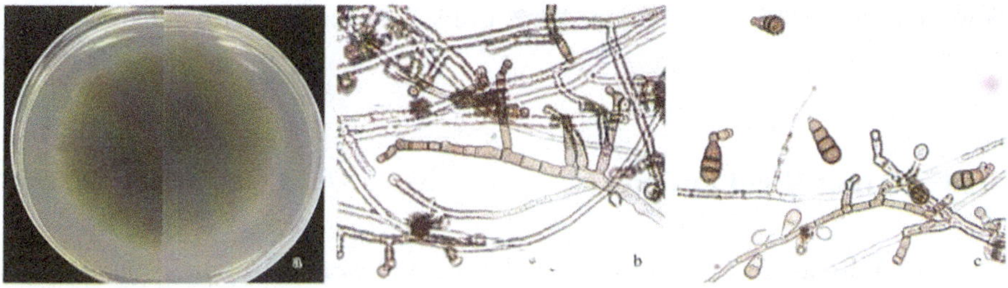

Fig. 35.6 Morphology of Woad Black Spot pathogen. (**a**) Culture characteristics. (**b**, **c**) Hyphae and Conidia

Pathogen

The pathogen causing woad black spot is *Alternaria napiformis* Purkayastha et Mallik, belonging to the asexual morph of Ascomycota in the genus *Alternaria*. The pathogen's conidiophores are either solitary or clustered, erect, or slightly knee-bent, with or without branches. They range in color from brown to light brown, with septations, measuring (31.0–70.0) μm × (3.0–5.5) μm. The conidia are club-shaped, brown, either single or in short chains, with sizes of (31.5–54.5) μm × (8.0–14.0) μm. They contain one to six transverse septa and zero to two longitudinal or oblique septa, with slight constrictions at the septa and a true beak (Fig. 35.6). Additionally, *A. brassicae* Sacc. can also infect woad.

Occurrence Pattern

The pathogen overwinters as conidia on infected plant debris, serving as the initial source of infection for the following year. The disease starts to occur in May and can persist until October, with peak incidence occurring during the hot and rainy months of July and August. High temperatures and rainfall favor disease development.

Control Measures

1. Scientific Site Selection.

   Choose loose, well-drained, fertile sandy loam for planting; poorly drained lowlands or clay soils are not conducive to woad growth.

2. Enhanced Field Management.

   After woad harvest, collect and destroy infected plants in the field to eliminate overwintering inoculum sources. Maintain appropriate plant spacing, fertilize rationally by reducing chemical fertilizers and increasing organic fertilizers, improve the soil microbial environment, enhance soil fertility, and promote plant growth to improve disease resistance. Ensure timely drainage after rainfall to reduce field humidity.

3. Chemical Control.

   At the initial stage of the disease, spray with a 1:1:100 Bordeaux mixture, 75% chlorothalonil wettable powder at a 600x dilution, or 65% mancozeb wettable powder at a 600–800-fold dilution. Apply every 7–10 days, with two to three applications in succession.

35.4 Woad Viral Disease

Woad viral disease is a newly discovered disease of woad. Overall, its incidence is mild and it is a systemic disease. Currently, it is reported in Beijing, Anhui, Shanxi, and Xinjiang, with an average incidence rate of 5–10%.

Symptoms
In the early stages of the disease, systemic mosaic patterns and mottling appear on the leaf surfaces. In severe cases, plants exhibit stunted growth, leaf curling, and reduced root size. Symptoms may become latent under high-temperature conditions (Fig. 35.7).

Pathogen
The pathogens causing woad viral disease are the Cucumber mosaic virus (CMV) and Broad bean wilt virus 2 (BBWV2).

CMV is a representative member of the genus Cucumovirus in the family Bromoviridae. The virus particles are isometric icosahedral, spherical structures without envelopes, approximately 29 nm in diameter. It is a positive-sense single-stranded RNA virus. CMV is a tripartite virus, having three genomic RNAs (RNA1, RNA2, RNA3), which together encode five proteins. The virions of CMV are each composed of 180 capsid protein subunits enclosing a corresponding RNA molecule. The content of single-stranded RNA is 18% and the protein content is 82%, with no lipids or other substances. The virus particles of CMV retains infectivity for 72–96 hours in vitro. The thermal inactivation temperature of the virus particles in the sap of diseased tissue is 55–70 °C.

Fig. 35.7 Woad Viral Disease Leaf symptoms

BBWV2, a member of the genus *Fabavirus* in the family Secoviridae. Its virus particles are spherical, with a diameter of approximately 25 nm. The genome consists of two single-stranded positive-sense RNA molecules. The thermal inactivation temperature of the virus particles is is 58°C. The dilution endpoint is between 10^{-5} and 10^{-4}, with an in vitro survival period of 2–3 days at 25°C. BBWV2 has a broad host range, infecting over 40 plant families and more than 300 plant species.

Occurrence Pattern

Both CMV and BBWV2 have widespread hosts in nature. They are transmitted primarily by various aphid species in a nonpersistent manner and can also spread mechanically through sap contact. High temperatures and drought favor aphid activity and virus transmission, leading to more severe outbreaks.

Control Measures

1. Insect Control for Disease Prevention: During aphid infestations, apply insecticides such as 10% imidacloprid wettable powder at 1,500-fold dilution, 5% acetamiprid emulsifiable concentrate at 2,500-fold dilution, 40% fenvalerate emulsifiable concentrate at 6,000-fold dilution, or 25% thiamethoxam emulsifiable concentrate at 5,000-fold dilution to control aphids and prevent virus spread.
2. Chemical Control: At the onset of symptoms, spray an 8% ningnanmycin aqueous solution at 800-fold dilution. Repeat every 5–7 days for a total of two to four applications.

35.5 Woad Fungal Diseases (Table 35.1)

Table 35.1 Characteristics of Woad fungal diseases

Disease (Pathogen)	Symptoms	Occurrence pattern
Anthracnose (*Colletotrichum higginsianum*)	This disease primarily affects leaves. The lesions are round, with a white and semi-transparent center and a reddish-brown edge. Small black dots, representing the acervuli, are visible on the lesions. As the disease advances, the lesions are prone to perforation. Sometimes, dense clusters of small, round spots may form on the leaves, though they typically do not cause the leaves to wither. When stems, flower stalks, and seed pods are infected, they show fusiform or strip-shaped reddish-brown sunken spots	The pathogen overwinters on infected plant tissues and prefers high temperatures. The disease typically occurs in late summer and early autumn; rainy and high-humidity conditions are conducive to its occurrence

(continued)

Table 35.1 (continued)

Disease (Pathogen)	Symptoms	Occurrence pattern
White rust (*Albugo candida*)	In the early stages, small yellow-green spots appear on the upper leaf surface, while the underside develops raised, glossy white blister-like pustules. These pustules rupture to release powdery white masses, which contain the sporangia of the pathogen. In the later stage of the disease, irregular necrotic spots are formed. When the petioles and young stems are infected, numerous white blister-like spots also develop on the diseased parts, causing the petioles and young stems to twist, deform, and eventually die	The pathogen overwinters as oospores in soil or infected plant tissues, serving as the primary source of infection for the following year. During the growing season, sporangia produced on the diseased parts are dispersed by air currents, wind, and rain, causing secondary infections and further spread of the disease. The disease thrives in cool, moist conditions. It occurs from mid-April to May, with a relatively short damage period
Stem rot (*Sclerotinia sclerotiorum*)	Roots, stems, leaves, and seed pods can all be infected, with stems being the most severely damaged. Infected areas become water-soaked and yellow-brown, later turning grayish-white; the tissues become soft and rotten, making the plants prone to lodging and causing large-area seedling death. White cotton-like mycelial layers and black dung-like sclerotia grow inside and outside the stems. In the late stage, the dried stem bark fibers resemble hemp threads. After stems and leaves are infected, they wilt and gradually die. When flower stalks and seed pods are infected, grayish-white spots also appear, resulting in poor fruiting or shrunken seeds	The pathogen overwinters as sclerotia or mycelia in diseased residual tissues, in the soil where sclerotia fall, or mixed with seeds, serving as the primary infection source for the next year. Under favorable conditions, sclerotia germinate to produce apothecia and ascospores, which are transmitted to the host surface by air currents, wind, and rain. The disease occurs in seed fields from March to April, with the peak incidence period from late April to May

(continued)

Table 35.1 (continued)

Disease (Pathogen)	Symptoms	Occurrence pattern
Leaf spot (*Cercospora* sp.)	The pathogen primarily infects the leaves, producing small, round, slightly sunken spots with brown edges and grayish-white centers. These spots become thin and brittle over time, easily cracking or perforating. Spots typically range from 2–6 mm in diameter. The upper leaf surface often displays brown mold-like substances, which are the conidiophores and conidia of the pathogen. The disease first occurs on old leaves and spreads from the bottom to the top. In the late stage, spots may coalesce, and the diseased leaves turn yellow and die	The pathogen overwinters in plant debris, serving as the primary infection source for the following year. Seeds may also carry the pathogen. Symptoms usually appear in early June, with peak infections occurring from late June to early September. When the daily average temperature is 23–25 °C, it is conducive to the occurrence of the disease, and the spread is rapid
Root knot nematode (*Meloidogyne hapla*, *Meloidogyne incognita*)	This pathogen primarily affects roots, causing the formation of small to large galls or knots on infected root tissues. In early stages, above-ground symptoms are not typical; however, as the root damage becomes severe, the above-ground plants show stunted growth, weakness, yellowing leaves, dullness, and leaf margin curling, which are similar to symptoms of nutrient deficiency	The root knots in the soil or the egg sac masses exposed on the root knots are the primary infection sources for the next season. In spring, when the soil temperature rises, the second-stage juveniles break out of the eggs and invade the young roots of the host to parasitize. The nematode disease is severe in areas with relatively high air temperatures, sort and not too cold winters, and in moist sandy

Chapter 36
The Diseases of Prince Ginseng

Tai Zi Shen (*Pseudostellaria heterophylla* (Mig.) Pax), also known as "Hai Er Shen" (children's ginseng), is a perennial herbaceous medicinal plant belonging to the genus *Pseudostellaria* in the family Caryophyllaceae. The root is used medicinally and has the effects of replenishing lung qi, strengthening the spleen and stomach, and promoting the production of body fluids. It is cultivated on a relatively large scale in Guizhou, Fujian, Shandong, Anhui, and Jiangsu provinces. Major diseases affecting Tai Zi Shen include leaf spot, root rot, violet root rot, viral diseases, and southern blight, with leaf spot, viral diseases, and root rot being particularly severe.

36.1 Leaf Spot Disease of *Pseudostellaria heterophylla*

Leaf spot disease, also known as "leaf blight," is a widespread fungal disease in Tai Zi Shen production areas, primarily damaging the leaves, with an incidence rate of 12–78%, and a yield loss of 8–60%.

Symptoms

In the initial stages, lesions appear as brown, circular or irregular spots. Over time, the center of the lesion turns grayish-white with a yellow halo around it, and small black dots form in concentric rings within the lesion. As the disease progresses, lesions merge into larger irregular patches. The center of the old lesion perforates, leading to leaf desiccation and decay. In severe cases, entire plants may wither and die (Fig. 36.1).

Pathogen

Several pathogens can cause leaf spot disease of Tai Zi Shen. The main ones include fungi of the asexual morph of ascomycetes, such as *Septoria* spp., *Phoma* sp., *Ascochyta versabilis* (Boerema Loer et Hamers) Q. Chen et L. Cai (synonym: *Phoma versabilis* Boerema Loer et Hamers), and *Colletotrichum fructicola*. On PDA medium, the colony of *Colletotrichum fructicola* is white to gray, with a raised center and a neat edge. The aerial hyphae are fluffy. In later stages, orange granules develop at the colony center, which contain conidia. There is melanin deposition on the back of the colony. Conidia are long elliptical, measuring (12.1–21.1) μm x (5.5–8.2) μm (Fig. 36.2).

© China Agriculture Press Co., Ltd. 2026

T. Wang et al., *Atlas of Chinese Medicinal Plant Diseases*,
https://doi.org/10.1007/978-981-95-6326-5_36

Fig. 36.1 Symptoms of Leaf Spot Disease in Tai Zi Shen (*Pseudostellaria heterophylla*) (**a**) Field symptoms, (**b**, **c**) Leaf symptoms

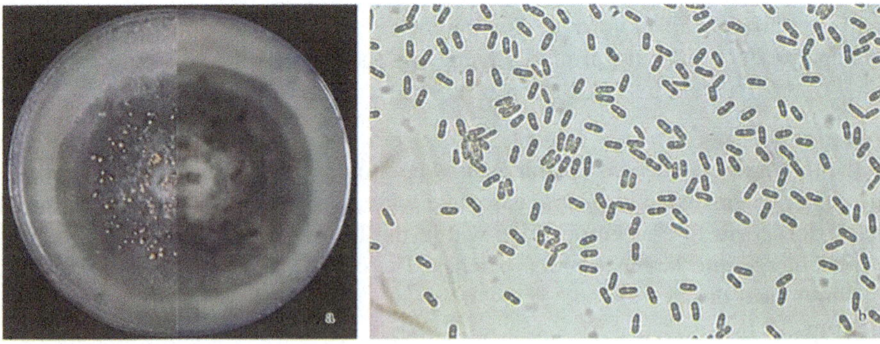

Fig. 36.2 Morphology of *Colletotrichum fructicola*. (**a**) Culture characteristics, (**b**) Conidia

Occurrence Pattern

The pathogen overwinters in the diseased residues in the form of pycnidia. In favorable conditions the following year, the pycnidia produce conidia to initiate primary infection. After the disease occurs, conidia are produced to cause secondary infections. Conidia are dispersed by air currents and rainwater, entering leaves through wounds and stomata. Symptoms begin in late March, with a peak infection period in early May. The disease is more severe in continuous cropping fields.

Control Measures

1. Seed Selection and Disinfection: Choose vigorous, virus-free Tai Zi Shen plants with good adaptability and high purity for seed stock. Before planting, disinfect the seeds by soaking them in a solution of 50% carbendazim WP (500x dilution) or 70% thiophanate-methyl WP (800x dilution) for 30 min, then rinse thoroughly before planting.
2. Field Management: Establish effective drainage and aeration channels; avoid continuous planting of Tai Zi Shen, and instead practice crop rotation with Poaceae family crops for at least 3 years or alternate with wet-dry crops for at least 2 years.
3. Chemical Control: Once seedlings reach 50% emergence, control leaf spot disease by promptly removing infected plant debris and diseased leaves, followed by localized pesticide applications to contain infection centers. Spray surrounding plants with 50% carbendazim wettable powder (600x dilution) every 7–10 days, for a total of two applications. During high temperatures and rapid disease progression, use 10% difenoconazole water-dispersible granules (1000x dilution) or 25% prochloraz emulsifiable concentrate (1500x dilution), spraying every 7–10 days for two to three applications.

36.2 Tai Zi Shen Root Rot Disease

Root rot is a significant soil-borne disease affecting Tai Zi Shen production and is a primary reason for continuous cropping challenges. This disease typically results in a 10–30% yield loss, with severe cases potentially leading to complete crop failure. It occurs in all major production areas across the country.

Symptoms
The disease primarily damages the basal stem and tuberous roots of Tai Zi Shen. Initially, infected areas develop water-soaked circular lesions. Later, white cottony mycelium forms on the affected tissues, and some of the mycelium turns pink in advanced stages. In severe cases, the connection between the basal stem and tuberous root breaks, and the tuberous root tissues rot and necrose (Fig. 36.3).

Pathogen
The pathogens causing Tai Zi Shen root rot include fungi of the asexual morph of ascomycetes, such as species of the *Fusarium* genus: *Fusarium oxysporum* (Schltdl. ex Snyder et Hansen), *Fusarium solani* [(Mart.) Appel et Wollenw. ex Snyder et Hansen], *Fusarium equiseti* [(Corda) Sacc.], and *Verticillium* species. *Fusarium oxysporum* on PDA medium appears cottony, ranging from white to light purple. The sporogenous cells are short with single phialides; chlamydospores are terminal or in chains, round or oblong. Microconidia are elliptical or oval, with zero to two septa, measuring (6.5–22.8) μm × (2.2–3.8) μm; macroconidia are sickle- or

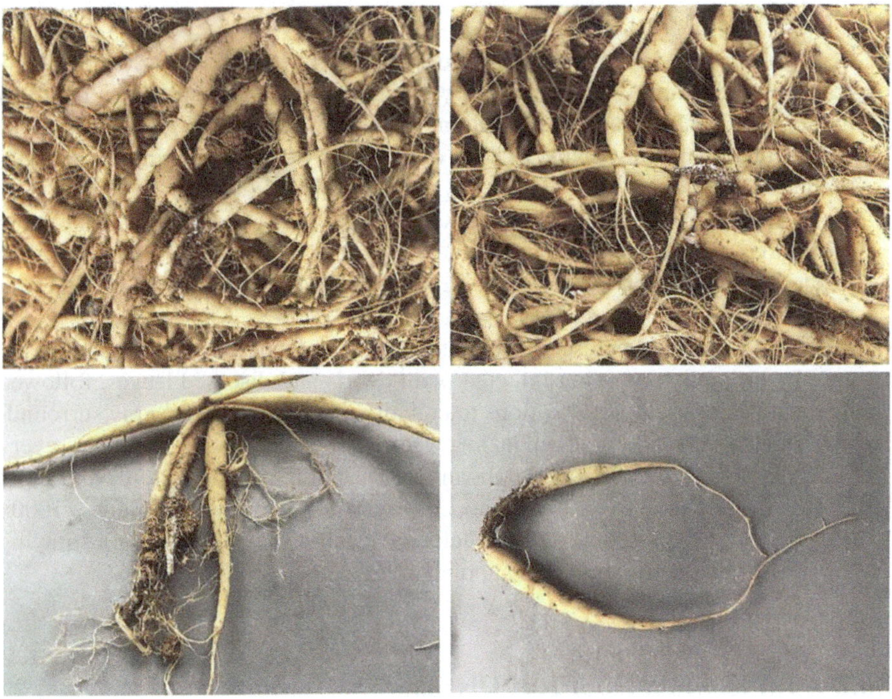

Fig. 36.3 Symptoms of Tai Zi Shen Root Rot

spindle-shaped, slightly curved, with foot cells and one to four septa, mostly with three septa, measuring (23.6–42.9) μm × (3.0–5.7) μm.

Fusarium solani colonies on PDA medium are initially light white and later develop a concentric ring pattern, with the central ring area turning yellow-brown in later stages, and the reverse side of the colony appearing light yellow. Macroconidia are sickle-shaped with zero to three septa, measuring (24.7–46.0) μm × (4.5–7.4) μm; microconidia are oval or kidney-shaped with zero to one septum, measuring (8.4–20.6) μm × (3.2–6.5) μm (Fig. 36.4).

Occurrence Pattern

The pathogen overwinters in the soil or is spread through infected tuberous roots. It can enter plants through wounds or directly penetrate them. The disease typically appears in late April, with severe outbreaks occurring from mid-May through the harvest period. Infections are most destructive during the summer after transplanting tuberous roots, causing significant yield losses. Soil pests (such as grubs, cutworms, mole crickets, and wireworms) aggravate the disease. High soil moisture and excessive rainfall further increase its severity.

Control Measures

1. Scientific Field Selection and Crop Rotation

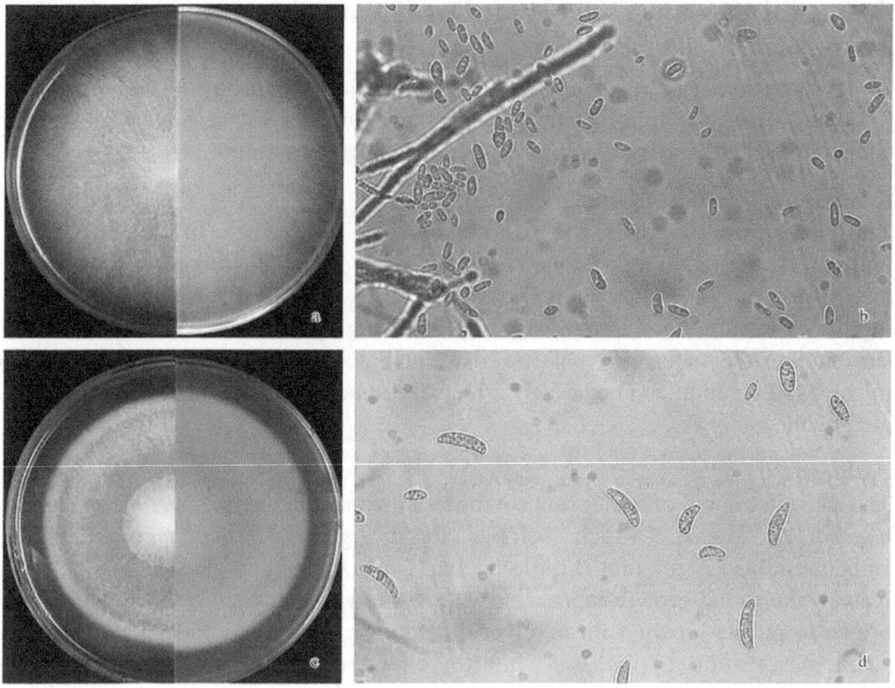

Fig. 36.4 Pathogen morphology of tuber fleeceflower root rot. (**a**) *Fusarium oxysporum* colony morphology, (**b**) *F. oxysporum* conidia, (**c**) *F. solani* colony morphology, (**d**) *F. solani* conidia

Choose well-drained fields rich in organic matter, and practice crop rotation with Poaceae crops.

2. Soil Disinfection

Before planting, apply 30–45 kg/hm^2 of 50% carbendazim wettable powder mixed with fine soil and evenly spread it across the field, or spray the soil with a 1000x solution of 30% hymexazol and mix thoroughly.

3. Tuber Disinfection

Disinfect seed tubers before sowing. Agents such as 6.6% azoxystrobin, 1.1% fludioxonil, 3.3% mefenoxam suspension seed coating agent, or 25% pyraclostrobin emulsifiable concentrate effectively inhibit the pathogen.

4. Biological Control

Reduce chemical fertilizers by 25% and increase the use of bio-organic fertilizers to improve soil conditions and promote healthy growth of Tai Zi Shen. Bio-organic fertilizers containing *Bacillus* spp. produce antagonistic substances that inhibit pathogen activity and induce stress-related enzyme in the plants to participate in the disease defense responses, enhancing disease resistance.

36.3 Tai Zi Shen Violet Root Rot Disease

Violet root rot disease, also known as "Purple Skin Disease," primarily affects Tai Zi Shen during the mid-to-late growth stages and harvest season. This disease occurs in production areas such as Guizhou, Fujian, and Anhui.

Symptoms
This disease mainly affects the tuberous roots of Tai Zi Shen. In the early stages, purple-red rope-like fungal strands form and wind around the tuberous roots. As the disease progresses, these strands expand into a network-like layer, eventually encasing the entire tuber and thickening over time (Fig. 36.5). In severe cases, the soil surrounding the tuberous roots are also entangled by the fungal strands, which impairs soil aeration and negatively affects the appearance and quality of the entangled by the.

Pathogen
The causative agent of violet root rot disease in entangled by the is *Helicobasidium mompa* N. Tanaka, which belongs to the genus *Helicobasidium*, family Helicobasidiaceae, phylum Basidiomycota. On PDA medium, the pathogen shows feather-like radial growth, appearing yellow-brown on the front side and yellow-brown to purple-brown with a velvety texture on the reverse side. Under microscopic observation, hyphae are white to yellow-brown, measuring 0.3–0.6 μm in diameter, with a small number of basidia and basidiospores.

Occurrence Pattern
The disease begins to occur in early May and reaches its peak during the harvest season. It often occurs in slightly acidic soils and areas with poor drainage, often spreading outward from a central point of infection. The disease can be transmitted

Fig. 36.5 Symptoms of entangled by the Violet Root Rot Disease

through rainwater and fertilizers. The occurrence of this disease is related to the previous crops. Fields where sweet potatoes or peanuts were planted as the previous crop have more severe occurrences.

Control Measures

1. Scientific Field Selection and Crop Rotation
   Select fields with good drainage and high organic matter content. Use fields that have not previously grown Tai Zi Shen or implement crop rotation with gramineous crops such as rice, corn, or wheat, ensuring a rotation period of at least 3 years.
2. Soil Disinfection
   Disinfecting the soil before transplanting Tai Zi Shen is a key method for controlling soil-borne diseases and eliminating residual pathogens (including those affecting the aerial parts). Before transplanting, apply 30–45 kg/hm^2 of 50% carbendazim wettable powder mixed with fine soil, evenly spread across the field, or spray the soil with a 1000x solution of 30% hymexazol aqueous solution and mix thoroughly.
3. Seed Tuber Disinfection
   Select Tai Zi Shen plants with good adaptability, high variety, vigorous growth, and free from virus infection for use as seed tubers. Before sowing, soak the seed tubers in a 600x solution of 50% carbendazim wettable powder for 20–30 min, or use a solution of biocontrol *Trichoderma* spp. at a concentration of 1×10^9 CFU/g, soaking 750 g/hm^2 of seed roots, and allow the surface water to dry before planting.
4. Cultivation Management
   Prepare the soil by deep tilling ahead of time to improve the soil ecosystem, which helps to control soil pests and reduce pathogen levels. Use raised beds and deep furrows for planting to prevent waterlogging. Ensure timely drainage in the rainy season to reduce humidity and prevent diseases. Regular tilling and weeding help aerate the soil, promoting a healthy root microbiome and inhibiting pathogenic bacteria. Reduce chemical fertilizer usage by 20–25% and increase the application of bio-organic fertilizers to improve soil conditions and enhance the plant's resistance to disease.

36.4 Tai Zi Shen Virus Disease

Tai Zi Shen virus disease occurs in all production regions in China, with an incidence rate of 20%–75%. In severely affected areas, the incidence rate can reach 100%, leading to yield losses of 15%–35%.

Symptoms
This disease is one of the major threats to Tai Zi Shen production, as infected plants remain virus carriers for life. In the early stages with mild infection, the leaf veins become lighter and yellow, forming a yellow-green mosaic symptom. Severe

Fig. 36.6 Field symptoms of Tai Zi Shen virus disease

infections result in leaf wrinkling, mottling, and leaf margin curling (Fig. 36.6). Infected seedlings exhibit stunted growth, necrosis of apical buds, restricted leaf expansion, reduced root tuber size, and reduced number of tuberous roots. Under high temperatures, the symptoms may become latent.

Pathogen

Ten viruses have been identified as pathogens of Tai Zi Shen virus disease, including *Turnip mosaic virus* (TuMV), *Broad bean wilt virus 2* (BBWV2), *Tobacco mosaic virus* (TMV), *Cucumber mosaic virus* (CMV), *Pseudostellaria heterophylla carlavirus 1* (PhCV1), *Pseudostellaria heterophylla carlavirus 2* (PhCV2), *Pseudostellaria heterophylla carlavirus 3*, (PhCV3), *Jasmine virus C* (JVC), *Stevia carlavirus 1* (StCV1), and *Pseudostellaria heterophylla amalgavirus 1* (PhAV1).

TuMV: A member of the Potyvirus genus, TuMV is a positive-sense single-stranded RNA virus with a genome length of approximately 10 kb, encoding a single polyprotein. Its particles are filamentous, measuring 680–780 nm in length and 11–13 nm in width. The host range is broad, mostly affecting Brassicaceae crops, and transmission occurs via sap and aphids in a nonpersistent manner.

BBWV2: A typical member of the *Fabavirus* genus, BBWV2 has a genome composed of two positive-sense RNA molecules, approximately 6 kb (RNA-1) and 4 kb (RNA-2). The spherical virus particles, approximately 25 nm in diameter, form crystalline inclusions in infected cells. It has a broad host range and infects economically important crops, spreading via aphids in a nonpersistent manner and through mechanical transmission (sap friction). TMV: A typical member of the *Tobamovirus* genus, TMV is a positive-sense single-stranded RNA virus with a genome length of approximately 6.3 kb. The virus particles are rod-shaped, about 300 nm in length and 15 nm in width. It has a wide host range, causing significant damage and strong stability, mainly transmitted through plant-to-plant contact.

CMV: A typical member of the *Cucumovirus* genus, CMV has a genome consisting of three positive-sense RNA segments. Subgenomic RNA4, transcribed from RNA3, encodes the coat protein (CP), facilitating particle packaging, aphid transmission, and long-distance movement. The virus particle is an icosahedral sphere, 20–30 nm in diameter, and infects over 1200 species across more than 100 plant families (both monocots and dicots), spreading through aphids in a nonpersistent manner.

PhCV1, PhCV2, PhCV3, JVC, and StCV1: These members of the *Carlavirus* genus are positive-sense single-stranded RNA viruses, with genome lengths of 8–9 kb and six open reading frames (ORFs). These ORFs encode a replicase (Rep), three triple gene block proteins (TGBs), a coat protein (CP), and a protein rich in leucine (CRP). PhCV1, PhCV2, and PhCV3 are newly identified viruses in Tai Zi Shen, with their field transmission vectors yet to be determined.

PhAV1: A new member of the *Amalgavirus* genus, PhAV1 is a positive-sense double-stranded RNA virus with a genome length of 3430 bp. It contains two partially overlapping ORFs, with ORF1 encoding CP, and ORF1 fused with ORF2 through a + 1 programmed ribosomal frameshift, encoding RdRp. PhAV1 is a newly identified virus in Tai Zi Shen, with its field transmission mode still unclear.

Occurrence Pattern

Virus-carrying tubers used for vegetative propagation are the primary cause of the disease. The virus can also spread through insect vectors and sap transmission. Epidemics occur when aphid populations peak during the growth season, It also occurs severely in drought years. The disease onset varies across regions; in Guizhou, symptoms generally appear from late March to mid-April, peaking from May to June. In Fujian, infections begin in mid-to-late February, peaking after mid-March.

Control Measures

1. Selection of Seed Tubers

Choose high-yield, disease-resistant varieties with plump, uniform seed tubers free from damage and disease, with healthy buds.

2. Planting Virus-Free Seedlings

Use seed propagation to cultivate virus-free seedlings. Virus-free Tai Zi Shen plantlets can be obtained through heat treatment to eliminate shoot-tip viruses, followed by rapid propagation in virus-free nurseries for field planting.

3. Field Management

Remove weeds and infected plant debris in a timely manner to reduce sources of infection. Establish efficient drainage systems and avoid continuous cropping of Tai Zi Shen, opting for rotations with gramineous crops for at least 3 years.

4. Aphid Control

Use yellow sticky traps to lure and kill aphids during the growing season, and apply 10% imidacloprid WP or 10% etofenprox suspension to kill aphids and prevent virus transmission.

5. Chemical Control

At the onset of the disease, apply a 40% olefin·hydroxy·morpholine guanidine soluble powder or an 8% ningnanmycin aqueous solution at 800x dilution. Spray every 5–7 days, repeating for two to four cycles.

36.5 Occasional Diseases of Tai Zi Shen (Table 36.1)

Table 36.1 Characteristics of Occasional Diseases of Tai Zi Shen

Disease (pathogen)	Symptoms	Occurrence pattern
White Mold (*Sclerotinia rolfsii*)	Cottony white mycelium develops at the stem base and on tuberous roots. In severe cases, the affected area rots into a "hemp-like" decay. In advanced stages, brown sclerotia resembling rapeseed seeds form on the diseased area	The pathogen overwinters as mycelium and sclerotia in infected plant residues and soil, serving as the primary infection source for the following year. High temperature and humidity create favorable conditions for disease occurrence
Early blight (*Stemphylium lycopersici*; synonym: *S. Solani*)	Primarily affects leaves. Initial symptoms appear as small, numerous circular or irregular white leaf spots. In severe cases, the leaf tips or margins of diseased leaves turn yellow-brown, and the leaf tips show withering and necrosis	Occasionally occurs in Tai Zi Shen production areas. Serious in fields with heavy disease or old seedbeds without disinfection. High temperature and humidity favor disease development. The disease spreads through conidia, and soil disinfection can reduce its occurrence
Damping-off (*Rhizoctonia solani, Alternaria alternata*)	A seedling disease of Tai Zi Shen, affecting the stem base of seedlings. At the initial stage, brown water-soaked lesions appear at the stem base, which then quickly expand and become sunken, spreading rapidly around with the diseased area as the center, girdling the young stem and causing the seedling to lodge and die	Damping-off typically occurs from February to May. It initially manifests as lesions on stems and spreads rapidly, causing a large number of seedling to wither and die

Chapter 37
The Diseases of *Pinellia ternata*

Pinellia ternata (Thunb.) Breit. is a perennial herbaceous plant in the Araceae family, with its tuber used as medicine. This plant is widely cultivated across China, with major production areas in Hubei, Gansu, Henan, Hebei, Guizhou, Shandong, and Sichuan. *P. ternata* has a mild aroma and a spicy, tingling taste that can irritate the throat. Internally, it is used to reduce dampness and phlegm, relieve nausea and vomiting, and dissolve nodules; externally, it is known for its anti-swelling and analgesic effects. Major diseases affecting *P. ternata* include leaf spot, southern blight, soft rot, and viral infections.

37.1 Leaf Spot Disease of *Pinellia ternata*

Leaf spot disease is the most common fungal leaf disease in *P. ternata* cultivation. The pathogens are diverse and complex, sometimes causing coinfections. This disease is prevalent across all major *P. ternata*-producing regions in China, spreads rapidly, causes significant damage, and is challenging to control. Infection rates range from 25% to 40%, with increased outbreaks in early summer and during hot, rainy seasons, which can significantly impact *P. ternata* yields.

Symptoms

Field investigations indicate that *P. ternata* leaf spot disease primarily appears in two types: large lesions and small lesions, both irregularly shaped, either circular or near-circular. For small leaf spot disease in *P. ternata*, lesions show a grayish-white center with brown edges, typically starting at the leaf edge and progressively expanding toward the center, eventually causing the entire plant to yellow. In large leaf spot disease, lesions have a brown center with grayish-white edges. As the disease progresses, lesions expand and merge, leading to leaf wilting and tearing; ultimately, the aboveground portions of the plant yellow and wither completely (Fig. 37.1).

Pathogen

The pathogen causing small leaf spot disease in *P. ternata* is *Stagonosporopsis cucurbitacearum*, a fungus belonging to the class Dothideomycetes and order Pleosporales, Didymellaceae family, while large leaf spot disease is caused by *Alternaria solani* of the genus *Alternaria*. *S. cucurbitacearum* produces white to

© China Agriculture Press Co., Ltd. 2026

T. Wang et al., *Atlas of Chinese Medicinal Plant Diseases*,
https://doi.org/10.1007/978-981-95-6326-5_37

Fig. 37.1 Symptoms of *Pinellia ternata* leaf spot disease. (**a**, **b**) Small leaf spot symptoms. (**c**, **d**) Large leaf spot symptoms

light gray mycelium with concentric rings; mycelium is short and sparse. Over time, the colony on PDA medium darkens, transitioning to gray-black and finally to brown-black after 20 days. Conidia are elliptical or short rod-shaped, measuring (4.6–8.7) μm × (1.2–2.4) μm, with most having transverse septa and containing small oil droplets. Chlamydospores are unicellular, spherical to oval, measuring (6.3–15) μm × (6–11) μm, and typically appear in chains of 4–13. *A. solani* forms sparse, radiating mycelium. Colonies on PDA medium display an orange-yellow center, transitioning to white at the edges, with an orange-brown pigment at the colony center on the reverse side. Conidia are colorless, elliptical or clavate, straight or slightly curved, with four to eight transverse septa and zero to three longitudinal septa, with a constriction at the septa, measuring (4.5–15.7) μm × (5.6–19) μm (Fig. 37.2).

Occurrence Pattern

The severity of *P. ternata* leaf spot disease varies with environmental conditions. Warm temperatures and a humid, rainy microclimate create ideal conditions for

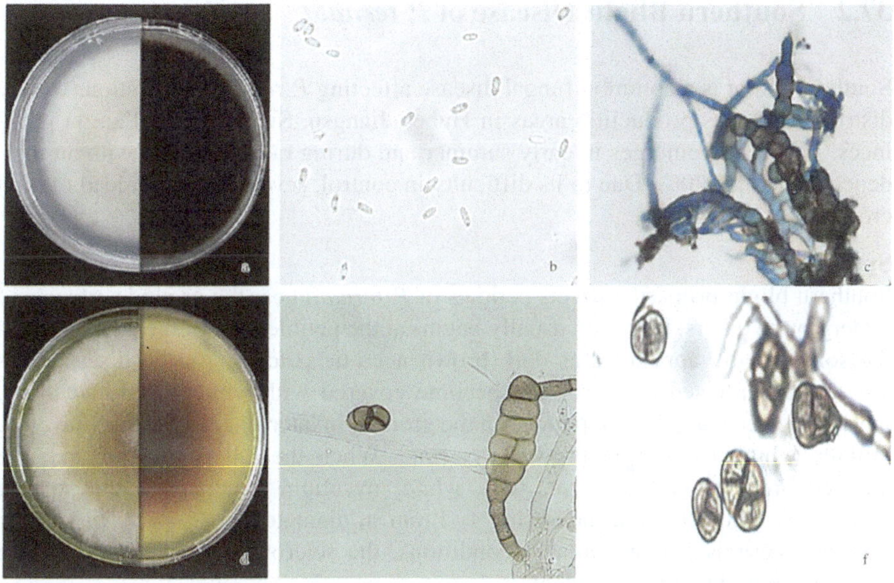

Fig. 37.2 Morphology of *Pinellia ternata* leaf spot pathogens. (**a**) *Stagonosporopsis cucurbitacearum* colony characteristics. (**b**, **c**) Conidia of *S. cucurbitacearum*. (**d**) *Alternaria solani* colony characteristics. (**e**, **f**) Conidia of *A. solani*

outbreaks, typically beginning in early June, with peak occurrence in July and August. The mycelia and conidia on plant residues, seed surfaces, and soil overwinter, serve as primary infection sources for the following season. These spores are spread by wind and rain, infecting through wounds or natural openings. Optimal conditions for spore germination are 15–25 °C with relative humidity above 95%. Long-term continuous cropping also promotes disease occurrence.

Control Measures

1. Seed Disinfection: Disinfect tubers before planting by soaking them in a 50% carbendazim wettable powder solution at 800× dilution for 3–5 min. Allow to dry before sowing to reduce pathogen load.
2. Site Selection: Choose loamy or sandy soil with a pH of 6.5–7.0 for *P. ternata* cultivation. In regions with high summer and autumn rainfall, ensure adequate drainage depth and ridge width to facilitate water runoff.
3. Enhanced Field Management: Improve field practices, moderately increase phosphorus and potassium fertilizer, weed in a timely manner, and ensure plant density is appropriate. Promptly remove diseased leaves.
4. Chemical Control: In cases of severe disease, spray with 50 mg/L carbendazim-thiram WP, 20 mg/L oxine-copper SC, or 10% difenoconazole WG. These fungicides effectively control the disease. Additionally, an 800–1,000-fold dilution of 80% allium essential oil can provide effective plant-based control.

37.2 Southern Blight Disease of *P. ternata*

Southern blight is a common fungal disease affecting *P. ternata* cultivation, widely distributed across production areas in Hubei, Jiangsu, Sichuan, and Jiangxi provinces. It typically emerges in early summer and during rainy seasons, with an incidence rate of 15–20%. Due to its difficulty in control, severe cases can lead to total crop loss.

Symptoms

Southern blight primarily affects petioles of *P. ternata* near the ground surface and underground tubers. Infection usually begins at the petiole base or the parts close to the soil, initially appearing as dark brown necrotic patches. As the disease progresses, the infected areas rot and become covered with white, silky mycelium, which spreads in a radiating pattern on the ground. In later stages, sclerotia develop, initially white but turning brown over time. When the tubers are infected, the affected areas shrivel and rot, with white mycelium attached to the surface (Fig. 37.3). The sclerotia, measuring 1–4 mm in diameter, can survive in soil for extended periods. Under suitable conditions, the sclerotia germinate, producing white mycelium that infects plants.

Pathogen

The pathogen responsible for *P. ternata* southern blight is *Sclerotium rolfsii* Sacc., an anamorphic fungus from the genus *Sclerotium*, with its teleomorph as *Athelia rolfsii* (Curzi) C.C. Tu et Kimbr. On PDA medium, this fungus grows rapidly, forming fluffy white mycelium, and later produces brown, glossy sclerotia that resemble rapeseeds (Fig. 37.4).

Occurrence Pattern

This pathogen has a broad host range and strong environmental adaptability, posing a significant risk if not managed effectively.

The pathogen overwinters in the soil or on plant debris as mature sclerotia, mycelial strands, or hyphae. It spreads through soil and water and proliferates within the soil. Disease typically begins in June, peaks in July and August, and decreases in September, ceasing entirely as temperatures drop.

Control Measures

1. Scientific Site Selection: Select moist, fertile soils with good water retention, fertility, and drainage, preferably loose, neutral sandy loam, for *P. ternata* cultivation. Slightly shaded, sloping land is also suitable. Special attention should be given to the depth of drainage ditches and field width to facilitate drainage.
2. Enhanced Cultivation Management: During prolonged rainy periods, promptly till and loosen the soil to prevent compaction and improve aeration. Avoid dense planting and limit irrigation frequency. If infected plants are found, remove them immediately and treat the surrounding soil.
3. Chemical Control: A 300-500-fold dilution of 29% lime-sulfur solution or an 8,000-10,000-fold dilution of 10% fluazinam SC effectively inhibits the mycelial

Fig. 37.3 Symptoms of *Pinellia ternata* southern blight disease

growth of the white mold pathogen. Botanical fungicides, including 1% osthole SL, 20% eugenol EW, and 0.5% berberine soluble formulation, have inhibitory effects on the pathogen, inducing hyphal deformation and reducing pathogen virulence. Additionally, microbial agents like *Trichoderma harzianum* and *Bacillus subtilis* effectively suppress mycelial growth of the southern blight pathogen.

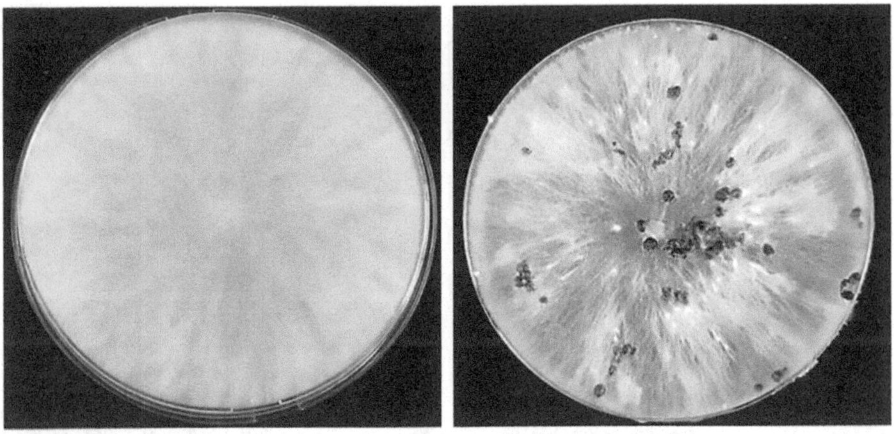

Fig. 37.4 Colony and Sclerotia morphology of the *Pinellia ternata* southern blight pathogen

37.3 Soft Rot Disease of *Pinellia ternata*

Soft rot is the most common soil-borne bacterial disease affecting *P. ternata* cultivation. It spreads rapidly, causes severe damage, and is difficult to control, occurring widely across major *P. ternata* production areas. Due to its high infectivity, once an outbreak begins, it quickly forms a disease center that spreads radially through soil and rainwater. The incidence rate of *P. ternata* soft rot ranges from 20% to 40%, with some key production areas experiencing rates as high as 80%, leading to total crop loss in extreme cases.

Symptoms

This disease primarily affects the leaf petioles and underground tubers of *P. ternata*. Initially, infected leaves exhibit water-soaked, irregular soft rot spots that gradually expand. In severe cases, the entire leaf becomes water-soaked and rots. The affected stems appear decayed and emit a foul odor. Exposure to sunlight causes infected stems and leaves to quickly dry out and die. Underground tubers rapidly undergo soft rot, with internal tissues decaying into a chalky, powdery mass that emits an offensive odor (Fig. 37.5). In the field, soft rot often appears in isolated patches, radiating from a diseased plant and causing clusters of seedlings to collapse and die.

Pathogen

The pathogen responsible for *P. ternata* soft rot is *Pectobacterium carotovorum* subsp. carotovorum, a member of the *Pectobacterium* genus within the Proteobacteria phylum. On LB medium, this pathogen forms slightly raised, glossy, translucent yellow-white colonies. It is a gram-negative, short rod-shaped bacterium with rounded ends, measuring approximately 0.5–0.7 μm × 1.0–2.0 μm, and has four to six peritrichous flagella (Fig. 37.6). The optimal growth temperature is 25–30 °C, with a maximum of 38–39 °C and a minimum of 4 °C. The lethal temperature is 48–51 °C.

Fig. 37.5 Symptoms of *Pinellia ternata* soft rot disease. (**a–c**) Above-ground symptoms. (**d–f**) Underground tuber symptoms

Occurrence Pattern

The onset and severity of *P. ternata* soft rot vary by region and field conditions. The disease usually begins in mid-May, although some areas report cases as early as April, and it peaks between June and September. The disease tends to spread widely following the rainy season under high temperature and humidity, especially in

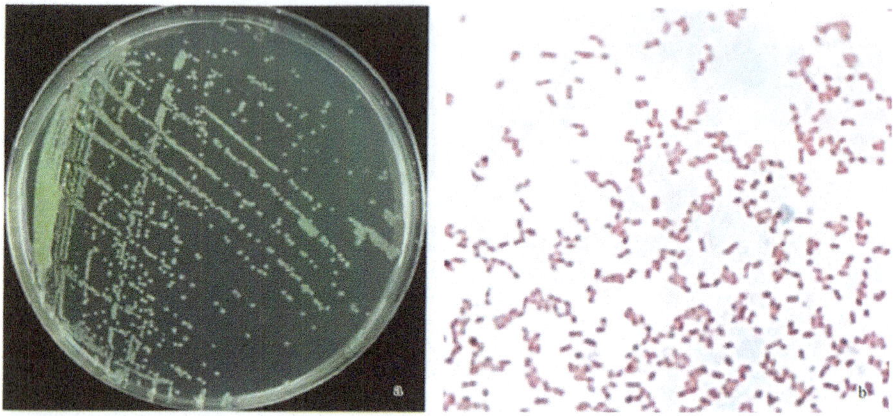

Fig. 37.6 Morphology of *Pinellia ternata* soft rot pathogen. (**a**) Colony characteristics. (**b**) Gram stain results

poorly drained soils. The pathogen overwinters in soil and plant debris, with initial and secondary infections occurring through wounds or natural openings in *P. ternata* under suitable environmental conditions. The disease spreads through rain splash or insect vectors. High temperature, high humidity, and poor ventilation are primary conditions that promote its spread. Pre-planting disinfection of tubers is closely related to the control of soft rot.

Control Measures

1. Scientific Site Selection: Choose elevated or sloped fields for *P. ternata* cultivation. Plant in deep furrows and high ridges; avoid clayey, poorly drained soils as they are more susceptible to disease.
2. Enhanced Cultivation Management: Avoid continuous cropping and high planting density, as extended monoculture increases disease severity. Newly planted or rotated fields have lower infection rates. After 1 year of *P. ternata* cultivation, rotate the crop for 5–8 years before replanting. Avoid crops susceptible to similar diseases, such as carrots, potatoes, tomatoes, cucumbers, and cabbage. Rotation with legumes or grasses is beneficial, and intercropping with green manure during corn rotation can improve soil fertility. During intense sunlight or high temperatures, plant tall crops like corn or sugarcane along ridges for shade or use shade nets. Promptly remove diseased plants and control underground pests.
3. Soil Disinfection: Before planting, disinfect the soil using fungicides such as 0.1% metalaxyl-fosetyl aluminum solution at 25–30 kg per mu or 50% carbendazim wettable powder at a 500–1,000-fold dilution. In high summer temperatures, turn the soil, irrigate, cover with mulch, and seal for 1–2 weeks to raise soil surface temperatures to over 70 °C, effectively killing soil-borne pathogens and insect eggs.

4. Seedling Disinfection: Soak seedlings in an 800-fold dilution of 80% garlicin SC, 40% sodium dichloroisocyanurate SP, or 0.4% formaldehyde solution for 20–30 min. Allow seedlings to air dry before planting.
5. Chemical Control: Various agents have shown good efficacy in controlling soft rot, including 40% sodium dichloroisocyanurate SP (900-1,200 g/hm^2), 80% garlicin SC (800-1,000-fold), 50% carbendazim WP (500-fold), 10% polyoxin AS (500-fold), 70% thiophanate-methyl WP (500-fold), 75% chlorothalonil WP (500-fold), and Bordeaux mixture at 200-300-fold dilution.

37.4 Viral Disease of Pinellia ternata

Viral disease is a widespread leaf disease affecting *P. ternata*, with an incidence rate of 30–60%. The disease is particularly severe in Hubei, where incidence rates can reach as high as 70% in severe cases.

Symptoms
P. ternata viral disease manifests through symptoms such as mosaic patterns, vein clearing, chlorosis, ring spots, yellowing, stunting, and leaf wrinkling. Affected leaves often display downward curling and significant wrinkling, with blister-like protrusions on the leaf surface and mottled yellow and green patches. Veins appear brown and contorted. In early stages, leaves may exhibit mild deformation and slight wrinkling, while advanced stages show increased curling and eventual vein necrosis, resulting in stunted growth and severe leaf curling. In the field, *P. ternata* plants often suffer from multiple viral infections simultaneously, which intensifies symptom complexity (Fig. 37.7).

Pathogen
The primary viruses causing *P. ternata* viral disease include Cucumber mosaic virus (CMV), Dasheen mosaic virus (DsMV), and Soybean mosaic virus (SMV). CMV, a single-stranded RNA virus in the Cucumovirus genus, has icosahedral, spherical particles approximately 22–28 nm in diameter. It remains viable in vitro at room temperature for 2–6 days and for 9–12 months at −5 °C. In shredded tissues at -5 °C or lower, CMV remains infectious for at least 3 years and can be preserved via freeze-drying or vacuum storage at −18 °C for over 11 years.

DsMV, a member of the Potyvirus genus, was initially identified in the 1960s in Florida and is now found in countries like China, Japan, India, and Vietnam. SMV, also in the Potyvirus genus, has filamentous particles located in the cytoplasm and nucleus, measuring 630–750 nm in length and 13–19 nm in width. SMV forms distinctive pinwheel-like inclusions within host cells (Fig. 37.8). The virus is inactivated at 55–65 °C within 10 min and remains viable outside the host for 3–4 days at 25 °C or up to 120 days below freezing; lower temperatures extend its survival.

Fig. 37.7 Symptoms of *Pinellia ternata* viral disease

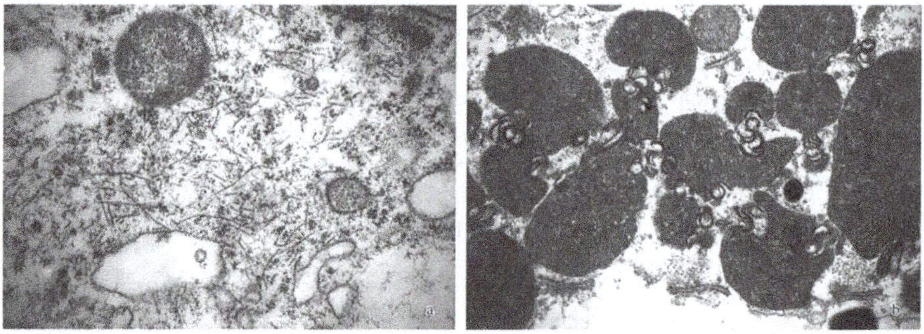

Fig. 37.8 Soybean Mosaic Virus. (**a**) Virus particle. (**b**) Pinwheel inclusion body

Occurrence Pattern

The virus spreads mainly through aphids, thrips, leafhoppers, and mites, as well as through mechanical injury. Infected mother plants retained for seed can harbor the virus through winter. High temperatures and humidity, combined with early aphid and thrip activity, can lead to severe outbreaks of *P. ternata* viral disease.

Control Measures

1. Scientific Seed Selection: Use virus-free mother plants as propagation material to cultivate clean seedlings. Virus-free plants can be obtained through tissue culture or seed propagation.
2. Enhanced Cultivation Management: Diseased *P. ternata* plants act as virus reservoirs. During introduction, apply strict quarantine to prevent spread to disease-free areas. Establish disease-free sites by selecting healthy plants for seed retention, increasing phosphorus and potassium fertilizers to improve plant resistance.

3. Pest Control: In the early stages, employ protective measures like insect-proof nets and sticky traps. In later stages, spray 10–20 g per 666.7 m^2 of 50% pymetrozine WP to control virus-transmitting aphids.
4. Chemical Control: At early infection stages, spray 20% morpholine hydrochloride-copper WP at a rate of 166–250 g/ha, repeating every 7–10 days, with three consecutive treatments to ensure effective control.

37.5 Occasional Diseases of *Pinellia ternata* (Table 37.1)

Table 37.1 Characteristics of occasional diseases in *Pinellia ternata*

Disease (Pathogen)	Symptoms	Occurrence pattern
Red Rot Disease (*Erwinia persicina*)	Leaves exhibit inward curling, wilting, and soft rot in stems, which decay without emitting an odor. Tissues near the nodes turn red, and eventually, the entire plant may rot, resembling ginger rot	Peak occurrence is from April to May. The disease outbreaks are frequent during the rainy season with high temperatures and humidity, especially in poorly drained fields, leading to widespread infection
Blight (*Phytophthora parasitica*)	Leaves develop greenish, water-soaked lesions that spread, causing yellowing and wilting. In advanced stages, leaves exhibit scorch marks, and stems collapse; the disease spreads easily	Peak occurrence is from April to May. Outbreaks are frequent during the rainy season with high temperatures and humidity, particularly in fields with poor drainage, leading to extensive infection
Fusarium Wilt (*Fusarium oxysporum*)	Stem bases develop shallow, water-soaked lesions that gradually expand, encircling the stem and forming brown spots. In severe cases, seedlings collapse, and high humidity promotes the growth of white or pinkish fungal mycelium	Begins in April, with peak occurrence from May to June. The disease is prevalent during the rainy season with high temperatures and humidity, especially in poorly drained soil, resulting in widespread infection

Chapter 38
The Diseases of Eleuthero (*Eleutherococcus senticosus*)

Eleuthero (*Eleutherococcus senticosus* [Rupr. & Maxim.] Maxim.), a shrub belonging to the Araliaceae family, has roots, stems, and leaves that can all be used medicinally. Its taste is slightly spicy and mildly bitter; it has a slightly warm nature, is nontoxic, and is associated with the spleen, kidney, and heart meridians. Modern medical research has shown that Eleuthero contains key components with notable antiaging and anti-fatigue properties, capable of enhancing both physical and mental stamina, while also regulating the nervous system. In recent years, Eleuthero has been used in China as a medicinal herb for liver protection, anti-cancer applications, as a tonic, and to strengthen the body.

Eleuthero grows naturally in forests or thickets and can thrive at altitudes ranging from several hundred meters to 2000 m. It prefers warm, humid climates, is cold-resistant, and can tolerate partial shade. The optimal cultivation environment includes sunny areas with deep layers of humus and slightly acidic, sandy loam soils.

In China, Eleuthero is primarily found in Heilongjiang (Xiaoxing'an Mountains, Dailing in Yichun City), Jilin (Jilin, Tonghua, Antu, Changbai), Liaoning (Shenyang), Hebei (Wuling Mountain, Chengde, Baihua Mountain, Xiaowutai Mountain, Neiqiu), and Shanxi (Huoxian, Zhongyang, Xingxian). The main diseases affecting Eleuthero include black spot, gray mold, seedling damping-off, and root rot.

38.1 Eleuthero Gray Mold

Gray mold has recently become a common foliar disease in Eleuthero cultivation. It tends to occur under dense planting conditions or in low lying areas. In mild cases, it causes leaf wilting and premature leaf drop. In severe cases, it impacts overall plant health. In high-density seedling beds, the disease may cause wilting and death, resulting in sparse stands. Severe gray mold infections have been reported in Jilin Province.

Symptoms

This disease primarily affects the leaves of Eleuthero. It typically occurs from June to August, initially appearing at the leaf margins or tips and gradually spreading to form irregular lesions. Later, the lesions dry out and may crack. As the disease progresses, lesions turn gray-brown, and in humid conditions, a layer of gray mold

© China Agriculture Press Co., Ltd. 2026

T. Wang et al., *Atlas of Chinese Medicinal Plant Diseases*,
https://doi.org/10.1007/978-981-95-6326-5_38

appears. In severe cases, lesions may cover half or the entire leaf. Under drier conditions, the leaves become desiccated and brittle (Fig. 38.1).

Pathogen

The causative agent of gray mold in Eleuthero is *Botrytis cinerea* Pers. Fr., of the genus *Botrytis* in the Sclerotiniaceae family (teleomorph: *Botryotinia fuckeliana* [de Bary] Whetzel, synonym: *Sclerotinia fuckeliana* [de Bary] Fuckel). The gray mold on infected areas consists of conidiophores and conidia. On PDA medium, the pathogen grows at 25 °C, initially producing gray-white mycelium that darkens to deep gray or gray-green and forms a velvety layer. After 4–6 days, it covers the plate, and by the seventh day, dark brown sclerotia form around and at the base of the medium, with abundant spore production by day ten. Conidiophores are long, mostly tufted, upright, brown, and septate, measuring (100–150) μm × (5–10) μm, with one or two branches at the tips. The branch tips are enlarged, resembling grape clusters, with dense spore-bearing structures. Conidia are spherical to ovoid, single-celled, colorless, with a smooth surface, measuring (6–10) μm × (7–13) μm (Fig. 38.2).

Fig. 38.1 Symptoms of Eleuthero gray mold

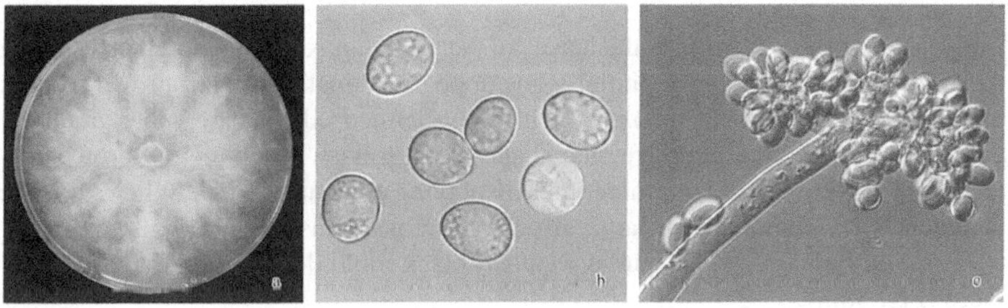

Fig. 38.2 Morphology of Eleuthero gray mold pathogen. (**a**) Culture characteristics. (**b**) Conidia. (**c**) Conidiophores

Occurrence Pattern

The pathogen primarily overwinters as mycelium in plant debris and soil. In the following year, spores or mycelium germinate and infect leaves through wounds or directly, producing large quantities of conidia for further spread. Wind, rain splash, and agricultural activities are the main means of disease transmission. During Eleuthero's growth period, the pathogen can cause multiple reinfections, spreading quickly. Prolonged cool, rainy conditions with high humidity favor disease development, with peak occurrence from mid-June to mid-August in northeastern regions.

Control Measures

1. Agricultural Control: Select high-quality disease-resistant varieties; ensure appropriate planting density to create a microclimate with good air circulation and light penetration, avoiding overcrowding. Enhance cultivation management by promptly draining water after rain; maintain field sanitation by clearing fallen leaves in autumn, collecting and deep-burying or destroying them to reduce primary infection sources.
2. Biological Control: Use biological agents such as 30 billion CFU/g *Bacillus subtilis* WP, 100 billion CFU/g *Pseudomonas fluorescens* WP, 10% polyoxin WP, 1% jinggangmycin SC, 1% matrine SL, 0.3% eugenol EW, and 5% carvacrol solution for control.
3. Chemical Control: At the onset of disease, apply 50% iprodione WP, 250 g/L propiconazole EC, 400 g/L fluazinam EC, or 30% carbendazim-thiram WP, repeating treatment every 7–10 days. Rotate chemicals to prevent resistance.

38.2 Acanthopanax Leaf Spot Disease

Leaf spot disease is the most common and severe foliar disease affecting Acanthopanax production in Jilin Province. This disease can result in extensive necrotic patches on leaves, which dry out and fall off, severely impacting the Acanthopanax industry.

Symptoms

This disease primarily affects the leaves of Acanthopanax, with young leaves being the first to show symptoms. Initial signs appear as small, circular, brown to dark brown spots, 1–2 mm in diameter, with well-defined edges. The spots gradually enlarge into near-circular or irregular shapes with grayish-white or gray-brown centers and dark brown borders, sometimes displaying concentric rings. When spots are abundant, they coalesce into large, irregular lesions, causing leaf blight, deformation, and early leaf drop. In humid weather, black mold develops on the lesion surfaces (Fig. 38.3).

Pathogen

The causative agents of this disease are *Alternaria tenuissima* [Nees.: Fr.] Wiltshire and *A. alternata* [Fr.] Keissl., both members of the *Alternaria* genus in the Ascomycota phylum. *A. tenuissima* grows densely on PDA medium, forming hairy, grayish-white aerial mycelium. Initially, colonies appear white, gradually turning light gray as they mature, with white edges. Colonies can cover the petri dish within 7 days. Conidia form in chains, appearing obclavate, ovoid, or near-elliptical, with one to six transverse septa and one to three longitudinal or oblique septa, with slight constrictions at the septa, measuring (6.65–12.43) μm × (16.87–46.32) μm (average 9.89 μm × 29.53 μm) (Fig. 38.4a–c). *A. alternata* forms abundant aerial mycelium on PDA medium. Colonies begin white, gradually turning gray with a black underside. Conidiophores may be solitary or clustered, straight or curved, with visible septa and slight branching, light brown to brown. Conidia are solitary or in short chains, ovoid or elliptical, light brown to brown, with three to seven transverse septa and one to four longitudinal or oblique septa, measuring (10–30) μm × (5–12) μm. Beaks or pseudobeaks are cylindrical and light brown (Fig. 38.4d–f).

Occurrence Pattern

The pathogens primarily overwinter as conidia and mycelia on infected leaves and twigs. In the following spring, when temperature and humidity are suitable, they produce abundant conidia that spread through wind and rain, leading to frequent

Fig. 38.3 Leaf symptoms of Acanthopanax leaf spot disease

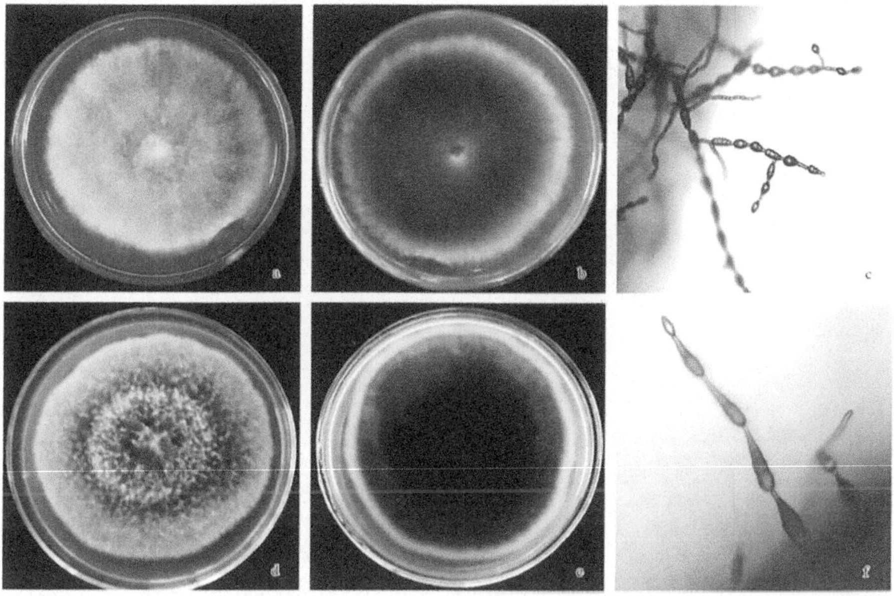

Fig. 38.4 Pathogenic characteristics of Acanthopanax leaf spot disease. (**a**, **b**) *Alternaria tenuissima* culture characteristics. (**c**) *A. tenuissima* Conidial chain. (**d**, **e**) *A. alternata* culture characteristics. (**f**) *A. alternata* Conidial chain

reinfections. In northeastern regions, initial symptoms appear around May 20, with peak infection occurring from June to August. The severity of leaf spot disease correlates with climate and plant vigor. Typical temperatures of 24–28 °C with continuous rain promote disease onset and spread. Healthy plants experience milder symptoms, while weaker plants are more severely affected. Inadequate fertilization and excessive nitrogen application also encourage disease development.

Control Measures

1. Agricultural Control: Enhance cultivation management by promptly pruning diseased and excess branches to improve ventilation and light penetration, which alters the microclimate in favor of the plants. Drain fields promptly after rain to prevent moisture retention. This practice supports both Acanthopanax growth and disease prevention. After leaf fall in autumn, clean the field of dead branches and leaves, and either burn or deeply bury them to reduce field inoculum sources.
2. Biological Control: At the onset of the disease, apply biocontrol agents such as 1000 billion CFU/g *Bacillus subtilis* WP, 10% polyoxin WP, 80% allicin EC, or 4% nucleoside pyrimidine antibiotic AS.
3. Chemical Control: In early infection (around early June), apply fungicides such as 70% thiophanate-methyl WP, 10% difenoconazole WG, 50% iprodione WP, 80% mancozeb WP, 25% azoxystrobin SC, or 25% propiconazole EC. Apply every 7–10 days for two to three consecutive treatments. Rotate fungicides to prevent resistance development.

Chapter 39
The Diseases of Ginseng

Ginseng (*Panax ginseng* C.A. Meyer) is a perennial herbaceous plant in the Araliaceae family, distributed across China, Russia, and Democratic People's Republic of Korea. In China, it is found mainly in eastern Liaoning, eastern Jilin, and eastern Heilongjiang. The fleshy root of ginseng serves as a potent tonic, beneficial for regulating blood pressure, restoring cardiac function, treating neurasthenia, and alleviating physical weakness. It also has expectorant, stomachic, diuretic, and stimulative properties. The stem, leaves, flowers, fruits, and by-products of ginseng are widely used in light industry to produce products containing ginseng extracts, such as cigarettes, wines, teas, and balms. Ginseng generally grows in deciduous broadleaf or mixed needle-broadleaf forests at elevations of several hundred meters. It thrives in loose, well-aerated, well-drained and nutrient-rich sandy loam rich in nutrients; it favors shaded, cool, and moist climates, tolerates low temperatures, and requires diffuse light, avoiding direct sunlight.

The primary diseases affecting ginseng include gray mold, black spot, anthracnose, damping-off, wilt, rust rot, sclerotinia, blight, and bacterial soft rot.

39.1 Ginseng Gray Mold

Gray mold primarily affects the leaves, petioles, flowers, and fruits of ginseng, causing stems and leaves to wither and die. In severe cases, it can also damage the stems and roots. The incidence rate in the field is generally 20–30%, and in severe cases, it can exceed 80%, bringing great troubles to growers.

Symptoms

The disease initially appears as water-soaked, grayish-brown spots on the leaves, typically beginning at the leaf tips or edges. A typical characteristic of infection sign is the inverted V-shaped lesions that expand rapidly, with both the upper and lower leaf surfaces covered in gray mold. In advanced stages, lesions merge, tissue necroses, and they are prone to rupture and fall off, and the leaves are easy to perforate. Stem infections begin as small, water-soaked spots that expand into light brown, long-elliptical or irregular-shaped lesions. In severe cases, the stems and leaves above the infected part wither and die, producing extensive gray mold layers. When

© China Agriculture Press Co., Ltd. 2026

T. Wang et al., *Atlas of Chinese Medicinal Plant Diseases*,
https://doi.org/10.1007/978-981-95-6326-5_39

the stigmas or petals are infected, the disease can spread to the fruits or fruit stalks, and the infected fruits cannot produce seeds normally (Fig. 39.1).

Pathogen

The causative agent of ginseng gray mold is *Botrytis cinerea* Pers.: Fr., an asexual morph of ascomycetes in the genus *Botrytis*. On PDA medium, the mycelia of *B. cinerea* grow relatively fast in a radial pattern, starting from white and gradually darkening to shades of gray-brown. The aerial mycelium is rather lush, with brown conidiophores sparsely or densely clustered throughout the colony, measuring (63 7.21–1213.68) μm × (6.72–16.23) μm. Conidiophores branch acutely or remain unbranched at positions above two-thirds of the conidiophores. Conidia are ellipsoidal to ovoid, with one end slightly protruding. Some retain a small stalk when the sporogenous cell detaches, while most remain on the main stem. Conidia measure (5.67–10.25) μm × (7.49–14.33) μm, with a length-to-width ratio of 1.00–1.74. Sclerotia are 2.58–3.16 μm in diameter, either solitary (with smooth surfaces) or clustered (with surfaces having warts or ring patterns); and have relatively diverse shapes; When the sclerotia mature, new mycelia can germinate from their surfaces (Fig. 39.2).

Occurrence Pattern

The onset time and severity of ginseng gray mold vary across different regions and under different field management practices. Typically, it begins to occur in late June and enters the peak incidence period from mid-July to late August. The occurrence of ginseng gray mold is closely related to environmental temperature and humidity; high air humidity, consecutive cloudy or rainy periods, and low-lying waterlogged fields all contribute to the occurrence of this disease.

Control Measures

1. Seed and Seedling Treatment: Before sowing, The onset time and severity of ginseng gray mold vary across different regions and under different field

Fig. 39.1 Field Symptoms of Ginseng Gray mold field symptoms

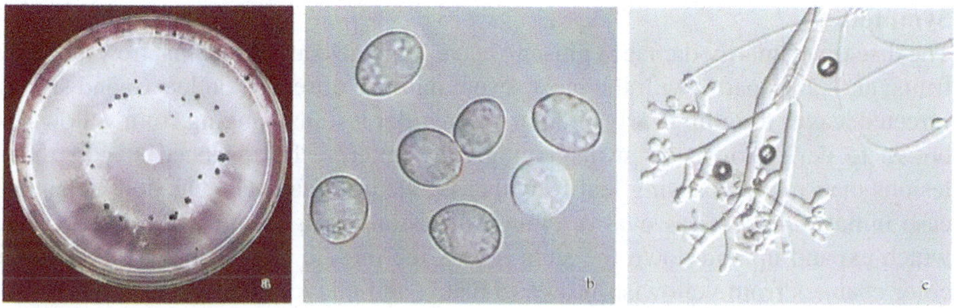

Fig. 39.2 *Botrytis cinerea* morphology. (**a**) Colony characteristics; (**b**) Conidia; (**c**) Conidiophore

management practices. Typically, it begins to occur in late June and enters the peak incidence period from mid-July to late August. The occurrence of ginseng gray mold is closely related to environmental temperature and humidity; high air humidity, consecutive cloudy or rainy periods, and low-lying waterlogged fields all contribute to the occurrence of this disease, then air dry in a cool, dry place before sowing.

2. Soil Disinfection: After removing winter mulch, disinfect the bed surface with 450–750 mL/hm² of 40% procymidone suspension or 1500–2100 g/hm² *Trichoderma harzianum* WP (3×10^8 CFU/g) at 1500–2100 g/hm².
3. Field Management: Adjust shading in ginseng beds to create a well-ventilated and illuminated environment, reduce internal temperature and humidity inside the shed, and minimize disease spread and reinfection. Before winter, thoroughly clean the fields and beds, and destroy or deeply burying diseased residues from the previous year.
4. Chemical and Biological Control: After ginseng emergence during the leaf expansion stage, use 600–900 g/hm² of 50% fluopicolide-WG, 375–450 mL/hm² of 43% fluazinam + fluopyram SC, or 900–1200 mL/hm² of 100 billion CFU/g *Bacillus subtilis* WP for foliar spraying. Alternate these fungicides during the growing season, with intervals of 7–10 days.

39.2 Ginseng Black Spot Disease

Ginseng black spot disease is one of the most prevalent and destructive diseases affecting the aboveground parts of ginseng, causing significant losses and presenting major challenges for ginseng farmers in China. Under favorable conditions, black spot disease progresses rapidly and can spread across entire ginseng fields in a short time. The incidence rate typically ranges from 20% to 30%, but in severe cases, it can exceed 80%. This disease leads to early leaf drop, resulting in reduced yields and compromised quality of both ginseng roots and seeds.

Symptoms

The disease primarily damages ginseng leaves, and also affects stems, flower stalks, fruits, and other parts, with the most severe damage observed on leaves and stems. Infected leaves develop nearly circular or irregular lesions, ranging from yellowish-brown to dark brown, often displaying ring patterns. In advanced stages, these lesions may merge, causing leaf to wither or shed prematurely. On stems, the disease initially presents as oval or elongated lesions with a yellowish-brown color, which expand up and down the stem. Later, the middle part becomes sunken, the color changes from yellowish-brown to black, and a layer of black mold grows on them. In severe cases, it can cause the stems to lodge, and farmers commonly call it "scarred stalks". Flower stalk infections can lead to inflorescence dieback, and in later stages, affected fruits and seeds shrivel. When fruits are infected, brown spots initially appear on the surface. In the later stage, the fruits are gradually drained and shriveled, dropping prematurely, and farmers commonly call it "dried-hanging seeds" (Fig. 39.3).

Pathogen

The pathogen causing ginseng black spot disease is *Alternaria panax* Whetzel, an asexual morph of ascomycetes in the genus Alternaria. Conidiophores are clustered in groups of 2–16, brown to blackish-brown, with lighter coloring at the tips. They have slightly enlarged basal cells, are unbranched, straight or with one geniculate node, and measure (16–64) μm × (3–5) μm, with one to five septa. Conidia are produced singly or in chains, being long-elliptical or obclavate, yellow-brown, with transverse and vertical septa. There is a slight constriction at the septa, and the conidia have a slightly short to slender beak at the top, with a pale color (Fig. 39.4). This pathogen primarily infects plants of the Araliaceae family, such as ginseng.

Fig. 39.3 Symptoms of Ginseng Black Spot Disease

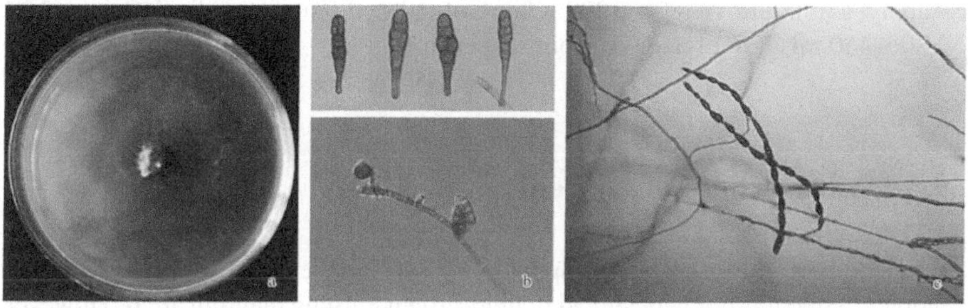

Fig. 39.4 *Morphology of Alternaria panax Causing Ginseng Black Spot* (**a**) Cultural characteristics; (**b**) Conidia; (**c**) Conidial chains

Occurrence Pattern

The pathogen overwinters as mycelium and conidia within infected plant debris, seeds, roots, and soil. In northeastern China, initial infections usually begin from mid-May to early June, with rapid disease progression in July and August. Abundant conidia can be produced on the lesions, which are dispersed by wind, rain, and air currents, causing repeated infections throughout the ginseng growing season until mid- to late September. Rainfall and humidity levels are critical factors for the occurrence, development, and epidemic spread of ginseng black spot disease. Based on long-term observations, high disease prevalence is associated with average field temperatures above 18 °C, rainfall around 40 mm in mid-June, and July–August temperatures between 18–27 °C with rainfall over 130 mm.

Control Measures

1. Field Management: Ensure proper ventilation within ginseng sheds. Reduce light exposure during summer and early autumn. Conduct thorough field cleaning after the harvest season. Remove and destroy or deeply bury diseased plant residues to prevent them from becoming initial infection sources for the following year.
2. Use of Disease-Free Seeds: Disinfect seeds and seedlings before planting. For every 100 kg of seeds, apply 200–400 mL of 25 g/L of fludioxonil suspension seed coating agent for seed dressing, or for every 100 kg of ginseng seedlings, soak them in 400 mL of 25 g/L of fludioxonil suspension seed coating agent for 5 minutes. Dry in the shade before planting for notable protective effects.
3. Chemical and Biological Control: After seedlings emerge, apply preventive sprays, especially in areas with a history of the disease. Before seedlings emerge, spray 800–1200 g/hm² of 10% difenoconazole WG at 800–1200 g/hm². During the leaf expansion stage, apply 300–450 mL/hm² of 25% azoxystrobin suspension, 2250–3000 g/hm² of 77% copper hydroxide WP, or 800–1200 g/hm² of 1×10^{11} CFU/g *Bacillus subtilis* WP. Repeat applications every 7–10 days, rotating fungicides to prevent resistance. During the growing season, promptly remove diseased plants for concentrated disposal. In severely affected areas,

apply a mixture of 800–1200 g/hm² of 10% difenoconazole WG and 25% 300–450 mL/hm² of azoxystrobin suspension.

39.3 Anthracnose of Ginseng

Anthracnose of ginseng is a widespread disease in ginseng cultivated under forest canopies, while it is not widespread in field-grown ginseng planting areas. However, in some individual years, it occurs severely, and the infection rates can reach 100%. Severe outbreaks lead to wilting and death of above-ground parts, impacting root growth and the formation of overwintering buds, ultimately reducing yield.

Symptoms
Anthracnose primarily affects the stems, leaves, and seeds of ginseng. Leaf lesions appear as round or nearly round spots, initially as small dark green dots that gradually expand, generally reaching diameters of 2–5 mm, with some lesions as large as 15–20 mm. Lesion edges are well-defined, showing yellow-brown or brown coloration with concentric rings. In later stages, the lesion centers turn yellow-white and become densely covered with black specks, which are the acervuli of the pathogen. Dry lesions may crack and perforate. In severe cases, the lesions mostly merge, resulting in leaf wilting and premature defoliation. On stems and flower stalks, lesions are long-elliptical, with dark brown edges and light yellow to light brown centers. Lesions on fruits and seeds are round and brown with distinct margins (Fig. 39.5).

Pathogen
The causative agent of ginseng anthracnose is *Colletotrichum panacicola* Uyeda et Takim, an asexual morph of ascomycetes in the genus Colletotrichum. Initially, acervuli develop within plant tissues and later break through the epidermis. Setae are sparsely distributed within acervuli, very few in number, dark brown with lighter tips, erect or slightly curved, broader at the base, and tapering at the apex, containing one to three septa and measuring (32–118) μm × (4–8) μm. Conidiophores are cylindrical, straight, unicellular, colorless, and measure (16–23) μm × (4–5) μm. Conidia are long cylindrical, colorless, unicellular, straight, with both ends rather rounded or one end blunt-rounded, having granular contents, and measuring (8–18) μm × (3–5) μm (Fig. 39.6).

Occurrence Pattern
The pathogen overwinters as mycelium and conidia in plant debris and seeds. In spring, under favorable conditions, the pathogen produces conidia, which are spread by wind and rain, causing primary infections. A large number of conidia can be continuously produced on the infected tissues, leading to repeated secondary infections. Conidia germinate easily under 100% relative humidity, forming germ tubes and appressoria. The pathogen typically infects directly but can also enter through wounds or natural openings. High rainfall and humidity facilitate disease

Fig. 39.5 Symptoms of *Ginseng Anthracnose*

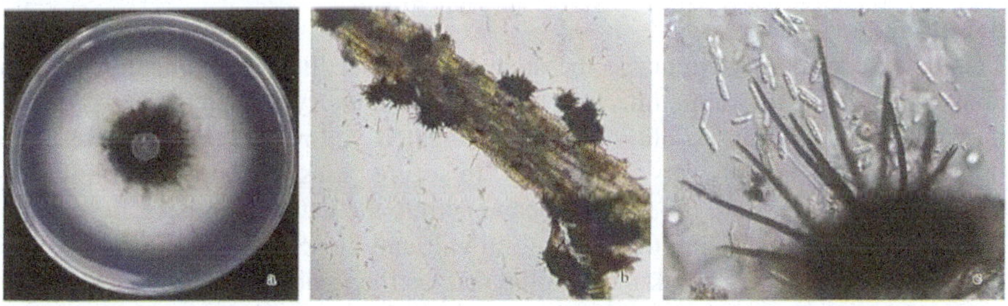

Fig. 39.6 Morphology of the Ginseng Anthracnose pathogen. (**a**) Culture characteristics; (**b**) Acervuli; (**c**) Setae

development and spread. The optimal temperature for the pathogen's growth and development is 24–25°C, while temperatures below 10°C or above 30°C inhibit its growth and spore germination. In northeast China, anthracnose typically appears in mid-to-late June, with peak incidence from early July to early August. As temperatures increase and the weather becomes hot, the occurrence of the disease gradually stops.

Control Measures

1. Seed and Seedling Treatment: For every 100 kg of seeds, use 200–400 mL of 25 g/L fludioxonil suspension seed coating agent for seed dressing; for seedlings, soak 100 kb of seedlings in 400 mL of 25 g/L fludioxonil suspension seed coating agent for 5 minutes. Allow seeds or seedlings to dry in a cool and ventilated place before planting to ensure effective control.

2. Bed Disinfection: After removing cold-protection soil, disinfect the bed surface with 300–450 mL/ha of 30% azoxystrobin·tebuconazole suspension.
3. Field Management: Optimize light exposure and ventilation within ginseng shade structures to reduce temperature and humidity, minimizing infection and reinfection opportunities.
4. Chemical Control: After ginseng emergence, during the leaf expansion stage, spray with 300–450 mL/ha of 30% azoxystrobin·tebuconazole suspension. Spray once every 7–10 days, and apply 2–3 times continuously for effective control.

39.4 Ginseng Damping-Off Disease

Damping-off is one of the primary diseases affecting ginseng during its seedling stage. This disease is widespread and common, occurring in nearly all ginseng-growing regions. The infection of ginseng plants is typically over 10%, and in severely affected areas, it can reach 60%, causing significant seedling mortality and substantial losses.

Symptoms
This disease primarily occurs at the stem base of seedlings, specifically 3–5 cm above the soil surface at the interface between wet and dry soil. In the early stages, the stem base develops slightly sunken, elongated yellowish-brown lesions. The infected tissue gradually rots and shrinks. In severe cases, lesions spread horizontally, penetrate deep into the stem and can encircle the entire stem base, destroying internal tissues and causing the seedling to wilt and die. Seedlings infected before emergence often fail to break through the soil, as the young buds rot underground. In the field, obvious disease foci can be observed, where infected plants rapidly spread the disease to surrounding plants, resulting in large patches of seedling death (Fig. 39.7). Under high humidity, brown mycelia may be visible on diseased parts and the surrounding soil.

Pathogen
The pathogen of ginseng damping-off is *Rhizoctonia solani* Kühn, an asexual morph of Basidiomycetes in the *Rhizoctonia* genus. On PDA culture medium, Rhizoctonia solani initially forms light gray colonies, later turning yellowish-brown to dark brown. The mycelium is septate, multinucleate, with a diameter of 7–13 μm. It branches at right angles, with constrictions at the branching points. The septa appear shortly after the branching points. In advanced stages, the mycelium turns to brownish-brown to brown and forms irregularly shaped sclerotia with a diameter of 1–3 mm, which are dark brown and usually several sclerotia are connected by mycelium strands (Fig. 39.8).

Fig. 39.7 Symptoms of Ginseng Damping-Off Disease

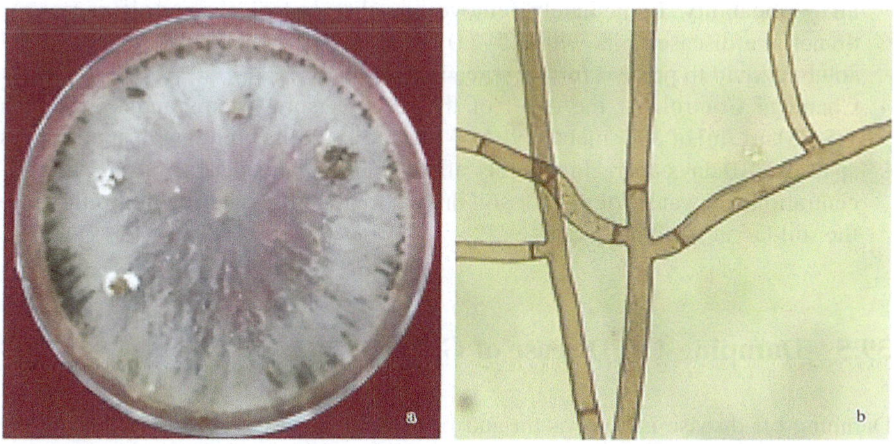

Fig. 39.8 *Rhizoctonia solani* morphology. (**a**) Colony characteristics; (**b**) Mycelium

Occurrence Pattern

The pathogen overwinters in plant debris or soil as mycelium and sclerotia, which serve as the initial infection source for the following year. Sclerotia can survive in the soil for 3 or more years. When temperature and humidity in the 5–6 cm soil layer are suitable, the mycelium rapidly spreads in the soil, infecting young ginseng stems through wounds or directly. Sclerotia are also dispersed by rainwater and farming operations. In northeastern regions, ginseng damping-off peaks from mid- to late June. In high-temperature, dry conditions, when soil temperatures exceed 16 °C, the pathogen enters a dormant stage, halting its growth and spread. Alternating cycles of rain, snow, freezing, and thawing in early spring often lead to outbreaks of damping-off disease. Heavier clay soils and low-lying fields are particularly susceptible, and high planting densities restrict airflow, increasing transmission between seedlings.

Control Measures

1. Seed Treatment: For every 100 kg of seeds, use 200-400 mL of 25 g/L fludioxonil suspension seed coating agent for seed pelleting. Allow to dry before planting for effective prevention.
2. Soil Treatment: Before planting, apply soil disinfectants such as 1.5–2.0 mL/m² of 30% metalaxyl-M + oxadixyl soluble liquid, 4–6 g/m² of 5 × 109 CFU/g *Paenibacillus polymyxa* WP, or 2–3 g/m² of 1 × 109 CFU/g Bacillus subtilis wettable powder. Ensure the treatment infiltrates approximately 5 cm into the soil. Alternatively, in early spring after ginseng seedlings emerge, use the above-mentioned dosage of soil disinfectants to irrigate the bed surface.
3. Improved Cultivation Practices: Use fertile, loose, well-aerated soil, preferably sandy loam, for seedbeds. Create raised beds to prevent waterlogging and ensure proper drainage during the rainy season. After seedlings emerge, frequently loosen the soil to increase soil temperature, make the soil loose, and ensure good air permeability. Immediately remove any diseased plants and, if necessary, drench the diseased pits with 1.5–2.0 mL/m² of 30% metalaxyl-M + oxadixyl soluble liquid to prevent further spread.
4. Chemical Control: At the onset of the disease, spray stems and leaves with 1.5–2.0 mL/m² of 30% metalaxyl-M + oxadixyl soluble liquid, with applications spaced 7–10 days apart. In severely affected areas, consider increasing the concentration and water volume for soil drench treatments, ensuring infiltration into the soil laryer by about 5 cm.

39.5 Damping-Off Disease of Ginseng

Damping-off disease is a prevalent and frequently encountered disease during the seedling stage of ginseng. Under severe conditions, it can lead to large-scale seedling die-off. This disease is widespread in ginseng-producing regions annually, with infection rates generally ranging from 1% to 10%; in severely affected fields, rates can exceed 30%, resulting in the patch-wise death of ginseng seedlings and relatively serious losses.

Symptoms
Early symptoms appear as water-soaked, dark lesions at the base of the young stems near ground level. When the humidity in the field is high, the lesions expand rapidly.

Fig. 39.9 Symptoms of Ginseng Damping-Off

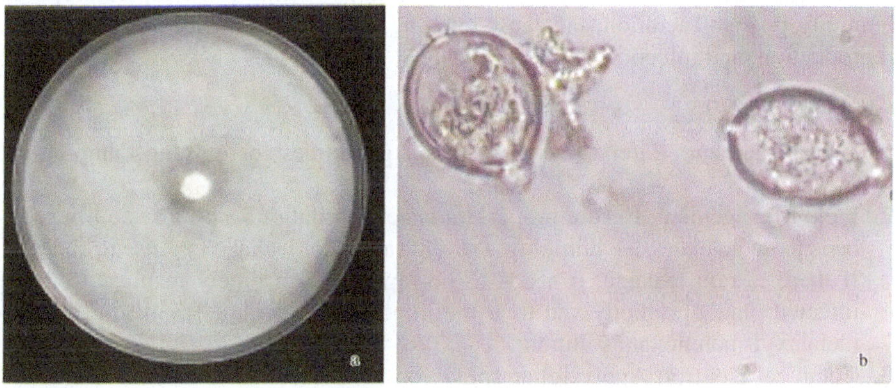

Fig. 39.10 Morphology of the Ginseng Damping-Off pathogen. (**a**) Culture characteristics; (**b**) Sporangia

The affected part becomes constricted and softened, ultimately the plants collapse and die while leaves remain green and unwilted (Fig. 39.9). When the humidity of the ginseng bed is high, a layer of gray-white mold-like substance often appears on the surface of the diseased part.

Pathogen

The causative agent of ginseng damping-off is *Pythium debaryanum* R. Hesse, a species of the genus Pythium in the Oomycota phylum. On PDA medium, the mycelium of *P. debaryanum* is fine, cottony, and white. The hyphae are thin, branched, non-septate, and measure 2–6 μm in diameter. Sporangia are either terminal or intercalary, spherical to nearly spherical or irregularly lobed, with a diameter of 15–25 μm. Mature sporangia generally remain attached and occasionally bear small papillae. These structures are colorless, with smooth surfaces and granular contents. When sporangia germinate, they produce germ tubes and the tips swell into vesicles. All the contents of sporangia are transferred into the vesicles through the germ tubes. Shortly afterward, zoospores develop within the vesicles, typically 30–38 in number. When the vesicles rupture, zoospores are released. Zoospores are kidney-shaped, colorless, measure (4–10) μm × (2–5) μm, and have two lateral flagella. They become immotile after a brief period. Oospores are spherical and pale yellow, with one oospore per oogonium, smooth-surfaced, and measure 10–22 μm in diameter (Fig. 39.10).

Occurrence Pattern

The pathogen P. debaryanum has extremely strong saprophytic properties, and can survive in the soil for a long time. Once inside the host, the pathogen spreads quickly within and between cells. It produces sporangia on infected tissues, releasing zoospores that drive secondary infections. Later, the pathogen forms oospores within infected tissues to overwinter. Oospores in soil can remain viable for more than one year. The pathogen primarily spreads by wind, rain, and water flow. Optimal infection occurs at temperatures of 15–16°C. Conditions such as low temperature, high humidity, poor soil aeration, and dense seedling beds are all conducive to the growth, reproduction, and infection of the pathogen.

Control Measures

1. Seed Treatment: Refer to seed treatment guidelines for ginseng damping-off prevention.
2. Field Management: Ensure proper drainage, ventilation, and loose soil in seedbeds to avoid excessive humidity. Prevent rainwater leakage in ginseng shelters. If there is rain leakage, it needs to be repaired in a timely manner. Remove infected plants promptly and treat the infected area with a 1.5 mL/m² of 30% metalaxyl-mefenoxam solution.
3. Chemical Control: Apply foliar sprays on seedbeds with a 1:1:180 Bordeaux mixture or 1.5–2 mL/m² of 30% metalaxyl-mefenoxam solution, 6–12 g/m² of 40% sodium dichloroisocyanurate soluble powder, or 4–6 g/m² of 5 billion CFU/g *Paenibacillus polymyxa* wettable powder at 4–6 g/m².

39.6 Ginseng Rust Root Disease

Rust root disease is one of the primary root diseases affecting ginseng, with field infection rates generally ranging from 20% to 30%. In severely affected areas, this rate can reach more than 70%. The disease can occur at any growth stages, from seedlings to mature plants, and the severity increases with plant age. This disease significantly reduces both the yield and quality of ginseng roots, negatively impacting market value and causing substantial economic losses for ginseng production.

Symptoms

This disease mainly affects ginseng roots, underground stems, and overwintering buds of ginseng. Infected roots initially develop yellow-brown spots at the infection

Fig. 39.11 Symptoms of Ginseng Rust Root Disease

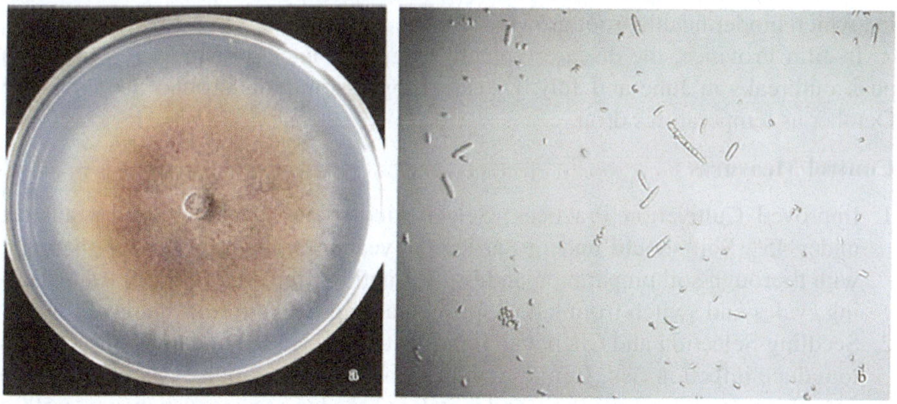

Fig. 39.12 Morphology of *Ilyonectria destructans.* (**a**) Colony characteristics; (**b**) Conidia

site, which gradually expand into near-circular or irregular brown lesions. In mild cases, the epidermis remains intact without affecting internal tissues, with the lesions limited to the subepidermal layers. Severe infections, however, penetrate deeper, damaging internal tissues and causing large accumulations of rust-colored powder at the lesions. Aboveground symptoms are generally absent in mild cases, but severe infections result in stunted growth, unexpanded leaves, which are reddish-brown, and, ultimately, wilting and plant death. When the pathogen infects the rhizome, it spreads up and down, causing the underground stems to become diseased and plant lodging and death. Overwintering buds display yellow-brown lesions that, in severe cases, lead to rot, preventing sprouting (Fig. 39.11).

Pathogen

The primary pathogen of ginseng rust root disease is *Ilyonectria destructans* (Zinssm.) Scholten, an asexual morph of ascomycetes in the genus *Ilyonectria.* The pathogen produces abundant aerial mycelium, initially gray-white, later turning brown. Conidia are single or clustered, cylindrical to nearly cylindrical, colorless, transparent, with one to three septa (occasionally up to four to six). Conidia may be straight or slightly curved. Chlamydospores are spherical, yellow-brown, and can be intercalary or in chains. The pathogen is a weak parasite, and is easily isolated from infected ginseng roots. Optimal growth occurs at temperatures between 22 and 24 °C, while growth declines noticeably below 15 °C or above 28 °C (Fig. 39.12).

Occurrence Pattern

This soil-borne pathogen can survive in the soil for extended periods as a soil-inhabiting organism. Ginseng roots are susceptible to infection throughout their growth period. The pathogen primarily overwinters as mycelium and chlamydospores in ginseng roots and soil. Under favorable conditions, it can enter ginseng roots through wounds. The disease spreads via infected seedlings, plant residues, contaminated soil, and human activities. Latent infections are common in ginseng roots, with infection rates increasing as the plants age. Rust root disease is more likely to occur in compacted, clay-heavy, waterlogged, acidic soils with low fertility, which hinder healthy root growth.

In Jilin Province, the disease typically begins in late April to early May, with peak outbreaks in June and July. Disease progression halts around September to October as temperatures drop.

Control Measures

1. Improved Cultivation Practices: Select wind-sheltered slopes with gradients under 45°. Soil should undergo at least 1 year of preparation before planting, with thorough soil preparation and bed formation. Implement 2-year transplanting cycles and switch from fall to spring planting to avoid frost damage.
2. Seedling Selection and Chemical Treatment: Use healthy, undamaged seedlings to reduce infection risk. Before planting, soak seedlings in a 500x dilution of a 25% thiazole · pyraclostrobin · and metalaxyl seed treatment suspension or 500× diluted 25g/L fludioxonil seed coat suspension for 30 minutes. This can reduce disease incidence. Additionally, apply seed coating treatment before sowing.
3. Soil Treatment: Disinfect the soil before planting or transplanting using 8–10 g/m^2 of 50% carbendazim wettable powder or 70% thiophanate-methyl wettable powder.
4. Removal of Infected Plants and Pit Disinfection: Remove infected plants promptly, and disinfect the surrounding soil in the affected pit using 70% thiophanate-methyl wettable powder. During the infection period, irrigate affected areas with a solution of 250 g/L of azoxystrobin suspension, 10% difenoconazole water-dispersible granules, or 500× diluted 70% thiophanate-methyl wettable powder to control disease spread.

Fig. 39.13 Symptoms of Phytophthora Blight on Ginseng

5. Biological Control: Apply biofertilizers and bioagents to prevent disease resistance and increase yield. During planting, apply *Trichoderma harzianum* wettable powder (3 billion CFU/g), *Bacillus subtilis* wettable powder (100 billion CFU/g), or *Bacillus amyloliquefaciens* WP (10 billion CFU/g), which has to effectively control on rust root disease.

39.7 Phytophthora Blight of Ginseng

Phytophthora blight is a severe disease affecting mature ginseng plants. It is prevalent across ginseng-producing regions in the United States, Canada, Russia, Democratic People's Republic of Korea, Japan, and northeast China. Generally, the plant infection rate is 10%–20%, and in severe cases, it can reach above 50%. When the disease is epidemic, the losses are relatively serious.

Symptoms

This disease primarily affects the leaves, petioles, stems, and ginseng roots, and occurs throughout the entire growth period. Early symptoms appear as dark green, water-soaked, irregular lesions on leaves. These lesions spread quickly, causing compound leaves to wilt and droop, which is commonly called "hand-holding towel" by ginseng farmers. Under high humidity, a yellowish-white mold layer appears on the diseased part. Infected petioles become water-soaked, soft, and rotten, causing the plant to wilt and droop, referred to as "hanging ghost" by ginseng farmers. Stems develop elongated, dark water-soaked lesions that quickly lead to

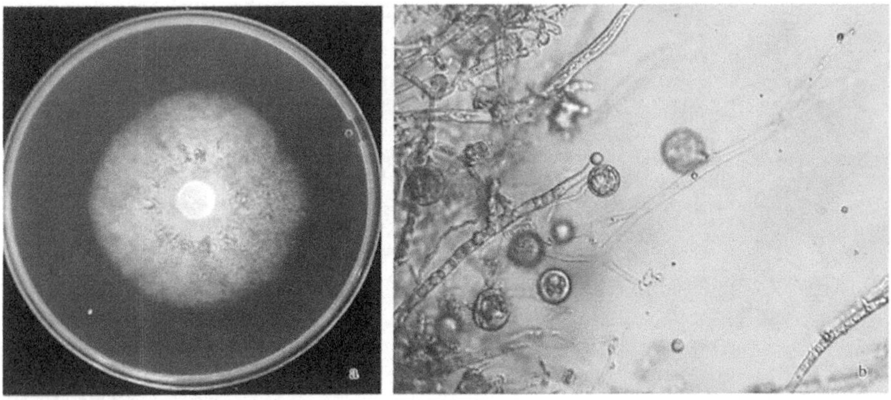

Fig. 39.14 Morphology of the Phytophthora Blight pathogen. (**a**) Culture characteristics; (**b**) Sporangiophores and Sporangia

rot, softening, and collapse (Fig. 39.13). Root infection initially appears as yellow-brown, moist rot, with easily detachable epidermis and yellow-brown internal tissue. Infected roots are often accompanied by combined infections of bacteria and *Fusarium* spp., and there are a large number of saprophytic nematodes. The rotten roots have a distinct fishy odor. In later stages, the outer skin of the rotting root is often surrounded by white mycelium intertwined with soil particles.

Pathogen
The pathogen causing ginseng Phytophthora blight is *Phytophthora cactorum* (Lebert et Cohn) J. Schröt., belonging to the genus *Phytophthora in the Oomycota.* The mycelium of this pathogen is white and cottony-like, with branched, non-septate, colorless hyphae. The sporangiophores are colorless, non-septate, and may be branched or unbranched, and measure 4–5 μm in width. Each sporangiophore bears one oval or pear-shaped, colorless sporangium, with a prominent papillate apical projection, measuring (32–54) μm × (19–30) μm. Upon germination, the sporangia release motile zoospores (Fig. 39.14).

Occurrence Pattern
The mycelium and sporangia of this pathogen cannot overwinter in soil; however, oospores can survive in soil for 4 years or more and serve as the primary infection source. Additionally, mycelium can overwinter on infected plant debris. The following year, when conditions are favorable, oospores germinate, producing sporangia that form germ tubes and appressoria. Infection pegs are generated to penetrate leaf tissue through leaf stomata. Sporangia can also germinate to release zoospores, which infect roots, leaves, and leaf axils directly. This disease primarily spreads through wind, rain, and agricultural operations, with root infection primarily occurring through direct contact. In northeastern China, initial infections typically occur from late May to early June, with peak incidence from late June to mid-July. The disease is favored by temperatures above 20 °C, relative humidity above 80%, and

soil moisture above 50%. Prolonged rainfall, compacted soil, dense plantings, and poor ventilation and light penetration can trigger widespread outbreaks.

Control Measures

1. Improved Shelter Management: Closely monitor ginseng shelters during rainy seasons to prevent water leakage and ensure proper drainage, maintaining optimal soil moisture levels.
2. Field Management: Maintain proper planting density to ensure good ventilation and light penetration. Regularly weed, loosen soil, and reduce moisture. Keep the field clean, and do not the use of unfermented fertilizers. In autumn, thoroughly remove plant debris and mulching. Remove infected plants from the field for disposal or deep burial, and disinfect infected areas with lime or an 80% dimethomorph WG.
3. Chemical Control: At the onset of disease symptoms, spray affected areas with 80–100 mL of 687.5 g/L oxathiapiprolin-ametoctradin suspension, 40–60 mL of 23.4% fluopicolide suspension, 80–120 g of 25% metalaxyl- propamocarb wettable powder, 25–35 mL of 500 g/L fluazinam suspension, or 100–170 g of 72% cymoxanil-mancozeb wettable powder per mu. Apply sprays every 7–10 days for a total of 2–3 applications.

Chapter 40
The Diseases of Rehmannia

Rehmannia glutinosa (Gaertn.) Libosch. ex Fisch. & C.A. Mey., a perennial herb of the Scrophulariaceae family, commonly grows on hillsides, along roadsides, and in wastelands at altitudes between 50 and 1100 meters. The root, known as Dihuang, derives its name from its yellow-white color and is used in traditional medicine.

Dihuang is cultivated across various regions domestically and internationally, with major domestic areas in Liaoning, Hebei, Henan, Shandong, Shanxi, Shaanxi, Gansu,, Jiangsu, Hubei province and Inner Mongolia Autonomous Region. Fresh Dihuang is used to cool blood and reduce heat, treating symptoms like heat-induced dehydration, tongue dryness, toxic rashes, and throat swelling. Dried Dihuang nourishes yin and promotes fluid production, addressing symptoms of heat in the blood, constipation due to fluid depletion, and yin deficiency-related fever. Cooked Dihuang serves as a tonic, treating blood deficiency, irregular menstruation, and liver and kidney deficiencies.

Dihuang is susceptible to a range of diseases including leaf spot, cyst nematode, leaf blight, viral infections, root rot, and wilt. Leaf spot and cyst nematode are particularly prevalent, significantly impacting the quality and yield of Dihuang, thus necessitating timely disease management.

40.1 Ring Spot Disease of Rehmannia

Ring spot disease is one of the significant diseases affecting Rehmannia, reported in the provinces of Henan, Shanxi, Shandong, Shaanxi, and Hebei. Early infection rates can reach up to 40%, severely impacting Rehmannia yields.

Symptoms
In the initial stages, circular or near-circular brown lesions appear on the leaves, displaying distinct concentric ring patterns. As the disease progresses, multiple lesions spread across the leaf surface, leading to necrosis and perforations (see Fig. 40.1). Under high humidity, extensive areas of leaf dieback may occur, severely affecting both the yield and quality of Rehmannia.

Pathogen
The causative agents of ring spot disease in Rehmannia include *Fusarium solani* (Mart.) Appelet Wollenweber ex Snyder et Hansen and *F. oxysporum* Schltdl. ex

© China Agriculture Press Co., Ltd. 2026

T. Wang et al., *Atlas of Chinese Medicinal Plant Diseases*,
https://doi.org/10.1007/978-981-95-6326-5_40

Snyder et Hansen, both within the genus *Fusarium*. On PDA medium at 25 °C, *F. solani* forms white, fluffy colonies. Macroconidia are sickle-shaped or cylindrical, while microconidia are oval or kidney-shaped. Under 20 °C, *F. oxysporum* initially forms dense, white colonies that gradually turn red. Conidiophores are flask-shaped, bearing unicellular, colorless, and oval microconidia clustered at the tips; macroconidia are sickle-shaped and slightly curved (see Fig. 40.2).

Occurrence Pattern

The pathogen overwinters as conidia in the soil, associated with plant debris. In the following year, spores spread via wind and rain, causing both primary and secondary infections. Initial symptoms appear in early May, with peak infection occurring in July and August in warm, humid conditions. After mid-August, the pathogen proliferates, producing a new generation of spores and hyphae, resulting in widespread infection during the harvest period from late October to November. High temperatures and humidity promote the germination, growth, and reproduction of overwintering spores in the soil, leading to disease outbreaks.

Control Measures

Resistant Varieties

Choose Rehmannia varieties with higher resistance to ring spot disease, such as Jinzhuangyuan, Jinbai No. 1, Hongjinhao, Beijing 11, Beijing No. 2, and Xiaoheiying.

Agricultural Control

Climate Monitoring and Forecasting

The incidence of ring spot disease is strongly influenced by weather conditions, making it essential to rely on weather forecasts for predicting outbreaks.

Field Management

Avoid excessive flooding and ensure proper drainage during rainy weather to reduce field humidity. Cultivate Rehmannia on raised beds to improve airflow.

Timely Removal of Infected Debris

Clear the field of plant debris after harvest and dispose of it properly.

Fertilization

Fig. 40.1 Symptoms of Rehmannia Ring Spot disease

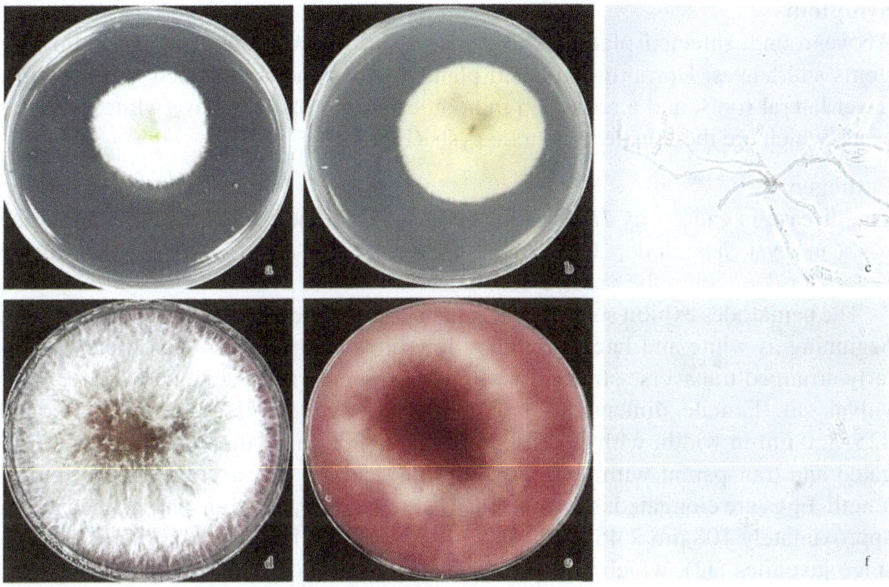

Fig. 40.2 Pathogens of Rehmannia Ring Spot disease: *Fusarium solani* and *F. oxysporum*. (**a, b**) Colony morphology of *F. solani*. (**c**) Macroconidia of *F. solani*. (**d, e**) Colony morphology of *F. oxysporum*. (**f**) Microconidia of *F. oxysporum*

Base fertilization should primarily consist of well-decomposed organic and compound fertilizers, supplemented with phosphorus and potassium fertilizers as necessary.

Crop Rotation

Rotate with nonhost crops for 3–4 years, and rotate nursery beds every 2–3 years to reduce soil-borne pathogens infection.

Chemical Control

Given that *Rehmannia glutinosa* is primarily propagated through root cuttings, it is necessary to disinfect the roots before planting. At the onset of the disease, treat with a Bordeaux mixture (copper sulfate: lime: water = 1:1:120–140. Subsequently, apply fungicides such as 70% Mancozeb WP at a 500-fold dilution, 50% chlorothalonil WP at a 500–700-fold dilution, or 25% triadimefon WP at a 1,000-fold dilution. Apply sprays every 7–10 days in rotation, with a total of 2–3 applications.

40.2 Rehmannia Root Cyst Nematode Disease

Root cyst nematode disease is a common affliction in Rehmannia cultivation, with widespread prevalence. In some regions, the infection rate can exceed 30%, affecting root systems, stunting plant growth, reducing yields, and impacting quality of Rehmannia.

Symptoms

Aboveground, infected plants exhibit inadequate growth, stunted, and yellowing stems and leaves. Uprooting affected plants reveals underdeveloped root systems, fewer lateral roots, and a reduction in nodules. The roots may have white, granular cysts, which are the female nematode cysts (Fig. 40.3).

Pathogen

The disease is caused by *Heterodera glycines* Ichinohe, commonly known as the soybean cyst nematode. It belongs to the genus *Heterodera* in the family Heteroderidae within the order Tylenchida.

The nematodes exhibit sexual dimorphism. Female nematodes are lemon-shaped, beginning as white and later turning yellow-brown. The surface has short, irregularly arranged transverse striations, with a prominent vulval cone and semicircular vulval slit. Female dimensions are approximately 480–550 μm in length and 425–520 μm in width, with stylet lengths of 28–30 μm. Male nematodes are elongated and transparent with a slightly ventrally curved tail, averaging 1.24 mm in length. Eggs are elongated, slightly concave on one side, and transparent, measuring approximately 108 μm × 45.7 μm. Juvenile nematodes hatch and molt into second-stage juveniles (J2), which are spindle-shaped with blunt heads and tapering tails (Fig. 40.4).

Occurrence Pattern

Soybean cyst nematodes have limited mobility and primarily spread through farm equipment, irrigation, flood or drainage water, or wind. They can also be dispersed over long distances in unprocessed compost or contaminated seeds. Overwintering occurs as eggs, embryo eggs, or juvenile nematodes encased in cysts within soil. The thick cyst cuticle allows them to survive in soil for over 10 years. Upon hatching, second-stage juveniles enter soil and infect Rehmannia roots. After mating,

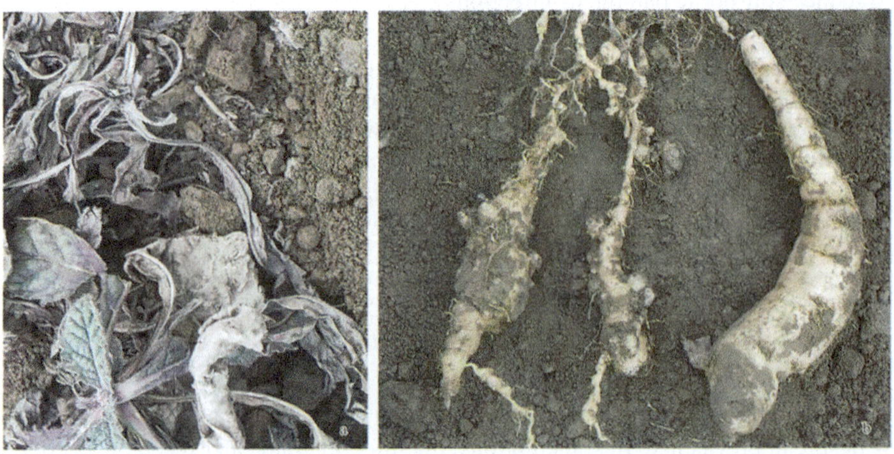

Fig. 40.3 Symptoms of Rehmannia Root Cyst Nematode disease. (**a**) Leaf symptoms. (**b**) Root symptoms

Fig. 40.4 Morphology of *Heterodera glycines*. (**a**) Second-stage juvenile. (**b**) Female. (**c**) Single egg mass. (**d**) Vulval pattern. (**e**) Cyst

males die, while females form cysts that fall into the soil. Juvenile nematodes only infect young roots. In autumn, as temperatures fall, eggs cease hatching and overwinter within cysts. Optimal temperatures of 20–22 °C complete the lifecycle within 22–35 days.

Control Measures

1. Protect Disease-Free Areas: The high risk of Rehmannia cyst nematode disease spreading through seed transportation necessitates prohibiting seed importation from infected regions.
2. Select Resistant Varieties: Use resistant varieties like Wen 85-5 and Beijing No. 3 to improve yield.
3. Agricultural Control

   (1) Crop Rotation: Rotate with nonhost crops like corn, wheat, cotton, or potatoes. Where feasible, alternate between paddy and dry fields. Introduce trap crops like snap beans, peas, or clover, which stimulate nematode egg hatching but prevent juvenile development, significantly reducing nematode numbers.

   (2) Water and Fertilizer Management: Apply fertilizers to enhance soil fertility, irrigate during dry periods, and promote Rehmannia growth to reduce infection rates.

4. Biological Control: Biological control agents like *Paecilomyces lilacinus* or *Bacillus subtilis* are used to protect plants from nematode infections and promote root development.
5. Chemical Control: Currently, effective nematicides include abamectin and dazomet, with an ongoing demand for environmentally safe alternatives.

Chapter 41
The Diseases of Corydalis

Corydalis yanhusuo [Corydalis yanhusuo (Y. H. Chou & Chun C. Hsu) W. T. Wang ex Z. Y. Su & C. Y. Wu], also known as "Yuanhu," is a perennial herbaceous plant of the Papaveraceae family, with its tubers used for medicinal purposes. It is mainly cultivated in Zhejiang, Jiangsu, Anhui, Hubei, and Henan, with the highest quality found in Jinyun County. The tuber has an irregular, flattened spherical shape, with a yellow to yellow-brown surface and irregular reticulate wrinkles. They have a slight odor and a bitter taste. It has the effects of promoting blood circulation, regulating Qi, and relieving pain, and can be used to treat chest and hypochondriac pain, epigastric pain, chest impediment and heart pain, amenorrhea, dysmenorrhea, postpartum stasis, and traumatic swelling. The main diseases affecting Corydalis yanhusuo are downy mildew, sclerotinia, rust, and damping-off, among which downy mildew causes the most severe damage.

41.1 Downy Mildew of *Corydalis yuanhusuo*

Downy mildew is one of the most important diseases in Corydalis yanhusuo production, often causing devastating damage. It typically occurs after rain, particularly in high-humidity conditions, and can destroy an entire field within ten days, hence the nickname "fire blight" or "plague". Yield losses can reach 50% to 70%.

Symptoms

The disease primarily affects leaves but can also infect stems. Affected leaves display irregular yellow-green to yellow-brown patches on the upper surface, with indistinct edges, while a dense, purplish-gray downy mildew layer appears on the lower surface. The bifurcations of stem petioles are also highly susceptible. Once the disease occurs in a field, it spreads rapidly, causing the stems of leaves of the entire plant to turn brown and rot (Fig. 41.1).

Pathogen

The pathogen causing Corydalis downy mildew is Peronospora corydalis de Bary, which belongs to Peronospora in Oomycota. The sorus is purplish-gray, with clustered sporangiophores. The main trunk of the sporangiophore measures (198–430) μm × (8–11) μm, with dichotomous to multiple branching in the upper part,, and the terminal branches are at right angles. Sporangia are solitary, lemon-shaped, nearly

© China Agriculture Press Co., Ltd. 2026

T. Wang et al., *Atlas of Chinese Medicinal Plant Diseases*,
https://doi.org/10.1007/978-981-95-6326-5_41

Fig. 41.1 Symptoms of *Corydalis* Downy Mildew. (**a**) Symptoms on the upper leaf surface. (**b**) Symptoms on the lower leaf surface

Fig. 41.2 Morphology of *Corydalis* Downy Mildew pathogen. (**a**) Hyphae. (**b**) Sporangia. (**c**) Hyphae and oospores

colorless, with a size of (16.9–23.7) μm × (13.5–16.9) μm. Oospores are yellow-brown, spherical, with a diameter of 33.8–37.14 μm (Fig. 41.2).

Occurrence Pattern

Oospores in the soil and infected plant debris serve as primary infection sources for the next season. Sporangia produced on the infected parts are spread by wind and rain, causing multiple secondary infections within the field. Downy mildew development is highly influenced by temperature and humidity, with cool, rainy, humid conditions favoring disease spread, while it occurs mildly in warm, dry weather with little rainfall. In early spring (March–April), when temperatures rises above 10°C, conditions such as frequent rain, sudden temperature changes, large day-night temperature differences, frequent dewdrops, or foggy weather promote its spread. Dense planting, excessive irrigation during early tuber enlargement period, and high field humidity also contribute to rapid disease expansion. The aboveground stems and leaves are prone to lodging and dying early. Conversely, the disease occurs mildly in plots with dry conditions, sandy loam soil, raised bed cultivation, and sufficient fertilization.

Control Measures

1. Scientific Site Selection

Select well-drained sandy loam in fields that have not previously grown *Corydalis*, such as rice-wheat rotation fields or mountainous areas, and avoid consecutive planting cycles.

2. Enhanced Cultivation Management

After harvest, thoroughly remove infected plant material, maintain proper planting density, and retain seeds only from disease-free fields. During cold, rainy spring seasons, ensure good drainage to reduce field humidity. Adjust irrigation based on weather conditions during the tuber enlargement stage.

3. Seed Selection and Breeding

Developing resistant cultivars through hybridization with wild *Corydalis* strains is an effective preventive approach.

4. Chemical Control

At the early stage of disease onset, apply a 1:1:300 Bordeaux mixture or a 1,000-fold dilution of 58% metalaxyl-mancozeb wettable powder every 7–10 days. Ensure even, thorough application for optimal control.

41.2 *Corydalis Sclerotium* Disease

Corydalis sclerotium disease occurs in the *Corydalis* cultivation regions of Jiangxi and Zhejiang provinces, with a relatively high incidence rate. The disease is most severe from mid-March to April. It spreads around, causing plants to collapse in a pattern resembling a "chicken nest," hence the local names "chicken nest blight" and "leaf rot." This disease seriously impacts the growth and quality of Corydalis.

Symptoms

Symptoms first appear as light brown to dark brown water-soaked streaks at the stem base near the soil surface, leading to soft rot and plant lodging. Affected leaves and petioles turn greenish-brown and rot; in severe cases, plants die in clusters. On the surface of infected plants near the soil, white, cotton-like mycelium and black, pellet-like sclerotia can be observed (Fig. 41.3). Additionally, the pathogen also infects the stem base and leaves near the ground. Under humid conditions, it can easily cause lodging and leaf rot, with visible cottony mycelium attached to small soil particle-like dark brown sclerotia.

Fig. 41.3 Symptoms of *Corydalis* Sclerotium disease

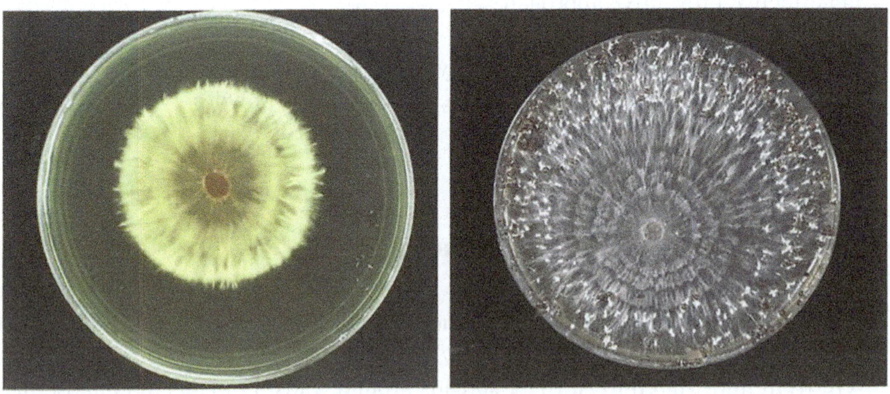

Fig. 41.4 Colony and Sclerotium Morphology of *Corydalis* Sclerotium Disease Pathogen

Pathogen

The pathogen causing *Corydalis* sclerotium disease is *Sclerotinia sclerotiorum* [(Lib.) de Bary], which belongs to the genus *Sclerotinia* in Ascomycota. The sclerotia are spherical, measuring (1.5–3) mm x (1–2) mm, and typically produce 4–5 stalked apothecia. The apothecia are disc-shaped, light reddish-brown, with a diameter of 0.4–1.0 mm. Asci are cylindrical, measuring (114–160) μm x (8.2–11) μm, and the ascospores are oval or fusiform, measuring (8–13) μm x (4–8) μm (Fig. 41.4).

Occurrence Pattern

The pathogen does not produce conidia and primarily overwinters and oversummers in the soil as sclerotia. In early spring, sclerotia germinate, and apothecia release ascospores, initiating primary infections. Sclerotia can also directly produce mycelium to infect nearby stems and leaves. Mycelium on infected stems and leaves may further spread to neighboring plants. Sclerotium disease favors low temperatures and high humidity. In early spring, when temperature reaches 15–18 °C, frequent rain and humid weather, poor drainage, dense planting, and tender foliage are most conducive to the disease occurrence. In Jiangsu and Zhejiang, the disease typically begins in mid-March, peaking from mid-to-late April.

Control Measures

1. Scientific Field Selection

Rotate *Corydalis* with plants of *Poaceae* or alternate wet-dry rotations with rice to avoid continuous cultivation of *Corydalis* for more than 2 years.

2. Enhanced Cultivation Management

Remove infected plant debris from the field, use appropriate planting density, and improve field ventilation and light conditions. As *Corydalis* is a fibrous-rooted crop that prefers moist conditions but is sensitive to waterlogging, ensure timely drainage after rainfall. Increase the application of phosphorus and potassium fertilizers to strengthen plant disease resistance.

3. Soil Disinfection

Use a mixture of 50% carbendazim WP and 50% metalaxyl WG with fine soil, spreading, and spread it evenly on the field surface or in planting furrows or holes before sowing.

4. Chemical Control

At the onset of the disease, spray the plants with 25% procymidone wettable powder or 50% carbendazim wettable powder every 5–7 days, repeating the treatment two to three times in succession.

41.3 *Corydalis* Rust Disease (Table 41.1)

Table 41.1 Characteristics of *Corydalis* rust disease

Disease (Pathogen)	Symptoms	Occurrence pattern
Rust Disease (*Puccinia brandegeei*)	The disease initially manifests as small, raised, yellowish-green spots on the upper leaf surface. On the lower surface, there are small round raised blisters. As the disease progresses, the epidermis ruptures to expose yellow-brown powder, which are uredinia and urediniospores of the pathogen. When the lesions are densely distributed at the leaf tips or margins, the local leaves curl and turn brown. When the stems and petioles are infected, rust spots also form, and they become deformed and curved. Black telia are rarely seen	The disease typically begins around early March, with peak infection occurring in April. When the uredinia rupture, a large number of urediniospores are released, causing secondary infections

Chapter 42
The Diseases of *Belamcanda chinensis*

Belamcanda chinensis (known as Shegan in traditional Chinese medicine) is a perennial herbaceous plant belonging to the genus Belamcanda in the family Iridaceae. Its rhizomes are used as medicine, which taste bitter, are cold in nature and slightly toxic. They can clear heat, detoxify, resolve masses and reduce inflammation, reduces swelling and alleviate pain, stop cough and resolve phlegm, and are used to treat tonsillitis and lumbago, etc. Belamcanda chinensis generally grows in forest edges or hillside grasslands, predominantly at relatively low elevations. However, in mountainous areas of southwestern China, it can grow at altitudes of 2000–2200 meters. Major cultivation areas in China include Hubei, Hebei, Hunan, and Guizhou.

The primary diseases affecting *B. chinensis* include southern blight, bacterial soft rot, leaf spot, and rust, etc. Among them, root diseases such as southern blight and bacterial soft rot are especially detrimental, significantly impacting the plant's growth, yield, and quality.

42.1 Southern Blight of *Belamcanda chinensis*

Southern blight, also known as sclerotial root rot, is an important disease affecting *B. chinensis* and occurs in all major production areas across China. The disease spreads rapidly after rainfall, forming infection centers, with an average infection rate of around 50%. Southern blight poses a severe threat to various medicinal plants; other affected medicinal plants include *Chrysanthemum morifolium*, *Vincetoxicum glaucescens*, *Macleaya cordata*, and *Dendrobium officinale*, etc.

Symptoms
In the early infection stage, white fungal mycelia appear at the base of the plant stems, gradually increasing and leading to water-soaked lesions at the stem base. These areas begin to turn yellow and soften. As the disease progresses, white sclerotia form on the mycelia, which later turn brown as the infection area expands. In the advanced stages, the stem base becomes completely water-soaked and rotted, leading to constriction. The aboveground parts of the plant die, and the underground rhizomes rot and become fibrous (Fig. 42.1).

© China Agriculture Press Co., Ltd. 2026

T. Wang et al., *Atlas of Chinese Medicinal Plant Diseases*,
https://doi.org/10.1007/978-981-95-6326-5_42

Pathogen

The pathogen causing southern blight in *B. chinensis* is *Sclerotium rolfsii* Sacc., which belongs to the genus Sclerotium. This pathogen grows rapidly on Potato Dextrose Agar (PDA) medium. Its mycelia are white and fluffy, and sclerotia form in later growth stage. The sclerotia are brown, shiny, and resemble rapeseed. Studies have shown that the pathogen can grow between 15–35 °C. The mycelia grow relatively fast at 25–30 °C, with 30 °C being the optimal temperature. The mycelia cannot grow normally when the temperature is below 5 °C or above 40 °C. Therefore, southern blight of *B. chinensis* rarely occurs in winter and high-temperature arid regions. On PDA medium, the pathogen grows at pH levels of 4.0–12.0, with optimal growth at pH 5.0. Under acidic conditions, the growth rate exceeds 2.4 cm per day, with fine and dense mycelia and more aerial mycelia (Fig. 42.2). Alkaline conditions, however, reduce the growth rate, and the mycelia become thicker and sparser, with morphological variations.

Occurrence Pattern

The pathogen overwinters in the field as mature sclerotia, rhizomorphs, and hyphae, or other structures. It persists in the soil or diseased plant debris, spreads through soil and water, and multiplies in the soil by means of hyphae. Initial infection typically begins in mid-to-late June, and spread rapidly after rain.

Peak infection occurs from July to August, as white hyphae encase the stem bases, accompanied by occasional brown, near-circular sclerotia, with water stains.

Fig. 42.1 Symptoms of Southern Blight on *Belamcanda chinensis*

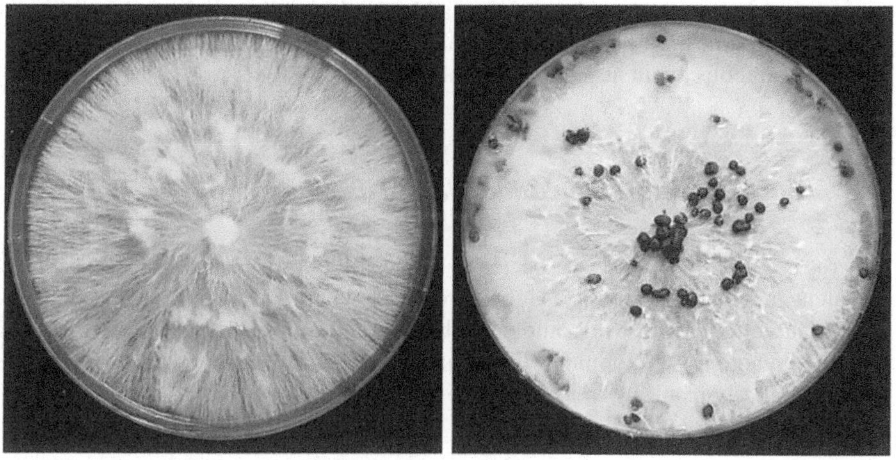

Fig. 42.2 Colony and Sclerotia morphology of Southern Blight pathogen

Infected plants are easy to break when pulled up and show symptoms of decay, with a field infection rate around 50%. After August, infection rates decline, and by September, the aboveground portions of the plants start to wither, with field density visibly decreasing. By late October, most of the above-ground parts have withered, with occasional new shoots emerging from rhizomes. The infection level stabilizes, though diseased plants with fungal hyphae and sclerotia at the stem base may still be observed after rainfall.

Control Measures

1. Scientific Site Selection

Cultivation should be conducted on well-drained sandy loam plains or sunny slopes, avoiding low-lying, waterlogged areas. Clay soils with poor drainage are prone to disease.

2. Field Management

Avoid continuous cropping and excessive planting density. The longerthe continuous cropping period is, the more severe the disease will be. *B. chinensis* should be rotated or intercropped with gramineous crops (e.g., maize) and should not follow crops that host the southern blight pathogen, such as legumes and *Solanaceae* plants. During the growing season, promptly remove diseased plant debris and apply ash or lime to surrounding areas to prevent spread.

3. Soil Disinfection

During ridge preparation, disinfect the soil with 50% carbendazim wettable powder at 7–8 g/m^2 or a mixed disinfectant powder (comprising wood ash, lime, and sulfur powder in a 50:50:2 ratio) at 7500 g/m^2.

4. Chemical Control

Upon detecting the disease, promptly apply fungicides to prevent the pathogen from spreading to healthy plants. Suitable treatments include 29% lime sulfur solution diluted 300–500 times, 10% fluazinam emulsion diluted 8000–10000 times, and various plant-derived fungicides such as 1% osthol emulsion diluted 800–1000 times, 20% clove phenol solution at 2250–3000 g/hm², and 0.5% berberine aqueous agent at 1290–1800 mL/hm².

42.2 Soft Rot in *Belamcanda chinensis*

Soft rot is a highly destructive bacterial disease affecting *B. chinensis*, especially prevalent in Tuanfeng County, Hubei Province, where it has led to multiple outbreaks. Infection can cause complete tuber rot and entire plant death, with rapid disease spread and development. Field infection rates are around 15%, with severe cases potentially resulting in total crop loss.

Symptoms
In the early stages, soft rot causes wilting of leaves with yellowing edges, and the affected areas of the rhizome soften, releasing a foul odor. As the infection progresses, the epidermal and woody layers of the rhizome separate, and the rhizome softens, turning into a mushy, decaying mass. The basal stem shows water-soaked, soft rot symptoms, and stems and leaves wither entirely, leading to full plant death (Fig. 42.3).

Pathogen
The causative agent of soft rot in *B. chinensis* is *Dickeya fangzhongdai*, a bacterium of the genus *Dickeya* within the Proteobacteria phylum (Fig. 42.4). Studies have shown that the bacterium grows between 10 and 45 °C, it multiplies and grows relatively fast at 23 and 33 °C, with 28 °C being the optimal temperature. Growth rate decreases significantly below 10 °C or above 45 °C. Therefore, soft rot of B. chinensis rarely occurs in winter and hot, arid regions. On LB medium, the pathogen forms round, slightly raised, glossy, transparent, yellowish-white colonies. It is a Gram-negative, short rod-shaped bacterium with blunt ends, measuring (0.5–0.7) μm × (1.0–2.0) μm, and has four to six peritrichous flagella. The bacterium grows across a pH range of 4.0–12.0, with optimal growth at pH 6.0. Growth rate slows as the pH becomes more alkaline, but the bacterium still grows at pH 12.0, while growth halts at pH 2.0.

Occurrence Pattern
Soft rot in *B. chinensis* tends to occur in the spring rainy season and spreads rapidly after rainfall. In summer, affected plants show basal stem softening with water-soaked spots and break easily when pulled from the ground, exhibiting rot symptoms. Field infection rates reach approximately 15%. In autumn, above-ground parts of infected plants all wither and die, leading to sparse field density. Partially rotted tubers may produce new sprouts, but the residual bacteria often continue to infect the tubers until full decay. Initial infection generally begins in mid-May, though some areas report earlier cases in April. Peak infection season spans June to September, especially following rainy seasons, with outbreaks common in

Fig. 42.3 Symptoms of Soft Rot disease of *Belamcanda chinensis*

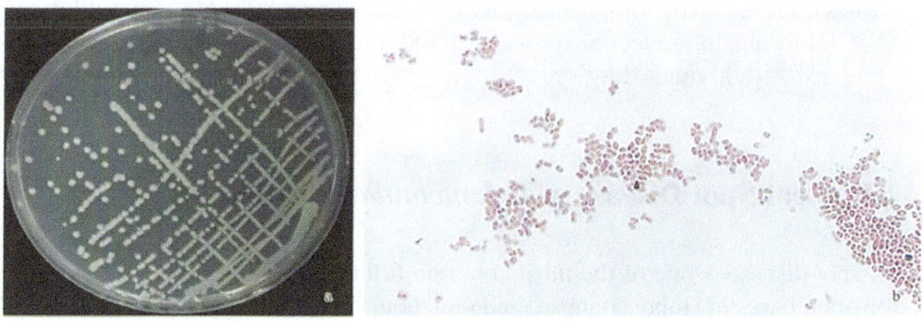

Fig. 42.4 Morphology of Soft Rot pathogen in *Belamcanda chinensis*. (**a**) Colony morphology. (**b**) Bacterial morphology

high-temperature, high-humidity conditions with poor soil drainage. The pathogen overwinters in soil and infected plant debris, re-infecting plants through wounds or natural openings under favorable conditions. Rain splash facilitates spread, and high humidity, inadequate ventilation, and high temperatures are the main conditions for the occurrence of this disease.

Control Measures

1. Scientific Site Selection

For *B. chinensis* cultivation, choose sandy loam soils on plains or sunny hillsides, avoiding low-lying, waterlogged areas. Clayey, poorly drained soils increase disease risk.

2. Field Management

Avoid consecutive cropping and excessive planting density, as prolonged consecutive cropping worsens disease severity. *B. chinensis* can be rotated or intercropped with gramineous crops such as corn. Avoid planting in fields where the previous crops were sweet potatoes, potatoes, and other crops that are hosts to the soft rot pathogen. During the growing period, promptly remove infected plant debris, applying ash or lime around the affected area to prevent disease spread.

3. Soil Disinfection

Before planting, disinfect soil with 0.1% metalaxyl-mancozeb mixture at 2.5–3 kg per mu or apply carbendazim or mancozeb. In summer, plough the land, water thoroughly, cover with plastic film, and seal for 1–2 weeks to raise the surface temperature of the soil above 70°C, which can kill the pathogens and pest eggs in the soil.

4. Chemical Control

Early treatment is critical to prevent pathogen spread to healthy plants. Effective agents include 40% sodium dichloroisocyanurate soluble powder at 900–1200 g/hm^2, 80% garlic extract suspension at 800–1000 times dilutions, 50% carbendazim wettable powder at 500 times dilutions, 10% polyoxin solution at 500 times dilutions, 70% thiophanate-methyl wettable powder at 500 times dilutions, 75% chlorothalonil wettable powder at 500 times dilutions, and Bordeaux mixture at 200–300 times dilutions.

42.3 Leaf Spot Disease of *Belamcanda chinensis*

Leaf spot disease is one of the most common foliar diseases affecting *B. chinensis* cultivation bases in Hubei, Jiangsu, and Guizhou. The disease spreads from the leaf tips down to the base of the stems, affecting the entire leaf and seriously impairing photosynthesis. This disease is widespread, challenging to control, and has an incidence rate of 30% to 50%. In some key production areas, the incidence rate can reach up to 90%, posing a severe threat to *B. chinensis* cultivation.

Symptoms

In the early stages of leaf spot disease, yellow or yellow-brown circular, oval, or ring-shaped spots appear on the leaves. The pathogen generally infects the young tips of leaves first, gradually spreading from the tender tips to the leaf base and stem as the disease progresses. In the late stages, the infected leaves wither completely (Fig. 42.5).

Fig. 42.5 Symptoms of Leaf Spot disease on *Belamcanda chinensis*

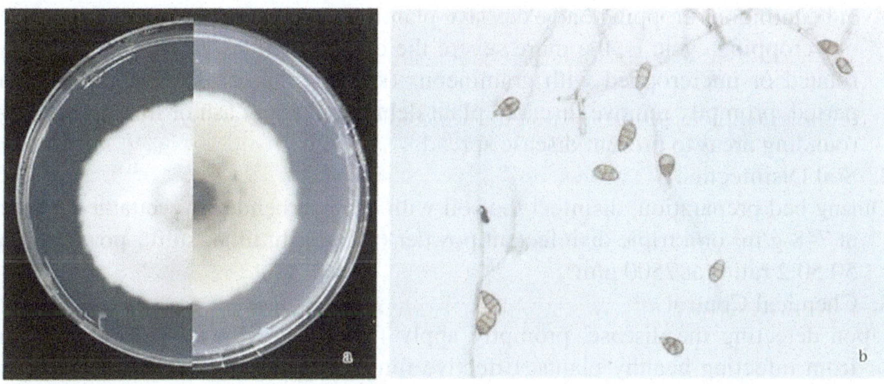

Fig. 42.6 Morphology of *Alternaria* spp. (**a**) Colony characteristics; (**b**) Conidia characteristics

Pathogen

Leaf spot disease in *B. chinensis* is caused by single or multiple infections of fungi in the genus *Alternaria*. Pathogenic species include *Alternaria solani*, *Alternaria alternata*, *Alternaria longipes*, *Alternaria tenuissima*, and *Alternaria alstroemeriae*, with *A. alternata* being particularly destructive. The aerial hyphae of *Alternaria* are dense, velvety, with concentric ringed patterns. Initially, the mycelia are milky white, but they gradually accumulate greenish-brown pigments over time. Conidiophores arise directly from the main hyphae, mostly single, upright or slightly curved, with septa, and range in color from light brown to yellow-brown. Mature conidia are also light brown to yellow-brown, in shapes of obclavate, obpyriform, ovate, nearly elliptical, or nearly round, and with transverse, longitudinal, and oblique septa, measuring (6.46–14.89) μm × (17.67–46.98) μm (Fig. 42.6).

Occurrence Pattern

Leaf spot disease typically starts in late March when new leaves sprout, with the peak infection period extending from May to October. Overwintering mycelia and conidia on plant debris, seed surfaces, and soil serve as primary infection sources for the following season. The pathogen spreads via wind and rain, entering plants through wounds or natural openings. Optimal conditions for spore germination include temperatures of 15–25 °C and relative humidity above 95%. Fields with extended continuous cropping cycles are conducive to the occurrence of this disease.

Control Measures

1. Scientific Site Selection

Cultivation should be conducted in sandy loam plains or sunny hillside slopes, avoiding low-lying, waterlogged areas. Clay soils with poor drainage are prone to the disease.

2. Field Management

Avoid continuous cropping and excessive planting density. The longer the continuous cropping cycle is, the more severe the disease will be. *B. chinensis* can be rotated or intercropped with gramineous (such as maize). During the growth period, promptly remove diseased plant debris and apply ash or lime to the surrounding areas to prevent disease spread.

3. Soil Disinfection

During bed preparation, disinfect the soil with 50% carbendazim wettable powder at 7–8 g/m^2 or a triple disinfectant powder (wood ash: lime: sulfur powder in a 50:50:2 ratio) at 7500 g/m^2.

4. Chemical Control

Upon detecting the disease, promptly apply fungicides to prevent the pathogen from infecting healthy plants. Effective fungicides include 12.5% triadimefon wettable powder at 450–900 g/hm^2 and 70% mancozeb wettable powder at 2.25 kg/hm^2.

42.4 Rust Disease of *Belamcanda chinensis*

Rust disease is a common foliar disease affecting *B. chinensis* throughout its growth period. It spreads rapidly and is highly infectious, making control challenging once the disease occurs. Typical infection rates of *B. chinensis* rust disease range from 20% to 40%, but in some key production areas, severe outbreaks have led to infection rates as high as 80%, which seriously restricts the development of the *B. chinensis* industry.

Symptoms

In the seedling stage, when infected by rust, multiple layers of yellow-brown spots (uredinia) arranged in concentric rings on the leaves. In mature plants, at the initial stage of leaf infection, the uredinia are long-strip, bright yellow, and elliptical,

Fig. 42.7 Symptoms of Rust disease in *Belamcanda chinensis*. (**a**, **b**) Field symptoms. (**c**) Leaf symptoms

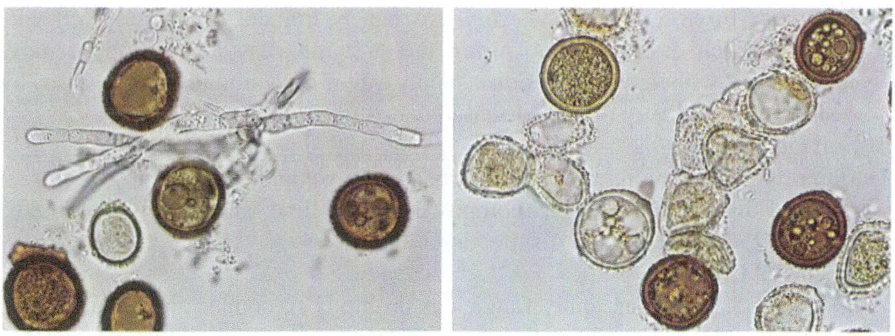

Fig. 42.8 Morphology of Urediniospores of *Puccinia iridis*

parallel to the veins and arranged in rows. The underside of the leaf is slightly raised. As the pustules expand, the color at the center of the lesions deepens to light brown or brownish-white, with a distinct yellow halo. The uredinia on the underside of the leaf are orange-yellow, grow on both sides of the leaf veins, protrude from the epidermis, and aggregate into large patches (Fig. 42.7). In the later stages, the pustules rupture, releasing large quantities of spores that infect nearby healthy plants. In addition to leaves, the flower receptacles and stems can also be infected.

Pathogen

Rust disease in *B. chinensis* is primarily caused by single or combined infections of fungi from the genus *Puccinia* in the phylum Basidiomycota phylum, with *Puccinia iridis* being the primary pathogen. The urediniospores of *P. iridis* are near-spherical, obovate, or elliptical, measuring (22–31) μm × (20–26) μm, with a wall thickness of

1.5–2.0 μm. The spores are reddish-brown, spiny, and feature two or three scattered pores, occasionally positioned medially (Fig. 42.8).

Occurrence Pattern
Rust disease typically starts to occur on leaves in late May, with severe outbreaks occurring from June to August. Puccinia spores are primarily dispersed by wind but can also spread through rain splash. The disease peaks in spring and autumn. The occurrence of the disease is closely related to humidity. Relative humidity levels exceeding 80% for consecutive days, especially when near saturation, significantly contribute to disease severity. The spores germinate only in the presence of water droplets or a moisture film, making rainy, dewy, or foggy weather conducive to disease outbreaks.

Control Measures

1. Scientific Site Selection

Select well-drained sandy loam plains or sunny slopes for cultivation, avoiding low-lying, waterlogged areas. Rust disease is more likely to occur in poorly drained clay soils.

2. Field Management

Avoid continuous cropping and high planting density, as prolonged cultivation intensifies disease severity. *B. chinensis* can be rotated or intercropped with legumes, such as soybeans. During the growing season, promptly remove infected and withered plant residues, apply ash or lime around affected areas for disinfection, and bury or burn diseased debris to prevent spread.

3. Soil Disinfection

During ridge preparation, apply 50% carbendazim wettable powder at 7–8 g/m^2 or a mixed disinfectant powder (composed of ash, lime, and sulfur in a 50:50:2 ratio) at 7500 g/m^2 for soil disinfection.

4. Chemical Control

At the onset of infection, apply fungicides promptly to contain the pathogen and protect healthy plants. In early infection stages, spray with a 15% triadimefon WP solution at a 1000× dilution or a 12.5% diniconazole WP solution at a 3000× dilution, applying every 7–10 days for one to two consecutive treatments. For mature plants, use a 25% triadimefon WP solution at a 250× dilution or a 25% propiconazole emulsion at a 3000× dilution, applying every 5–7 days for two to three consecutive treatments.

42.5 Occasional Diseases of *Belamcanda chinensis* (Table 42.1)

Table 42.1 Characteristics of occasional diseases in *Belamcanda chinensis*

Disease (Pathogen Unidentified)	Symptoms	Occurrence pattern
Leaf Blight (Pathogen Unidentified)	Infection starts from the leaf edges and tips, spreading from small to large irregularly shaped spots. Lesions begin as reddish-brown, eventually turning ash-gray. The spots coalesce, covering up to 1/3 to 1/2 of the leaf area. A dark necrotic border often surrounds the lesion, clearly marking the boundary between healthy and diseased tissue	Outbreaks begin in April, with peak incidence from May to September. During the rainy season, high temperatures and humidity, along with poor soil drainage, facilitate rapid spread
Flower Blight (Pathogen Unidentified)	Initially, angular greenish spots appear on the upper side of the leaves, gradually changing to brown. The infected leaves show a mosaic pattern of light green and evergreen. In severe cases, the leaves become deformed and yellow	Outbreaks begin in April, with peak incidence from May to September. During the rainy season, high temperatures and humidity, combined with inadequate soil drainage, facilitate rapid spread
Spot Blotch (Pathogen Unidentified)	Small circular brown spots initially appear on the upper leaf surface, gradually expanding into near-circular patches. The spots turn grayish-white or light brown, with dark brown edges. Small black dots are scattered or arranged in a ring within the spots	Outbreaks begin in April, with peak incidence from May to September. During the rainy season, high temperatures and humidity, coupled with poor soil drainage, contribute to rapid spread

References

Bai Jinkai, 2003. Fungi Flora of China. Beijing: Science Press.
Chen Jun, Ding Wanlong, Cheng Huizhen, 2019. Medicinal Plant Protection Science. Beijing: Electronic Industry Press.
Eisenback J D, Triantaphyllou A C, Hirschmann H, et al. (1981) A Guide to the Four Most Common Species of Root-Knot Nematodes (Meloidogyne spp.), With A Pictorial Key[M]. Raleigh: North Carolina State University.
Fu Junfan, 2007. Medicinal Plant Pathology. Beijing: China Agricultural Press.
National Pharmacopoeia Commission, 2020. Pharmacopoeia of the People's Republic of China (Volume I). Beijing: China Medical Science and Technology Press.
He Yunzhuan, Xie Xiaoliang, Liu Tinghui, et al., 2019. Color Atlas of Major Pests and Diseases of 35 Traditional Chinese Medicinal Herbs. Beijing: China Medical Science and Technology Press.
Hong Jian, Li Debao, Zhou Xueping, 2001. Atlas of Plant Virus Classification. Beijing: Science Press.
Xie Hui, 2005. Plant Nematology. Beijing: Higher Education Press.
Zeng Lingxiang, 2017. Pests and Diseases of Medicinal Plants. Guizhou: Guizhou Science and Technology Press.
Zhou Rujun, Fu Junfan, 2014. Color Atlas of Medicinal Plant Diseases. Beijing: China Agricultural Press.
Agrios, G.N., 2009. Plant Pathology. Beijing: China Agricultural University Press.

© China Agriculture Press Co., Ltd. 2026

T. Wang et al., *Atlas of Chinese Medicinal Plant Diseases*,
https://doi.org/10.1007/978-981-95-6326-5

The manufacturer's authorised representative in the EU is Springer Nature Customer Service Centre GmbH, Europaplatz 3, 69115 Heidelberg, Germany. If you have any concerns regarding our products, please contact ProductSafety@springernature.com

Printed and bound by CPI Group (UK) Ltd, Croydon, CR0 4YY
22/09/2026
02220954-0012